Advances in Intelligent Systems and Computing

Volume 1145

The series "Advances in Intelligent Systems and Computing" contains publications on theory, applications, and design methods of Intelligent Systems and Intelligent Computing. Virtually all disciplines such as engineering, natural sciences, computer and information science, ICT, economics, business, e-commerce, environment, healthcare, life science are covered. The list of topics spans all the areas of modern intelligent systems and computing such as: computational intelligence, soft computing including neural networks, fuzzy systems, evolutionary computing and the fusion of these paradigms, social intelligence, ambient intelligence, computational neuroscience, artificial life, virtual worlds and society, cognitive science and systems, Perception and Vision, DNA and immune based systems, self-organizing and adaptive systems, e-Learning and teaching, human-centered and human-centric computing, recommender systems, intelligent control, robotics and mechatronics including human-machine teaming, knowledge-based paradigms, learning paradigms, machine ethics, intelligent data analysis, knowledge management, intelligent agents, intelligent decision making and support, intelligent network security, trust management, interactive entertainment, Web intelligence and multimedia.

The publications within "Advances in Intelligent Systems and Computing" are primarily proceedings of important conferences, symposia and congresses. They cover significant recent developments in the field, both of a foundational and applicable character. An important characteristic feature of the series is the short publication time and world-wide distribution. This permits a rapid and broad dissemination of research results.

** Indexing: The books of this series are submitted to ISI Proceedings, EI-Compendex, DBLP, SCOPUS, Google Scholar and Springerlink **

More information about this series at http://www.springer.com/series/11156

Lakhmi C. Jain · Sheng-Lung Peng ·
Shiuh-Jeng Wang
Editors

Security with Intelligent Computing and Big-Data Services 2019

Proceedings of the 3rd International Conference on Security with Intelligent Computing and Big-data Services (SICBS), 4–6 December 2019, New Taipei City, Taiwan

 Springer

Editors
Lakhmi C. Jain
University of Technology Sydney
Broadway, NSW, Australia

Sheng-Lung Peng
CSIE Department
National Dong Hwa University
Hualien City, Taiwan

Shiuh-Jeng Wang
Central Police University
Taoyuan City, Taiwan

ISSN 2194-5357   ISSN 2194-5365 (electronic)
Advances in Intelligent Systems and Computing
ISBN 978-3-030-46827-9   ISBN 978-3-030-46828-6 (eBook)
https://doi.org/10.1007/978-3-030-46828-6

This Springer imprint is published by the registered company Springer Nature Switzerland AG
The registered company address is: Gewerbestrasse 11, 6330 Cham, Switzerland

Preface

It is our Third International Conference on Security with Intelligent Computing and Big-data Services, SICBS'19. The proceedings of the first two conferences, 2017 and 2018, were published in Springer AISC 733 and Springer AISC 895 series, respectively. The conference was held during December 4–6, 2019, in Chihlee University of Technology, New Taipei City, Taiwan.

SICBS'19 provides a platform for researchers, engineers, academicians as well as industrial professionals from all over the world to present their research results and development activities in Security with Intelligent Computing and Big-data Services. It aims at strengthening the international academic cooperation and communications, and exchanging research ideas.

We received a large number of papers from all around the world. After rigorous peer-review process, where each submission was reviewed by at least two TPC members, we selected 30 papers, covering the topics as follows: cryptography, multimedia security and artificial intelligence, Internet of things and big data, intelligent computing, intrusion detection, privacy and cryptography. Moreover, our conference is further enriched by two keynotes entitled "Secure Connected Car over 5G," by Mr. Koji Nakao, National Institute of Information and Communications Technology, Japan, and "Blockchain Applications in Financial Management," by Prof. Patrick S. Chen, Taiwan.

Organization of conferences is a hard work. We would like to thank all those who contributed to the Advisory Committee, the Technical Program Committee and the Organizing Committee for their efforts in the course of conference preparations. We are grateful to all authors and reviewers for their contributions. Our thanks are due to Springer for their assistance during the preparation phase of the manuscript.

Lakhmi C. Jain
Sheng-Lung Peng
Shiuh-Jeng Wang

Organization

Honorary Chairs

Lakhmi C. Jain	University of Technology Sydney, Australia
Chin-Chen Chang	Feng Chia University, Taiwan
Ju-Long Chen	Chihlee University of Technology, Taiwan

General Chairs

Shiuh-Jeng Wang	Central Police University, Taiwan
Sheng-Lung Peng	National Dong Hwa University, Taiwan

International Advisory Committee

Nakao Koji	NICT, Japan
Javier Lopez	University of Malaga, Spain
Heung Youl Youm	Soonchunhyang University, Korea
Hiroaki Kikuchi	Meiji University, Japan
Witold Pedrycz	University of Alberta, Canada
Sakurai Kouichi	Kyushu University, Japan
Sang-Soo Yeo	Mokwon University, Korea
Liudong Xing	University of Massachusetts, Dartmouth, USA
Laurence T. Yang	St. Francis Xavier University, Canada
Binod Vaidya	University of Ottawa, School of Electrical Engineering and Computer Science, Canada

Program Committee Chairs

Chih-Hung Wang	National Chiayi University, Taiwan
Kuo-Yu Tsai	Feng Chia University, Taiwan
Taeshik Shon	Ajou University, Korea

Damien Sauveron   University of Limoges, France
Ching-Nung Yang   National Dong Hwa University, Taiwan

Publicity Chairs

Vincent FC Lee   E-Security Analysis and Management
                                       Association, Taiwan
Kuo-Cheng Ho   E-Security Analysis and Management
                                       Association, Taiwan
Aneesh Sharma   University of California, Berkeley, USA
Seungmin Rho   Sungkyul University, Korea
Kuo-Jui Wei   AAA Security Technology Co., Ltd., Taiwan
Chin-I Lee   Ling Tung University, Taiwan

Publication Chairs

Chi-Yao Weng   National Pingtung University, Taiwan
Cheng-Hsing Yang   National Pingtung University, Taiwan
Chih-Ping Yen   Central Police University, Taiwan

Finance Chairs

Chien-Lung Chan   Yuan Ze University, Taiwan
Jen-Chun Chang   National Taipei University, Taiwan
Cheng-Ta Huang   Oriental Institute of Technology, Taiwan
Changpo Chiang   National Police Agency, Taiwan

Local Arrangement Chairs

Chih-Ho Chou   National Center for Cyber Security Technology,
                                       Taiwan
Jung-San Lee   Feng Chia University, Taiwan
Hao-Kuan Tso   Chien Hsin University of Science
                                       and Technology, Taiwan
Min-Yi Tsai   E-Security Analysis and Management
                                       Association, Taiwan
Sheng-Chih Ho   Department of Information Management,
                                       Management College, National Defense
                                       University, Taiwan
Chung-Nan Chen   Institute for Information Industry, Taiwan

Organizing Chairs

Chien-Lung Hsu	Chang Gung University, Taiwan
Chung-Fu Lu	Chihlee University of Technology, Taiwan
Jongsung Kim	Kookmin University, Korea
Chao-Lung Chou	Chung Cheng Institute of Technology, National Defense University, Taiwan

Program Committees

Winston Seah	Victoria University of Wellington, New Zealand
Jung-Shian Li	National Cheng Kung University, Taiwan
Geyong Min	University of Exeter, UK
Wen-Chuan Wu	Aletheia University, Taiwan
Athanasios V. Vasilakos	Lulea University of Technology, Sweden
Hsi-Chung Lin	Aletheia University, Taiwan
I-Ching Hsu	National Formosa University, Taiwan
Zhao Yao	Institute of Information Science, Beijing Jiaotong University, China
Bo-Chao Cheng	National Chung Cheng University, Taiwan
Hyunguk Yoo	University of New Orleans, USA
Chi-Chao Chang	Chang Jung Christian University, Taiwan
Liwei Chen	Zhengzhou University, China
Te-Yu Chen	National Tainan Junior College of Nursing, Taiwan
Ken Choi	Illinois Institute of Technology, USA
Chih-Ho Chou	National Center for Cyber Security Technology, Taiwan
Rashid Mehmood	King AbdulAziz University, Saudi Arabia
Da-Ren Chen	National Taichung University of Science and Technology, Taiwan
Yi-Chao Wu	Chihlee Institute of Technology, Taiwan
Yen-Wen Chen	National Central University, Taiwan
Richard Chbeir	Universite de Pau et des Pays de l'Adour, France
Cheng-Chi Lee	Fu Jen Catholic University, Taiwan
Hung-Yu Chien	National Chi Nan University, Taiwan
Donghoon Chang	IIIT Delhi, India
Narn-Yih Lee	Southern Taiwan University of Science and Technology, Taiwan
Yujie Wang	University of Electronic Science and Technology of China
Chu-Hsing Lin	Tunghai University, Taiwan
Ji-Han Jiang	National Formosa University, Taiwan
Wei Ping Chang	Central Police University, Taiwan
Taeshik Shon	Ajou University, Korea

Wei-Hua He	Soochow University, Taiwan
Gwoboa Horng	National Chung Hsing University, Taiwan
Yu-Chen Hu	Providence University, Taiwan
Wen-Shenq Juang	National Kaohsiung First University Science and Technology, Taiwan
Aneesh Sharma	University of California, Berkeley, USA
Jin-Cherng Lin	Tatung University, Taiwan
Jia-Ning Luo	Ming Chuan University, Taiwan
Ali Shahrabi	Glasgow Caledonian University, UK
Jau-Ji Shen	National Chung Hsing University, Taiwan
Wei-Liang Tai	Chinese Culture University, Taiwan
Seokjun Lee	Kennesaw State University, USA
Woei-Jiunn Tsaur	National Taipei University, Taiwan
Jonathan Kavalan	University of Florida, USA
Ming-Hour Yang	Chung Yuan Christian University, Taiwan
Ray-Lin Tso	National Chengchi University, Taiwan
Luis Javier Garcia Villalba	Complutense University of Madrid, Spain
Chung-Ming Ou	Kainan University, Taiwan
Hai-Cheng Chu	National Taichung University of Education, Taiwan
Bingwen Feng	School of Cyberspace Security, Jinan University, China
Yu-Chi Chen	Yuan Ze University, Taiwan
Yue-Shan Chang	National Taipei University, Taiwan
Marco Aiello	University of Groningen, The Netherlands
Kwok-Yan Lam	Tsinghua University, China
David Naccache	Gemplus, France
Xuejia Lai	Shanghai Jiao Tong University, China
Wei-Jen Wang	National Central University, Taiwan
Ryma Abassi	University of Carthage, Tunisia
Samia Bouzefrane	CEDRIC Lab, Conservatoire National des Arts et Métiers, France
Serge Chaumette	LaBRI, University of Bordeaux, France
Emmanuel Conchon	University of Limoges, France
Gerhard Hancke	City University of Hong Kong, Hong Kong
Jose Onieva	University of Malaga, Spain
Henrich C. Pöhls	Institute of IT-Security and Security Law, Universität Passau, Germany
Akka Zemmari	LaBRI, University of Bordeaux, France
Stefanos Gritzalis	University of the Aegean, Greece
Pierre-François Bonnefoi	University of Limoges, France
Andrew Kusiak	The University of Iowa, USA
Brahim Hamid	University of Toulouse, France
Yi-Chao Wu	National Taitung University, Taiwan

Contents

About the Editors

Lakhmi C. Jain, PhD, ME, BE(Hons) is with the University of Technology Sydney, Australia; Liverpool Hope University, UK; and KES International, UK. He is Fellow of the Institution of Engineers Australia.

He founded the KES International for providing a professional community the opportunities for publications, knowledge exchange, cooperation and teaming. Involving around 5,000 researchers drawn from universities and companies worldwide, KES facilitates international cooperation and generates synergy in teaching and research. KES regularly provides networking opportunities for professional community through one of the largest conferences of its kind in the area of KES. http://www.kesinternational.org/organisation.php.

Sheng-Lung Peng is Professor of the Department of Computer Science and Information Engineering at National Dong Hwa University, Hualien, Taiwan. He received the BS degree in mathematics from National Tsing Hua University and the MS and PhD degrees in computer science from the National Chung Cheng University and National Tsing Hua University, Taiwan, respectively. He is Honorary Professor of Beijing Information Science and Technology University of China and Visiting Professor of Ningxia Institute of Science and Technology of China. He serves Director of the ICPC Asia Taipei-Hsinchu Site and Director of Institute of Information and Computing Machinery, of Information Service Association of Chinese Colleges and of Taiwan Association of Cloud Computing. He is also Supervisor of Chinese Information Literacy Association and of Association of Algorithms and Computation Theory in Taiwan. His research interests are in designing and analyzing algorithms for bioinformatics, combinatorics, data mining and networks. He has edited several special issues at journals, such as soft computing, Journal of Real-Time Image Processing, Journal of Internet Technology, Journal of Computers, MDPI algorithms and so on. He published over 100 international conferences and journal papers.

Shiuh-Jeng Wang (http://hera.im.cpu.edu.tw/sjw_2006/) was Visiting Scholar in the Computer Science Department at Florida State University (FSU), USA, in 2002 and 2004. He also was Visiting Scholar in the Department of Computer and Information Science and Engineering at University of Florida (UF) in 2004, 2005, 2010 and 2011. He served Editor in Chief of the Journal of Communications of the CCISA in Taiwan from 2000 to 2006 and Director in Chinese Cryptology and Information Security Association (CCISA, http://www.ccisa.org.tw/) from 2000 to 2015. He was, then, with the Vice President of CCISA in 2012–2015. He academically toured the CyLab with School of Computer Science in Carnegie Mellon University, USA, in 2007 for international project collaboration inspection.

He has authored 15 books (in Chinese versions). The partial list as follows: Information Security, Cryptography and Network Security, State of the Art on Internet Security and Digital Forensics, Eyes of Privacy – Information Security and Computer Forensics, Information Multimedia Security, Computer Forensics and Digital Evidence, Computer Forensics and Security Systems, Computer and Network Security in Practice, Image Processing with Applications, Digital Forensics in Mobile Service with Evidence Investigations and Science Book in Fundamental in Cryptography Learning, all published in 2003, 2004, 2006, 2007, 2009, 2012, 2013, 2015, 2016 and 2019, respectively. He has published over 300 papers in referred journals/conference proceedings/technique reports so far. He is Full Professor and Member of the IEEE, ACM. He served a lot of academic and reputable journals in the position of guest-editor, including the lead guest editor at the issue of Advances in Digital Forensics for Communications and Networking - IEEE J-SAC (IEEE Journal on Selected Areas in Communications), at http://www.comsoc.org/jsac, http://ieeexplore.ieee.org/document/5963154/ and the issue of Security and Privacy Information Technologies and Applications for Wireless Pervasive Computing Environments—Information Sciences at http:s//www.sciencedirect.com/science/article/pii/S0020025515005332, http://www.sciencedirect.com/science/journal/002 00255/321

He is currently with the President of E-Security Analysis and Management Association at http://esam.nctu.me/esam/, the Editor-in-Chief of Journal at JITAS (http://jitas.im.cpu.edu.tw/) and the chairs with ICCL (Information Cryptology and Construction Lab., http://hera.im.cpu.edu.tw/) and SecForensics (Intelligence and Security Forensics Lab., http://www.secforensics.org/).

Cryptography, Multimedia Security and Artificial Intelligence

Embedding Advanced Persistent Threat in Steganographic Images

Hung-Jui Ko[1], Cheng-Ta Huang[2], Gwoboa Horng[1],
and Shiuh-Jeng Wang[3(✉)]

[1] Department of Computer Science and Engineering,
National Chung-Hsing University, Taichung City, Taiwan
[2] Department of Information Management, Oriental Institute of Technology,
New Taipei City, Taiwan
[3] Department of Information Management, Central Police University,
Taoyuan City, Taiwan
dopwang@gmail.com

Abstract. The main characteristic of Advanced Persistent Threats is the stealthy and long period of attack. These threats usually aim at stealing secure information and include six phases: (i) Reconnaissance, (ii) Delivery, (iii) Initial intrusion, (iv) Command and control, (v) Lateral movement, and (vi) Data exfiltration. This paper proposes an APT attack scheme that exploits image steganography, in which a malicious code is embedded into a digital image. The stego-image can be triggered by an extractor program. The most common methods of delivering malicious code to target include phishing, watering hole attack, removable drive, compromised control system, and provided system service; these can all be detected. In contrast, schemes using our steganography-based approach are less likely to be detected since there is no detecting system/software that analyzes pixel values of an image. The behavior and signature of the extractor program are very different with the other malwares. The steganography is always used to deliver secret message by building a covert channel, but this technique can also be used to hide malware. In our experiments, both stego-image analysis and extractor program are tested. We present the more efficient detection methods to countermeasure advanced persistent threat in this paper.

Keywords: Steganography · Stegosploit · Advanced Persistent Threats · Evidence investigations

1 Introduction

With the emergence of information age, the core functions of many organizations, such as financial transactions and secure information storage, are built and managed by information systems. But some information systems are vulnerable to hacking. Walker [22] considers hacking to include the following phases (as shown in Fig. 1):

© Springer Nature Switzerland AG 2020
L. C. Jain et al. (Eds.): SICBS 2019, AISC 1145, pp. 3–17, 2020.
https://doi.org/10.1007/978-3-030-46828-6_1

(i)   Reconnaissance: This involves the gathering of target information, and it includes Passive Reconnaissance and Active Reconnaissance, whereby both differ on the basis of direct contact with the target. Passive Reconnaissance involves publicly accessible information, such as information provided by social media, public websites, and search engines such as Google and the famous Google Hacking Database (GHDB) [26]. Active Reconnaissance is about methods that involve direct interaction with targets, such as social engineering.

(ii) Scan and Enumeration: This deals with deeply scanning the internal network of a target to find the open ports, operating system, service information, and vulnerabilities of each host. It also involves drawing the network diagrams and finally enumerating all the information to enable attackers design attack procedures.

(iii) Gaining Access: In this phase, hackers use the exploits to perform actual attacks on a target.

(iv) Privilege escalation: It involves elevating privileges of the system so that attackers can access protected resource or information, such as administrator in Windows OS or root in Unix-like OS. It depends on the demands of attackers, and not every attack includes this phase.

(v)   Maintain Access: It involves maintaining access of a system until an attacker achieves his goal, such as data exfiltration, and system destruction. The common methods are the installation of back-door program on a target and creation of a privileged account to enable an attacker access the system without exploiting vulnerabilities.

(vi) Covering tracks: This occurs after an attack, whereby an attacker clears related records such as system log or command history to stay stealthy.

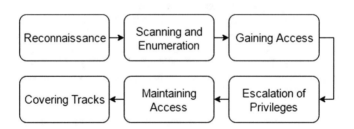

Fig. 1. Hacking phases

Furthermore, a new threat "stegosploit" appeared. "stegosploit" means that exploits vulnerability by steganography during an attack. Steganography is a technique of delivering secret message through building a covert channel, and it uses different carriers to accommodate data such as text, image, audio, and video information [10, 11]. Another application of steganography is watermarking, whereby owner copyright identification is embedded with the host image [17]. Image steganography can be briefly divided into two subfields: spatial domain and transform domain, which involve various algorithms [12, 14, 15, 20, 23, 24]. There are both lossy and lossless methods of each aforementioned subfield. The lossless data hiding technique is known as "Reversible Data Hiding". We can fully restore the host image after data extraction by this technique. The concept of spatial domain is to take the host image as a two-dimensional matrix with different elements (pixels). The frequency domain is performed by implementing transformation on spatial domain using different functions such as Discrete Fourier Transform, Discrete Cosine Transform, and Discrete Wavelet Transform domains.

We use steganography as a malware hosting and delivery method to avoid detection and examine its pattern. This concept is discussed deeper in the rest of this paper, which is organized as follows: Sect. 2 briefly introduces related works; Sect. 3 describes our Advanced Persistent Threat (APT) attack scheme based on steganography; Sect. 4 demonstrates the experimental results; and we concludes this paper in Sect. 5.

2 Related Work

To understand how the steganography is used as APT attack, we introduce the various phases of APT attack, and then we discuss common techniques of steganography and make a deep introduction of Stegosploit.

2.1 Advanced Persistent Threat

National Institute of Standards and Technology defines APT: "An adversary with sophisticated levels of expertise and significant resources, allowing it through the use of multiple different attack vectors (e.g., cyber, physical, and deception) to generate opportunities to achieve its objectives, which are typically to establish and extend footholds within the information technology infrastructure of organizations for purposes of continually exfiltrating information and/or to undermine or impede critical aspects of a mission, program, or organization, or place itself in a position to do so in the future; moreover, the advanced persistent threat pursues its objectives repeatedly over an extended period of time, adapting to a defender's efforts to resist it, and with determination to maintain the level of interaction needed to execute its objectives [16]."

Chen et al. [4] summarized four characteristics of APTs: (i) specific targets and clear objectives; (ii) highly organized and well-resourced attackers; (iii) a long-term campaign with repeated attempts; and (iv) stealthy and evasive attack techniques. In [4], Chen et al. also summarized a six-phase model of APT attacks from [9]:

(i) Reconnaissance and weaponization: This phase does information gathering and preparation using tools for different attack methods. It includes scanning and enumerating devices and services of a target and employing techniques such as Social Engineering [29] and Open Source Intelligence Techniques [3] to acquire useful information for penetration.

(ii) Delivery: This is the phase where attackers deliver their exploits to the targets. There are different types of delivery mechanisms, such as spear phishing [8] and watering hole attack. Each attack is discussed in the following section.

(iii) Initial intrusion: This is the first unauthorized access to a target's device/network, mainly to make sure the attacker can maintain his control of the targets, which is usually by placing a backdoor malware or creating accounts for access of long-term campaign.

(iv) Command and control: This is the phase of controlling the compromised devices and exploiting of the network. For example, the attackers implanted malicious code (e.g. TCP reverse shell) in the victim system. The malicious code would access attackers' command and control system on port 80. The traffic flow would be considered as normal access for firewall system. But attackers can exploit these malicious codes to create accounts and escalate privilege for long-term hacking.

(v) Lateral movement: Chen et al. [4] summarized main activities of this phase as (a) performing internal reconnaissance to map the network and acquire intelligence, (b) compromising additional systems in order to harvest credentials and gain escalated privileges, and (c) identifying and collecting valuable digital assets such as development plans.

(vi) Data exfiltration: It is one of the main purposes of APT attack, and it involves stealing sensitive information from a target, such as personal information and secure document of an enterprise or government.

In the delivery phase of APT attacks, there are many methods to distribute exploits to targets:

(i) Phishing: It is a social-engineering attack that uses spoofing email to lure people to share sensitive information or download malware. Spear phishing, which customizes content, is more selective on their target, so it looks more realistic and harder to detect. Phishing generally involves emails, URLs, and websites and works based on several factors, such as (a) user's lack of computer system knowledge, (b) visual deception, and (c) user's lack of attention to security indicators [5]. Nowadays, phishing uses different kinds of techniques, such as data mining, machine learning, and online learning, to improve the accuracy of detection methods [1, 2].

(ii) Watering Hole Attack: This involves compromising a website that is regularly visited by a target and abusing the trust of the target toward the websites. Attackers usually exploit browsers or browser plug-in related vulnerabilities.

(iii) Removable Device: In this method, malware is delivered to a target directly, and this only works with lack of proper physical protection of the target. Some serious APT attack such as Stuxnet [13] gets its initial intrusion by such method.

(iv) Compromised Control System: The control system is set to distribute update-patch to other host, but hackers utilize this feature to deliver their malicious codes to targets. This method is used in the APT attack of First Bank ATM heist [25] to deliver the malware for manipulating ATMs to dispense banknotes.

(v) Provided System Service: The malware is transmitted using services provided by targets. Vulnerabilities such as CVE-2017-7494 deliver arbitrary shared-object and execute using SMB service provided by the target server.

2.2   Simplest Image Steganography: Least Significant Bit (LSB) Embedding

In this paper, we focus on image steganography and try to embed a binary file into an image in order to avoid detection.

Spatial domain is the simplest and most used domain for data hiding. The simplest method of spatial domain is LSB embedding [18], whereby a color image can have three color channels (red, green, blue), and each channel is composed of one byte: an 8-bit value, which produces 256 discrete levels of color and a gray-scale image with one channel of color black. The 8-bit value can be represented as a bit-plane:

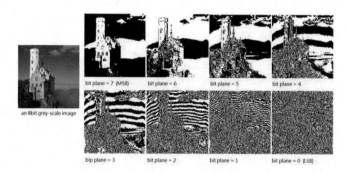

Fig. 2. Bit plane of an 8-bits grey-scale image

In Fig. 2, we have an original image and its bit-plane. In the bit plane 0 (the LSB), we embed the secret message by simply replacing it with the bit-stream message.

There are also surveys about how to detect steganography, which are called "Steganalysis." In spatial domain, there are some basic detection methods and tools such as chi-square attack [21] and RS-diagram Attack [6]. In our experiment, we implemented the chi-square attack using a tool provided by Guillermito [27].

2.3   Stegosploit

The term "stegosploit" was introduced by Saumil Shah in 2015 Black Hat Europe [30]. It means to use images as a host of malicious payload to attack and exploit vulnerabilities. He had shown how to exploit CVE-2014-0282 by stegosploit. In his definition, "stegosploit" includes two subfields: steganography and polyglot. We have already discussed steganography in Sect. 2.2, and Shah implemented his steganography method using a feature of HTML5 called CANVAS. Shah defined polyglot as a combination of HTML and image, which is implemented by JavaScript, HTML, and modifying of the image header. Both methods provided by Shah are based on browser vulnerabilities.

In Sect. 5 of Shah's demonstration he shows some brief introduction of the attack model using "stegosploit":

(i) Reaching the target browser, which is implemented by hosting a web server, exploiting on a URL shortener, and uploading the image on 3rd party websites.
(ii) Content sniffing, which mainly uses the behavior of browsers when recognizing data formats.
(iii) Time-shifted exploit delivery, which uses URL as a trigger to attack.

2.4 Malware Analysis

There are three main analyzing methods nowadays: static- analysis, dynamic-analysis, and machine learning [7]. Static- analysis examines the malware without executing; it usually focuses on string signature, byte-sequence n-grams, syntactic library-calls, control flow graph, opcode frequency distribution, etc. Dynamic-analysis focuses on the behavior of malware while executing it in a controlled environment. It is time-intensive and resource-consuming, and sometimes different virtual environments result in different malware performances. Some behaviors are only triggered under specific circumstance and cannot be detected in virtual environment. Many approaches have been proposed for machine learning to analyze malware: Association Rule, Support Vector Machine, Decision Tree, Random Forest, Naive Bayes, Clustering, etc. Some static features of malware are used for malware classification and detection: strings, byte sequence n-gram, DLL (list of DLLs used by the binary, DLL function-calls and system-calls used within each DLL).

In our proposed scheme, we exploit steganography to deliver malicious payload to avoid detection and analyze the host-image and extraction of binary in order to explore patterns of the scheme (Table 1).

Table 1. Optional approaches for different phases to implement the scheme.

Preparation (creating stego-image)	Preparation (setting trigger)	Implantation	Activation	Extraction
[19]	CVE-2014-8517	Phishing	URL pointing [17]	[19]
[27]	CVE-2017-12477	Watering Hole Attack	Init.d	[27]
[28]	CVE-2017-7494	Compromised Control System	System Task Scheduler	[28]
[29]	CVE-2017-12478	Removable Drive	-	[29]
[30]	-	Provided System Service	-	[30]

3   Proposed Scheme

Our APT attack scheme is constructed by the following five phases: preparation, implantation, activation, extraction, and execution.

A. Preparation phase

This phase includes two tasks:

(i) Creating the stego-image: Steganography algorithms are used to embed malicious payload into host image and obtain the stego-image.
(ii) Setting trigger on target: This is for the payload extraction. After the stego-image is delivered into a target's machine, it needs a extractor to extract payload from stego-image and execute the malicious payload.

B. Implantation phase

Here, we deliver the stego-image and extractor into the target. As we have discussed in Sect. 2.1, there are lots of approaches to achieve the implantation.

C. Activation phase

This phase includes detection and condition identification. The trigger sets in preparation phase, detects the existence of stego-image and other environmental setting to determine whether to extract the malicious payload.

D. Extraction phase

When the activating condition is matched, the extractor will start extracting payload from the stego-image. One who does not know the steganography algorithm and extractor cannot detect the existence of the malicious payload and its true functionality, which results in longer latency of the attack.

E. Execution phase

This phase executes the malicious payload, and its execution is related to the activation or extraction phases. The designer decides how to evoke/control execution.

4   Experiment and Analysis

4.1   Experiment Environment

The environment of this experiment is described as follows:

Both attacker and victim are hosted on ORACLE VirtualBox 5.1.26 r117224 (Qt5.6.2). In the attacker's side, the OS is kali-linux-2017.1-amd64, and we use binaries provided by Metasploit framework console 4.15.5-dev to exploit CVE- 2017-7494, and we use gcc version 6.4.0 to compile our source code: including P1, P2 mentioned in Sect. 3 and the steganography tools. In the victim's side, the OS is ubuntu- 16.04.3-server-amd64, and we have a samba server without patch, which can be acquired at [28].

10   H.-J. Ko et al.

4.2 Implementation of the Scheme

Figure 3 explains how we implement the proposed scheme, and its variables are explained as follows: S is the stego-image which contains malicious payload; E is the extractor, used for extracting malicious payload from S; P1 is the malicious payload used in the experiment, which is capable of performing connection request and providing a built-in shell for remote controller; and P2 is the binary used by attacker to communicate with P1 and get remote control of the target.

In preparation phase, we create the stego-image by LSB algorithm and exploit CVE-2017-7494 on the target to modify its system task scheduler (crontable) for setting trigger. In implantation phase, since the example target hosts samba server, we deliver the stego-image and extractor by uploading to share folder of the server. In activation phase, the trigger will detect whether the stego-image is in the machine at a specific time: if not, it would wait for the next time of detection; if yes, it would perform the extraction. In extraction phase, malicious payload will be extracted and then executed by task scheduler to perform connection request. After connection is built, P1 will send a message to P2 with information of the victim device, and then P1 resumes its built-in shell for P2 for further attack.

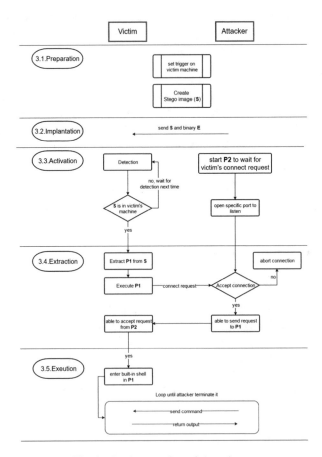

Fig. 3. Implementation of the scheme

4.3   Analysis

The two main files in the proposed scheme are the stego-image (image with malicious payload) and the extractor. In the stego-image analysis, we analyze chi-square attack and LSB average to explain the characteristics of stego-images. When analyzing the extraction of binary, we use tools to examine library-calls and system-calls to realize the characteristics of the extractor.

(a) The Stego-Image Analysis

Stego-image is the host of malicious payload. Performing common static-analysis on an image file is nothing suspicious. In spatial-domain steganography, such as LSB of pixel value is widely used in order to reduce image distortion, regardless of if LSB embedding algorithm is adopted [18]. Likewise, using high- frequent coefficient in frequency-domain steganography has the same effect. To explain the concept, we analyze the following 24-bits BMP color images (Fig. 4):

Fig. 4. Experiment image: (a) Emma (b) Wildflower (c) Lena

The distribution of LSBs in an image can be random or with some specific patterns that depend on the image. For example, the LSBs of a pure black image will definitely be 0, without randomness. An image with specific file embedded in LSBs will show some specific patterns, which destruct the original pattern of LSBs in the image, so that it becomes detectable.

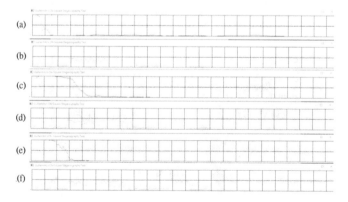

Fig. 5. Experiment result of using Chi-square attack tool: (a) original Emma (b) stego-Emma (c) original Lena (d) stego-Lena (e) original wildflowers (f) stego-wildflowers

In Fig. 5, stego-image is the image with binary embedded in LSB, and the original image is the unmodified one. The red lines in Fig. 5 indicate the level of randomness of LSBs. Each spot is calculated in 128-bytes blocks, which gives the green points. The green points indicate the average of LSBs in each 128-bytes block. In common situation, numbers of 1 and 0 should be similar, that is, an average value around 0.5.

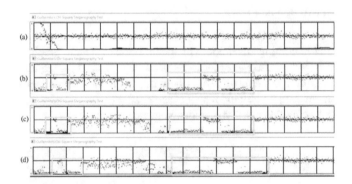

Fig. 6. (a) Original Emma (b)(c)(d) Emma with different binaries

To clarify the pattern, we separately embedded three different binaries into one image and the result is showed in Fig. 6. In the yellow rectangles, all three binaries show the same pattern, and this comes from the ELF header and section header table of Executable and Linkable File (ELF).

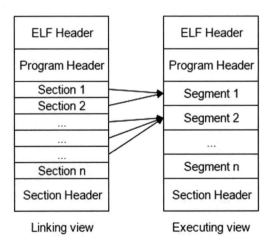

Fig. 7. ELF structure

As seen in Fig. 7, an ELF has a linking view and executing view. Linking view shows the structure of non-executed ELF, and executing view shows the structure of executed ELF. Section header records how sections are assigned to segments, and program header records relative address of segments in memory.

(b) **The Extractor Analysis**

In this section, we use two Linux tools, "strace" version 4.19 and "ltrace" version 0.7.3, to trace the system-calls and library-calls used by the extractor. The tool "strace" intercepts and records the system-calls and signals of the program. We can monitor the activities of system-calls. The tool "ltrace" can intercept and record the dynamic library-calls which are used by the program. We also can monitor the system-calls of the program by "ltrace".

In system-call analysis, we compare the output of the extractor with a normal program by "strace". We can see there are unique system-calls only used by the extractor, such as "lseek", "rt_sigaction", "rt_sigprocmask", "clone" and "wait4".

The system-call "lseek" is used to change the current file offset. The extractor uses "lseek" to locate the positions of the malicious payload in stego-image. Before explaining "rt_sigaction" and "rt_sigprocmask," we have to make a brief introduction of relation thread, process, and signal (Table 2):

Table 2. System-Calls statistics of extractor from strace

System-Call	Called times
read	6
write	24
Open	5
Close	5
fstat	6
lseek	1
mmap	8
mprotect	6
munmap	1
brk	3
rt_sigaction	4
rt_sigprocmask	2
access	4
clone	1
execve	1
wait4	1
arch_prctl	1

As Fig. 8 depicts, signal can change the action of the process. Each process has its own signal mask. When a process receives the signal, the signal mask would determine which thread should act. The system-call "rt_sigaction" is used to change the action taken by a process on receipt of a specific signal, and "rt_sigprocmask" is used to fetch and/or change the signal mask of the calling thread. The extractor uses these system-calls to change the action of the process.

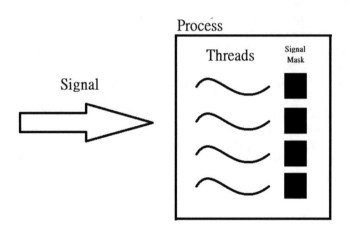

Fig. 8. Relation between signal and thread

In our experiment, we did not use multi-thread. We observed the system-calls forked a child process. The "rt_sigaction" and "rt_sigprocmask" is immediately called after "write" system-call. The system-call "write" had created a file, the "rt_sigaction" and "rt_sigprocmask" execute the file.

The sequence of system-calls to fork (clone) a process is as follows:

```
...
rt_sigaction(SIGINT, ...) rt_sigaction(SIGQUIT, ...)
rt_sigprocmask(SIG_BLOCK, ...) clone
wait4
rt_sigaction(SIGINT, ...) rt_sigaction(SIGQUIT, ...)
rt_sigprocmask(SIG_SETMASK, ...)
...
```

We can see the "rt_sigaction" and "rt_sigprocmask" are used to interrupt and block the current process, respectively, and then clones (forks) the child process with "clone" and waits for child process to finished with "wait4." Finally, it interrupts the process again and unblocks the original process with "rt_sigaction" and "rt_sigprocmask," respectively.

In library-calls analysis, we focus on the abnormal library-calls. When the extractor performs payload extraction, it would take times to arithmetic computing. The arithmetic computing would use the specific library-calls. The size of malicious payload be the cue of the count of the specific library-calls. In our experiment, the malicious payload size is 13504 bytes (108032 bits). The specific library-call is the power function (pow in Table 3), it is used by the extractor frequently.

Table 3. Library-Calls statistics of extractor from ltrace

% time	Library-call	Called times
98.99	POW	108032
0.18	System	1
0.02	Puts	7
0.01	Malloc	7
…	…	…

To summarize this section

(a) Stego-image analysis: We realized that embedded payload in an image will cause distortion to original image and destruct randomness of LSB-distribution. Furthermore, if we know the steganography algorithm implemented, we can still reconstruct patterns of the embedded payload.

(b) Extractor analysis: We realized the malicious payload extraction, the extractor would need to locate specific position of the image file and use power function to perform arithmetic computing so that it can reconstruct payload from bit-stream information.

5   Conclusion

In this paper, we introduce the characteristics and procedure of APT attack and the basic concepts of steganography and Stegosploit, and then we propose a 5-phase attack scheme based on steganography and enumerate the feasible approaches to achieve this scheme. Finally, we examine the stego-image and extractor to explore patterns. Attacks through this scheme have longer latency and are harder to detect because the characteristics of extractor differ from those of common malware. The malicious payload hide in the pixels of images is not easy to be examined. We could study more Stegosploits in terms of examinations and investigations, by using the representations of experiments, to be our next challenges.

Acknowledgement. We thank Jui-Tai Wang for performing these experiments. This research was partially supported by the Ministry of Science and Technology of the Republic of China under the Grant MOST 107-2221-E-015-001-MY2-.

References

1. Aburrousa, M., Hossaina, M.A., Dahala, K., Thabtahb, F.: Intelligent phishing detection system for e-banking using fuzzy data mining. Expert Syst. Appl. **37**(12), 7913–7921 (2010)
2. Blum, A., Wardman, B., Solorio, T., Warner, G.: Lexical feature based phishing URL detection using online learning. In: AISec '10 Proceedings of the 3rd ACM Workshop on Artificial Intelligence and Security, pp. 54–60, October 2010
3. Bazzell, M.: Open Source Intelligence Techniques: Resources for Searching and Analyzing Online Information, 5th edn. CCI Publishing, Aarhus (2016). ISBN-13 978-1530508907
4. Chen, P., Desmet, L., Huygens, C.: A study on advanced persistent threats. In: International Federation for Information Processing (IFIP) International Conference on Communications and Multimedia Security, pp. 63–72 (2014)
5. Dhamija, R., Tygar, J.D., Hearst, M.: Why phishing works. In: CHI '06 Proceedings of the SIGCHI Conference on Human Factors in Computing Systems, pp. 581–590, April 2006
6. Fridrich, J., Goljan, M., Du, R.: Detecting LSB steganography in color, and gray-scale images. IEEE MultiMed. **8**(4), 22–28 (2001)
7. Gandotra, E., Bansal, D., Sofat, S.: Malware analysis and classification: a survey. J. Inf. Secur. **5**(2), 56–64 (2014)
8. Hong, J.: The state of phishing attacks. Commun. ACM **55**(1), 74–81 (2012)
9. Hutchins, E.M., Cloppert, M.J., Amin, R.M.: Intelligence-driven computer network defense informed by analysis of adversary campaigns and intrusion kill chains. Lead. Issue Inf. Warf. Secur. Res. **1**, 80–106 (2011)
10. Kayarkar, H., Sanyal, S.: A survey on various data hiding techniques and their comparative analysis. ACTA Technica Corviniensis **5**(3), 35–40 (2012)
11. Kamila, S., Roy, R., Changder, S.: A DWT based steganography scheme with image block partitioning. In: International Conference on Signal Processing and Integrated Networks (SPIN) (2015)
12. Kittawi, N., Al-Haj, A.: Reversible data hiding in encrypted images. In: International Conference on Information Technology (ICIT) (2017)
13. Kushner, D.: The real story of stuxnet. IEEE Spectr. **50**(3), 48–53 (2013)
14. Li, X., Li, J., Li, B., Yang, B.: High-fidelity reversible data hiding scheme based on pixel-value-ordering and prediction-error- expansion. Signal Process. **93**(1), 198–205 (2013)
15. Mao, Q., Li, F., Chang, C.-C.: Reversible data hiding with oriented and minimized distortions using cascading trellis coding. Inf. Sci. **317**, 170–180 (2015)
16. Ross, R.S.: Managing information security risk: organization, mission, and information system view. National Institute of Standards and Technology (NIST) Report Number 800-39, March 2011
17. Saini, L.K., Shrivastava, V.: A survey of digital watermarking techniques and its applications. In: International Joint Conference On Science And Technology (IJCST), May–June 2014, vol. 2(3), pp. 70–73
18. Sutaone, M.S., Khandare, M.V.: Image based steganography using LSB insertion technique. In: IET Conference on Wireless, Mobile and Multimedia Networks, pp. 146–151 (2008)
19. Thakur, K., Kopecky, S., Nuseir, M., Ali, M.L., Qiu, M.: Advanced Persistent Threats: Behind the Scenes. In: IEEE 3rd International Conference on Cyber Security and Cloud Computing (2016)
20. Tian, J.: Reversible data embedding using a difference expansion. IEEE Trans. Circuits Syst. Video Technol. **13**(8), 890–896 (2003)

21. Westfeld, A., Pfitzmann, A.: Attacks on steganographic systems. In: Breaking the Steganographic Utilities EzStego Jsteg Steganos and S- Tools—and Some Lessons Learned. Lecture Notes in Computer Science (LNCS), vol. 1768 (1999)
22. Walker, M.: CEH Certified Ethical Hacker All-in-One Exam Guide, 2nd edn. McGraw-Hill Education, New York (2011). ISBN-13 978-0071772297
23. Yang, C.-H.: Reversible data hiding in encrypted image based on frequency domain. In: Cryptology and Information Security Conference (CISC) (2014)
24. Zhicheng, N., Shi, Y.Q., Ansari, N., Su, W.: Reversible data hiding. IEEE Trans. Circuits Syst. Video Technology **16**(3), 354–362 (2006)
25. iThome: First Bank ATM Heist of Taiwan. https://www.ithome.com.tw/news/107294 (2018). Accessed 17 Jan 2018
26. Exploit Database: Google Hacking Database. https://www.exploit-db.com/google-hacking-database (2018). Accessed 19 Jan 2018
27. Guillermito: Chi-square Steganography Test Tool. http://www.guillermito2.net/stegano/tools/ (2018). Accessed 19 Jan 2018
28. Samba.org.: Samba 4.5.9 Available for Download. https://www.samba.org/samba/history/samba-4.5.9.html (2018). Accessed 19 Jan 2018
29. SANS Institute InfoSec Reading Room: Social Engineering: A Means To Violate A Computer System. https://www.sans.org/reading-room/whitepapers/engineering/social-engineering-means-violate-computer-system-529 (2018). Accessed 19 Jan 2018
30. Shah, S.: Stegosploit Exploit-Delivery with Steganography and Polyglots. http://stegosploit.info/, https://www.blackhat.com/eu-15/briefings.html (2018). Accessed 19 Jan 2018

AI-Based Farm Survey Technique for Efficient Fruit Harvesting

Aditya Indla[1], Aneesh Sharma[2(✉)], and Chi-Yao Weng[3]

[1] Bellarmine College Preparatory, San Jose, USA
[2] University of California, Berkeley, USA
mail@asharma.info
[3] Department of Computer Science, National Pingtung University,
Pingtung, Taiwan

Abstract. Strawberry farming is labor intensive and time critical. Farmers need to harvest ripe fruit within 2–3 days, and the size of their farms makes it extremely difficult to accurately survey the farm and determine the optimal harvest time for strawberries. As a result, farmers employ fixed labor and pick their crop three times a week, often incurring losses due to this inefficient system. Image recognition service and machine learning can be used to solve this problem by analyzing the farm and more accurately determining the optimal time and labor requirements for an efficient harvest. However, off-the shelf image recognition models [1, 2] perform poorly with low accuracy and recall scores when identifying strawberries, as they have not been trained with datasets that reflect farm conditions such as daylight or terrain. The solution proposed here identifies strawberries through a custom Convolutional Neural Network (CNN) trained with images of an actual strawberry farm. The proposed solution provides farmers more information about the state of the farm and can be used in conjunction with a drone or robot to more effectively survey and analyze the farm to plan the harvest procedures. Future work includes using this solution with a fruit-picking robot to further decrease labor costs and increase efficiency.

Keywords: Machine learning · Image recognition · Tensorflow · CNN · Model training · Farm · Strawberries · Robot

1 Introduction

Strawberries have a very small window of ripeness where they should be picked. Today farmers typically pick strawberries with a fixed number of workers on alternate days, to maximize the yield. However, this process is inefficient, incurring high labor costs [3] and wasting over 30% of strawberries as they rot in the field. Rather than having a fixed schedule, if farmers had more information about the ripeness of their strawberries, they could dynamically adjust the size of the workforce and the time of picking to save labor costs and increase yield. Machine learning is uniquely suited to solve this problem, [4] through its ability to rapidly and accurately identify and classify strawberries as ripe or not, on a large scale.

© Springer Nature Switzerland AG 2020
L. C. Jain et al. (Eds.): SICBS 2019, AISC 1145, pp. 18–26, 2020.
https://doi.org/10.1007/978-3-030-46828-6_2

In the following, we will first describe the issues with off the shelf object detection models. Secondly, we will explain the advantages of the proposed solution. Third, we discuss the way the solution works, specifically, SSD (Single Shot Detector) and Inception followed by implementation details. Fourth, we discuss the model and prove its efficacy, most notably in relation with off the shelf models. Finally, we explain how the model can be improved upon in the future, and other possible applications.

2   Status of Existing Solutions

Strawberry detection is a difficult process, and in this paper, we have compared multiple different state of the art models to a custom model.

The models [1, 2] are trained on a generic dataset and they performed well over many classes, but failed to effectively identify strawberries, as they have not been trained on the specific fruit. The Table 1 below demonstrates the results of these models on images of a strawberry farm and the lack of accuracy and precision of the classifications.

Table 1. Incorrect recognition of strawberries

Image	Classification	Model Name
	Apple	ssd_inception_v2_coco_2018_01_28
	None	ssd_mobilenet_v2_coco_2018_03_29

(continued)

Table 1. (*continued*)

Image	Classification	Model Name
	Banana, Broccoli, Person, Elephant, Apple	ssd_mobilenet_v1_coco_11_17
	None	faster_rcnn_resnet101_kitti_2018_01_28

These models are unable to effectively identify the strawberries, instead classifying them as people or other fruits. These models are not trained with images of strawberries at all, making them ineffective.

Additionally, some research has been done into using current systems to identify strawberries. However, these studies often focus on multiple fruits at once, and are not specifically for strawberry farming. Additionally, these models are trained on images of strawberries taken in front of a black tarp, not on a farm. As a result, conditions on a farm, which vary drastically from in a house, may confuse the models [5].

3   Solution Overview

The proposed solution is to create a custom image recognition model that identifies strawberries and categorizes them based on their ripeness with high accuracy and precision. This model is used during or after a survey of the farm to identify the number of ripe strawberries, labeled as strawberry 100%; semi-ripe strawberries, labeled as strawberry 50%; and overripe strawberries, labeled as strawberry 100% + . The system then reports the percentage of ripe strawberries, allowing the farmers to determine the harvest time and labor required, based on this data.

3.1   Custom Model

The model developed here is built on the Single Shot Detector (SSD) Inception model [2, 6]. The Inception module is the base CNN, while SSD is the image detection model built on top of Inception. Like other CNN's, Inception performs convolutional transformations, be they with a 3×3 kernel or a 5×5 kernel. However, unlike other CNN's, Inception performs multiple transformations at the same time, concatenating all the results for the next layers to determine which features are most useful. However, this massively increases the computation costs. Thus, in order to reduce the computational load, Inception uses a 1×1 convolution to compress the information being filtered and make further transformations easier, creating a sort of transformation stack. Inception, as a result of this unique arrangement, Inception requires fewer parameters than other models, only around 5 million, as compared to the dozens of millions required by other models. As a result of this efficiency in transformation, Inception is one of the most used CNN models, especially for object detection [6].

SSD is the object detection model used in conjunction with the Inception framework. SSD works by, first, passing an image through a series of convolutional layers, creating feature maps of different sizes (5×5, 6×6, 3×3, etc.) Within these feature maps, a 3×3 convolutional filter is used to evaluate a set of bounding boxes placed around each object that is detected. For each box, the offset between the actual bounding box and the original box is predicted, as well as the class of each box. Then, during training, the ground truth boxes, or boxes added to the image used for training, will be compared to the bounding boxes produced by the model. All boxes with an Intersection over Union (IoU), or an overlap with the ground truth box, greater than 0.5 will be marked as positives, while most of the rest will be discarded. SSD is unique from most object detection models in that it does not identify areas of interest prior to classifying. Thus, SSD produces many more bounding boxes, a majority of which are negatives. To get around this, for areas with many overlapping boxes, SSD only considers the box with the highest confidence, discarding the rest in a technique known as non-maximum suppression. SSD also uses hard negative mining, which only uses a subset of negative examples, those with the highest training loss (false positives) in each iteration. These negative examples are always kept in a 3:1 ratio with positive examples. These methods allow SSD to be one of the fastest image recognition models while still maintaining a high degree of accuracy, making it extremely easy to create a simple model for farmers to use effectively [7].

3.2   Model-Building Process

The model-building process involved the collection of data from a strawberry farm, the preparation of said data for the model, and the training of the model using this data.

3.3   Collecting and Labeling Data

For this project, a custom model was created using the SSD inception Tensorflow framework. The test images are from videos of one strawberry farm, recorded with an Intel RealSense camera [8] under two configurations: 60FPS with 640×480

resolution and 30FPS with 1280 × 720 resolution. The videos were recorded using multiple angles and at multiple times in order to train the model with differing amounts of sunlight. These images were then labelled using the labelimg [9] tool program as unripe, ripe, or overripe to produce .xml files to be used to train the custom model (Fig. 1).

3.4 Preparing Data

However, Tensorflow requires a specific format of file to be used in training, the *tfrecord* (tensorflow records) filetype. Prior to converting the xml files, however, the raw data of the image labels and the distribution of these files were saved in a .csv file for later reference. Upon converting the xml to *tfrecord* files, the inception framework was trained with a set of training images, which represented ∼80% of the labeled images, or188 images. Test images represented the other 20% of the images, or47 images. The training process created checkpoint files after a certain number of intervals, but these checkpoint files are unusable for testing. Instead, they are transformed into *frozen_inference_graph.pb* files, which can then be used in the testing process (Fig. 2).

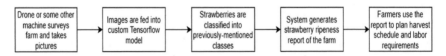

Fig. 1. High level solution overview

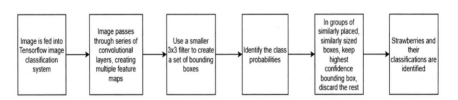

Fig. 2. SSD Inception model identification process

4 Results

In order to determine the accuracy of the model over multiple areas, it was tested on images from four different farms, taken at different times of day. If the model can be accurate over multiple farms, it should be effective for use across the country (Figs. 3 and 4).

The first model was trained with *3,000* intervals, but was still partially inaccurate, mostly due to the use of green unripe strawberries, labeled as strawberry *0%*. These strawberries confused the model, resulting in the classification of leaves as unripe strawberries (Figs. 5 and 6).

Fig. 3. First model test identifying strawberries, but including leaves

Fig. 4. First model test not able to identify strawberries accurately

Upon removing labels for completely unripe strawberries and consolidating the rest under the "strawberry" label, the model became far more accurate. With even more training, this time with *4,000* intervals, the model was able to accurately identify many strawberries. However, it still could not identify *100%* of the strawberries (Fig. 7).

Fig. 5. New model is able to accurately identify one strawberry, but cannot find the other

Fig. 6. Model accurately identifies most strawberries, some are still unlabeled

As the model was able to identify strawberries, the next step was for the model to categorize the strawberries as ripe, unripe or overripe. The *csv* and *tfrecord* files were updated accordingly and the model was trained, this time for a longer period of *25,000* intervals. As a result of this increase in training time, the model was able to far more accurately identify strawberries, classifying more of them than previous models, generating average precision values of *0.69* and average recall values of *0.55* (Fig. 8).

However, there were still some issues with identifying strawberry *+100%*, due to shadows and the black tarp, and identifying strawberry *50%*, due to a lack of data provided to the model. For instance, the model sometimes identified what should be a strawberry *100%* as a strawberry *+100%* due to the shadows making them look similar.

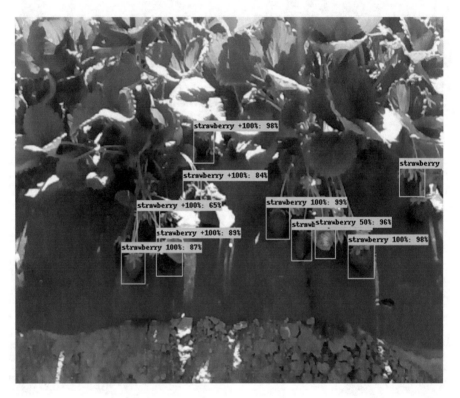

Fig. 7. Model correctly identifies most strawberries, misidentifies one as +100% and misses some

Fig. 8. Model identifies most strawberries accurately

5   Conclusion

Currently, the proposed custom machine learning model can identify ripe strawberries effectively. This would provide some aid to farmers by providing concrete data on the number of ripe strawberries to be picked. However, it struggles with unripe and overripe strawberries, mostly due to a lack of training data. By providing more data sources, such as ripe and unripe strawberries, data bias could be minimized, and the model could become more accurate. Additionally, the model could be trained longer, as increasing the number of intervals increases accuracy as well. The model serves as a good basic system for farmers to use but requires improvements to fully achieve its stated goal. Were it to count the number of strawberries and determine the percentage that are ripe, it would be more useful.

In the future, by training the model with diseased leaves and weeds, thus providing farmers with more information on the health of the farm, the solution will be improved upon. The proposed model could also be used effectively by a fruit picking robot to further help farmers with labor costs. The robot could find ripe strawberries and pick them, and the farmers could pick the few stragglers, providing significant labor savings, making strawberry farming more profitable.

Acknowledge. This work was partially supported by the Ministry of Science and Technology of the Republic of China under the Grant No. 108-2221-E-153-004-MY2 and the Taiwan Information Security Center at Tsing Hua University (TWISC@NTHU).

References

1. Huang, J., et al.: Speed/accuracy trade-offs for modern convolutional object detectors. In: Proceedings of the IEEE conference on computer vision and pattern recognition (2017)
2. Liu, W., et al.: SSD: single shot multibox detector. In: Leibe, B., Matas, J., Sebe, N. (eds) Computer Vision-ECCV (2016)
3. Estabrook, B.: Farmers can't find enough workers to harvest crops-and fruits and vegetables are literally rotting in fields. EatingWell, 29 March 2019. www.eatingwell.com/article/291645/farmers-cant-find-enough-workers-to-harvest-cropsand-fruits-and-vegetables-are-literally-rotting-in-fields/
4. TensorFlow: TensorFlow. https://www.tensorflow.org/
5. Sa, I., et al.: DeepFruits: a fruit detection system using deep neural networks. Sens. (2016). (Basel, Switzerland), MDPI. https://www.ncbi.nlm.nih.gov/pmc/articles/PMC5017387/
6. Xu, J.: An Intuitive Guide to Deep Network Architectures. Medium, To-wards Data Science, 15 August 2017. https://towardsdatascience.com/an-intuitive-guide-to-deep-network-architectures-65fdc477db41
7. Xu, J.: Deep Learning for Object Detection: A Comprehensive Review. Medium, Towards Data Science, 11 September 2017. https://towardsdatascience.com/deep-learning-for-object-detection-a-comprehensive-review-73930816d8d9
8. Intel® RealSense™ Depth Camera D415: "Intel® RealSense™ Depth Camera D415", Intel RealSense Store. http://store.intelrealsense.com/buy-intel-realsense-depth-camera-d415.html
9. Tzutalin. LabelImg: Git code (2015). https://github.com/tzutalin/labelImg

A Lightweight Authentication Stream Cypher Mechanism for Industrial Internet of Things

Shih-Hao Chang[1](✉) and Ping-Tsai Chung[2]

[1] Department of Computer Science and Information Engineering,
Tamkang University, New Taipei City 25137, Taiwan
shhchang@mail.tku.edu.tw
[2] Long Island University, 1 University Plaza, Brooklyn, NY 11201, USA
pchung@liu.edu

Abstract. Cyber security is becoming an increasingly critical issue in auto-mated manufacturing process and particularly for production control engineer-ing. Cyber security in manufacturing process has a massive impact on production management and control systems with the potential effects ranging from production failure, performance degradation to the loss of confidential data. Since smart manufacturing competencies to predicate on levels of technical sophistication, automation and integration far beyond conventional manufac-turing processes. There will be new vulnerable points and the lack of cyber security protection need to be concern. Moreover, as networking in process automation grows, there is a greater risk of cyber attacks from network channels. This paper proposed a lightweight method of industrial data encryption trans-mission and device authentication by applied WG-8 stream cypher mechanism. By improving its pseudorandom number to protect the manufacturing facilities' communication transmission. In order to comply with the actual manufacturing facilities, we utilize the Raspberry Pi3 B$^+$ as experimental test bed for power consuming, real-time data transmission evaluation. The proposed mechanism evaluation result shows that the WG-8 encryption algorithm will lead Raspberry Pi3 central processing unit (CPU) utilization increased 40%, but it empower secure communication and facilities authentication in industrial Internet of Things (IIoT).

Keywords: Industrial Internet of Things · Stream cypher · WG-8 · Power consuming · Real-Time

1 Introduction

Many companies need industrial production management to integrate intercessors of production channels into their supply chain networks. The main reason for automated supply chain networks is to improve the performance of product production, trans-portation, and distribution. This is an end-to-end process related to the movement of products from suppliers to manufacturers, distributors, retailers, and ultimately to end consumers [1, 2]. However, the production supply chain is a complex combination of parties that requires collaboration, coordination and information exchange between them. On the other hand, Industrial Internet of Things (IIoT) is a network of devices

© Springer Nature Switzerland AG 2020
L. C. Jain et al. (Eds.): SICBS 2019, AISC 1145, pp. 27–34, 2020.
https://doi.org/10.1007/978-3-030-46828-6_3

connected via wired/wireless communications technologies to form systems that collect, monitor, exchange and analyze production line data, then deliver valuable insights that enable industrial companies to make smarter business decisions faster. The driving philosophy behind IIoT is those production line machines are better than human at accurately and reliably capturing and communicating real-time data. This data enables companies to pick up on inefficiencies and problems sooner, saving time and money and supporting business intelligence efforts.

To develop an IIoT system in a smart factory will lead thousands of sensors and it is necessary to track down the location of parts in real-time while collecting and analyzing data before arranged to the automation process. As shown in Fig. 1, complex and precise industrial process management systems remain vulnerable to malicious attacks by hackers. Malfunctions of certain facility parts can lead to production safety issues and crash down the quality of products. Among these attack issues, data transmission is very important in smart factories. Due to once the sensor data was modified or stolen, the process of making goods at the factory may cause the factory production system to stop, or to be tampering. An attack was a risk and they could not be ignored after providing production services, since security is required for industrial communications between gear and monitoring servers. An example case is in the sudden occurrence of equipment failure at a petroleum plant in the Middle East at the end of 2017. Resulting in a forced interruption of the production process. Finally, the investigation found that the safety process system was planted. Although the information security algorithms currently in use can solve most of the problems, there is no information security algorithm that can meet the various requirements of smart factories.

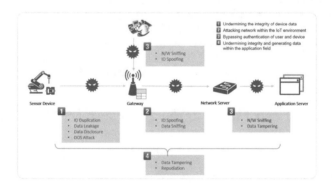

Fig. 1. Cyber security issues in smart factory

The paper is mainly aimed at the needs of the factory, and the encryption algorithm is designed to ensure that the information of the smart factory will not be stolen or falsified, which will lead to the outflow of the factory secrets, and even more serious, the factory will be shut down. Therefore, this paper focuses on how to protect data transmission in IIoT and initiatives like the German Industry 4.0 by means of encryption and verification. The rest of this chapter is organized as follows. In the second section, the background and related work will be briefly introduced. In the third section, we introduced the Link Guard scheme (LGS) and the implementation results will be described in Sect. 4. Finally, Sect. 5 explores our conclusions and future work.

2 Background and Related Works

In the traditional industrial communication system [1], it is based on a wired fieldbus system and must meet the needs of the applications it serves. These fieldbus communication systems are usually different from IT networks and connect multiple industrial gears (slave) with a control device called a master. Recently, as real-time Ethernet based protocols, such as EtherCAT have become more widely used in various fields such as automation systems and motion control, many studies on their design have been conducted [2, 3]. However, these automation systems and motion control will face same threats and almost weakly since their design goals in most cases do not include any reinforcing against security attacks. Representative threats have been appears to manipulate process automation or shutdown by cyber attacks [4], loss of know-how, insufficient product protection or a disclosure of sensitive personal data. For example, there is a high risk that security attacks might cause the severe failures of production operation leading to intense consequences on humans, the environment and machines.

As fieldbus become more integrated on the IT base networks, it seems applicable to reuse the tools designed for web applications in order to controls the robots in the production line. However, the problems that the fieldbus will face are similar to those the computer revolution faced with the widespread of the Internet decades ago. Among the common attacks computers may suffer including denial-of-service, eavesdropping, privilege escalation, spoofing, tampering, or information disclosure, etc. [5]. To face these problems, fieldbus robots add the additional factor of cyber security. Due to taking the control of a desktop computer or a server may result in loss of information, taking the control of a master device may endanger whatever or whoever is near. There are somewhat differences between regular computers communicating through the network, and fieldbus robots. Fieldbus robotics are stateful, multiprocessor, and require a bi-directional communication with the client. These fundamental changes may lead to different tradeoffs and design choices and may ultimately result in different software solutions for web and robotics applications. To these differences, we could also add the real-time constraints that characterize robotics applications. Despite other issues, like software bugs or vulnerabilities such as command injection, buffer overflow, etc. [6], we consider that communications currently are one of the main vulnerabilities in robotics.

Encryption [7] is a mathematical algorithm that utilizing in cyber security for information protection.

Typically there are text, keys and cipher texts. The interception is just a cipher text that cannot be comprehended. Information encryption technology is the use of mathematical or physical means to protect electronic information during transmission and storage to prevent leakage. Due to the leakage of communications, the threat to data security, the problem of decryption and piracy is becoming increasingly serious, and even international disputes have arisen. If the keys are same and used by both parties can be encrypted and decrypted, this encryption method is called symmetric encryption [8, 9]. If keys are different, a pair of keys consists of a public key and a private key. Private key decrypts public key encrypted data and public key decrypts private key encrypted data, this is called an asymmetric encryption algorithm [10, 11]. In cryptography, a block cipher [12, 13] is a deterministic algorithm that operates on a fixed-length group of bits (called a block) and has an invariant transformation specified by a symmetric key. Stream ciphers are symmetric key ciphers in which plaintext digits are combined with a pseudo-random cipher digital stream [14, 15].

In block cipher, to encrypt variable-length data, the data will be separate into blocks. In most cases, the last piece of data also needs to be filled in appropriately way to expand the same length as the fixed cipher block size. The working model describes the process of encrypting each block of data, and normally uses commonly called initial vector as additional inputs. For security reasons, the initial vector values are random. Research on encryption methods [16–18] includes data integrity protection, that is, the error propagation characteristics of passwords caused by certain data being reformatted. Subsequent [19] identified integrity protection as a completely different encryption target that has nothing to do with encryption. Part of the modern working model combines encryption and authentication in an effective way, known as authentication encryption mode. Although the working mode is usually applied to symmetric encryption, it can also be applied to public key encryption.

Stream encryption is different from block ciphers, which are bit-level encryption algorithms. Both encryption and decryption use a constant pseudo-random number as the key. The pseudo-random code [21] represents the randomness of the numbers. However, because the number of cycles is too large, the sequence size of a pseudo-random column in a cycle is greater than 1040, so the cycle value cannot be seen [22]. Each time the plaintext data is encrypted with the key data stream, a ciphertext data stream is obtained. Although block encryption is more widely used than stream encryption, but in the smart factory cases, they require real time encryption and decryption. Stream encryption is faster than block encryption and consumes less power. Because the data transmission time in the smart factory is short, hence, it is not suitable for block cipher algorithms. We believe the bit-level stream cipher method is more suitable for use in smart factories.

3   Mechanism Design

Stream cipher is a real-time and high-precision encryption algorithm, which is usually used for multimedia data encryption. However, most of them lack validation. Therefore, the proposed mechanism modifies the stream cipher by integrating with a pseudo-random code so that the stream cipher is embedded in the authentication function. As provided by the EU eSTREAM project, the proposed mechanism will utilize a WG-8 encryption algorithm that matches the low-power encryption algorithm. These low-power consumption encryption algorithms included WG-8 [20], Grain, MICKEY, and Trivium. For the best confidentiality, WG-8 has the lowest power consumption compared with rest of three encryption algorithms. The formulations of WG-8 encryption algorithms can be list below:

$$S_0 = \left(K_3, K_2, K_1, K_0, IV_3, IV_2, IV_1, IV_0\right) \tag{1}$$

$$S_1 = \left(K_7, K_6, K_5, K_4, IV_7, IV_6, IV_5, IV_4\right) \tag{2}$$

$$S_2 = \left(K_{11}, K_{10}, K_9, K_8, IV_{11}, IV_{10}, IV_9, IV_8\right) \tag{3}$$

$$S_3 = \left(K_{15}, K_{14}, K_{13}, K_{12}, IV_{15}, IV_{14}, IV_{13}, IV_{12}\right) \tag{4}$$

In the experiment, we utilize Raspberry PI 3^+ board as the experimental test bed. As shown in Fig. 3, the reason why Raspberry Pi 3^+ will be used as the experimental test bed is because the encryption mechanism proposed in this paper can be implemented by the embedded system and have been implemented by many factories. The console decrypts, when the master station needs to upload the data, it is decrypted by the host station machine encryption cloud, and the console station is the only networked external machine, because of its cost considerations, Most of them are cheap computers, so if the encryption system of this paper can be executed on the Raspberry Pi 3^+. It can be executed on a cheap computer, and its power consumption is not high, because the Raspberry Pi3 is more than a computer. Verify that the CPU usage of the encryption system is up to 49%, and when the computer is going to do data transmission, it is up to 98%, as shown in Fig. 3. The horizontal axis coordinates are in seconds. When the two line graphs are overlapped, it can be found that although the data encryption verification system will increase the CPU usage, it can still be executed on the Raspberry Pi 3^+, and the intermediate computing performance is about 40%, as shown in Fig. 3, the blue line has The encryption verification system is implemented, and the yellow line is a non-execution encryption verification system.

$$S_{k+20} = (\omega \otimes S_k) \oplus S_{k+1} \oplus S_{k+2} \oplus S_{k+4} \oplus S_{k+7} \oplus S_{k+8} \oplus S_{k+9} \tag{5}$$

However, it will not return to the loop after conversion as formula 6.

$$\text{WGT} - 8\left(x^d\right) = \text{Tr}\left(Q\left(S_{K+19}^{19}\right)\right) = \text{Tr}\left(x^9 + x^{37} + x^{53} + x^{63} + x^{127}\right),$$
$$\text{Tr}(x) = x + x^2 + x^{2^2} + \cdots + x^{2^7}, x^d = S_{K+19}^{19} \tag{6}$$

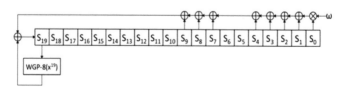

Fig. 2. WG-8 encryption subsequent loop

All operations of WG-8 are handled for the key, so that the key can have pseudo-randomness. The WG-8 key stream has a period of $2^{160} - 1$. The number of 0 s in a period of the key stream is only one less than the number of ones. The linear range of the key stream is $2^{33.32}$. In the subsequent steps will use WG-8 pseudo-random key for encryption and verification. Through the calculation of WG-8, a set of pseudo-random key S can be obtained, and its cycle period is $2^{160} - 1$ is larger than [10]^40, so it can be proved that the pseudo-random number of WG-8 is used as encryption. As shown in Fig. 2, its encryption level is based on subsequent loop S and XOR to generate cipher text C, every 11744 bit to do encryption. Due to EtherCAT packet data capacity can store up to 1470 bytes is equal to 11760 bit, the remaining 2 bytes is the verification code. The packet can be transferred immediately after receiving it. The verification system takes $S_0 \sim S_7$ converts to 10-bit x^0, and inserts 1 in C_{x_0} to replenish it later, then takes $S_{x_0} \sim S_{x_0+7}$ and converts to 10-digit x_1 Insert 1 in C_{x_1} and repeat 16 times. Before decryption, check whether the position of $C_{x_0} \sim C_{x_{16}}$ is 1. If it is 1, the data has not been modified. If it is 0, the data has been modified, so it will be discarded directly. If the verification is correct, it will be The XOR of C and S changes back to this article.

4   Experiment Results

In the experiment, we utilize Raspberry PI 3^+ board as the experimental test bed. As shown in Fig. 3, the reason why Raspberry Pi 3^+ will be used as the experimental test bed is because the encryption mechanism proposed in this paper can be implemented by the embedded system and have been implemented by many factories. The console decrypts, when the master station needs to upload the data, it is decrypted by the host station machine encryption cloud, and the console station is the only networked external machine, because of its cost considerations, Most of them are cheap computers, so if the encryption system of this paper can be executed on the Raspberry Pi 3^+. It can be executed on a cheap computer, and its power consumption is not high, because the Raspberry Pi3 is more than a computer. Verify that the CPU usage of the encryption system is up to 49%, and when the computer is going to do data transmission, it is up to 98%, as shown in Fig. 3. The horizontal axis coordinates are in seconds. When the two line graphs are overlapped, it can be found that although the data encryption verification system will increase the CPU usage, it can still be executed on the Raspberry Pi 3^+, and the intermediate computing performance is about 40%, as shown in Fig. 3, the blue line has The encryption verification system is implemented, and the yellow line is a non-execution encryption verification system.

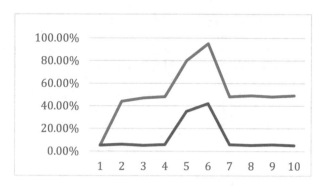

Fig. 3. Verify that the CPU utilization rate by proposed encryption system utilize Raspberry PI 3$^+$ test bed

5 Conclusion

The communication security is important but somewhat not integrated in current industrial 4.0 facilities. Due to complex and precise industrial management systems remain vulnerable to malicious attacks by hackers. Therefore, this paper proposed a lightweight method of industrial data encryption transmission and device authentication by applied WG-8 stream cypher mechanism. By improving its pseudorandom number to protect the manufacturing facilities' communication transmission. In order to comply with the actual manufacturing facilities, we utilize the Raspberry Pi3 B$^+$ as experimental test bed for power consuming, real-time data transmission evaluation. The proposed mechanism evaluation result shows that the WG-8 encryption algorithm will lead Raspberry Pi3 central processing unit (CPU) utilization increased 40%. In the near future, we hope to continue to study the network attack mode for smart factory and provide system intrusion detection analysis and machine learning method to improve the security of smart communication in smart factory in the future.

References

1. Sadeghi, A.-R., Wachsmann, C., Waidner, M.: Security and privacy challenges in industrial internet of things. In: Proceedings of the 52nd Annual Design Automation Conference, Article No. 54, pp. 1–6 (2015)
2. Gong, G., Youssef, A.M.: Cryptographic properties of the Welch–Gong transformation sequence generators. IEEE Trans. Inf. Theory **48**(11), 2837–2846 (2002)
3. Xiong, G., Ji, T., Zhang, X., Zhu, F., Liu, W.: Cloud operating system for industrial application. In: Proceedings of the IEEE International Conference on Service Operations and Logistics, and Informatics (SOLI)
4. http://www.twgreatdaily.com/cat77/node1562986. October 2017
5. https://www.digitimes.com.tw/iot/article.asp?cat=130&cat1=45&cat2=25&id=0000424643_ee45qu335sxgr42fqo53o. June 2018
6. Diedrich, L., Jattke, P., Murati, L., Senker, M., Wiesmaier, A.: Comparison of Lightweight Stream Ciphers: MICKEY 2.0, WG-8, Grain and Trivium

7. Naor, M.: Bit commitment using pseudo-randomness. In: Proceedings of the IEEE CRYPTO Conference, pp. 128–136 (1989)
8. Hamann, M., Krause, M., Meier, W.: LIZARD – a lightweight stream cipher for power-constrained devices. In: Proceedings of ACR Transactions on Symmetric Cryptology, vol. 2017, no. 1, pp. 45–47. ISSN 2519-173X
9. Usman, M., Ahmed, I., Aslam, M.I., Khan, S., Shah, U.A.: SIT: a lightweight encryption algorithm for secure internet of things. Proc. Int. J. Adv. Comput. Sci. Appl. 8(1) (2017)
10. Cheminod, M., Durante, L., Valenzano, A.: Review of security issues in industrial networks. Proc. IEEE Trans. Ind. Inform. 9(1), 277–293 (2013)
11. Bhuiyan, Md.Z.A., Wang, G., Choo, K.K.R.: Secured data collection for a cloud-enabled structural health monitoring system. In: Proceedings of the IEEE 18th International Conference on High Performance Computing and Communications (2016)
12. Wattanapongsakorn, N., Sangkatsanee, P., Srakaew, S., Charnsripinyo, C.: Classifying network attack types with machine learning approach. In: Proceedings of 7th International Conference on Networked Computing (2011)
13. Misra, S., Hashmi, S., Maheswaran, M.: Security challenges and approaches in internet of things. In: Proceedings of the International Workshop on Adaptive Security, ASPI 2013, pp. 2:1–2:4 (2013)
14. Raj, S., Rajesh, R.: Descriptive analysis of hash table based intrusion detection systems. In: Proceedings of 2016 International Conference on Data Mining and Advanced Computing (SAPIENCE) (2016)
15. Xu, P., He, S., Wang, W., Susilo, W., Jin, H.: Lightweight searchable public-key encryption for cloud-assisted wireless sensor networks. Proc. IEEE Trans. Ind. Inform. 14(8), 3712–3723 (2017)
16. Bonnerji, R., Sarkar, S., Rarhi, K., Bhattacharya, A.: COZMO - A New Lightweight Stream Cipher (2017)
17. McKay, K., Bassham, L., Sönmez Turan, M., Mouha, N.: Report on Lightweight Cryptography (2016)
18. Blocher, U., Dichtl, M.: Fish: a fast software stream cipher. In: Proceedings of FSE 1993: Fast Software Encryption, pp. 41–44 (1993)
19. Wu, W., Wu, S., Zhang, L., Zou, J., Dong, L.: LHash: a lightweight hash function. In: Proceedings of Information Security and Cryptology (Inscrypt), pp. 291–308 (2013)
20. Fan, X., Mandal, K., Gong, G.: WG-8: a lightweight stream cipher for resource-constrained smart devices. In: Proceedings of Quality, Reliability, Security and Robustness in Heterogeneous Networks (QShine 2013), pp. 617–632 (2013)
21. Yin, C., Zhu, Y., Fei, J., He, X.: A deep learning approach for intrusion detection using recurrent neural networks. Proc. IEEE Access 5, 21954–21961 (2017)
22. Li, Z., Gong, G.: On the node clone detection in wireless sensor networks. Proc. IEEE/ACM Trans. Netw. (TON) Arch. 21(6), 1799–1811 (2013)

The Research of Attacking TOR Network Users

Yu-Wei Huang, Chia-Hao Lee$^{(\boxtimes)}$, and Fu-Hau Hsu

Department of Computer Science and Information Engineering,
National Central University, Taoyuan City, Taiwan, ROC
diolee@csie.ncu.edu.tw

Abstract. Tor anonymous network is to protect the privacies of the users in the Internet. In this paper, our goal is to research what potential security risks and vulnerabilities we will encounter when using Tor anonymous network. We try to find some determined results from the perspective of an attacker. Using the system architecture of Tor anonymous network and combining the concept of man-in-the-middle (MITM) attack, we design a system for experiments to instigate the attacks in practice. Finally, we discuss and analyze the potential security threats and risks of using Tor anonymous network based on our results of experiments.

Keywords: Tor · Anonymity network · Man-in-the-middle attack · Botnet

1 Introduction

In some network security threats cases recently, the attackers cover their tracks by using the high anonymity of the tor network, avoid the defenders to take reverse tracing, and attack securely [2–4]. This paper is as a reinforcement of a thesis of a preliminary research [1]; we will use contrary thinking to discuss the role of the aforementioned attackers, the tor network users whether they surfer opposite direction attacks, and the any kinds of attacking methods to the tor network anonymous users widely. We will use the concept of man-in-the-middle attacks, prompt one kind of the man-in-the-middle attack models with high anonymity and difficult detection, and do real experimental attacks to the tor network users. According to this experimental model, we prove that our designed man-in-the-middle attack model can make diverse attacks to the tor network users and cause some degrees of information security threats with more and more time. Through the researching process and experimental result of this thesis, we will discuss the affected range of the experimental model, the security of using the Tor network and the potential information security threats. Finally, we discuss the defense methods and future work, and make a conclusion.

2 System Architecture and Function Design

2.1 Scenario

We assume that the users download tor client software from the tor official website (www. torproject.org), install it, and connect to tor network by the standard of procedure in Fig. 1.

© Springer Nature Switzerland AG 2020
L. C. Jain et al. (Eds.): SICBS 2019, AISC 1145, pp. 35–38, 2020.
https://doi.org/10.1007/978-3-030-46828-6_4

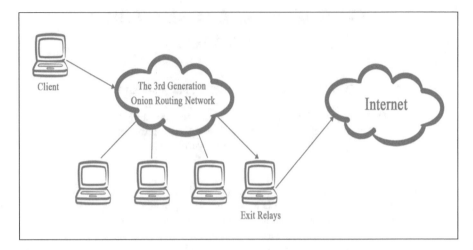

Fig. 1. The architecture of using tor network

When the clients connect to tor network, all network service data packages are encrypted and passed through to a virtual circuit of tor network by the tor client software (onion proxy). Finally, the data packages are decrypted gradually in the virtual circuit to the exit relay, and the original data packages of service send to the destination server. Data packages send back vice versa.

2.2   System Architecture

We use the concept of man-in-the-middle attack to design the system architecture of our experiment. We use four computers to do our experiment with tor network architecture. They show in Fig. 2 and make with star (*) symbol.

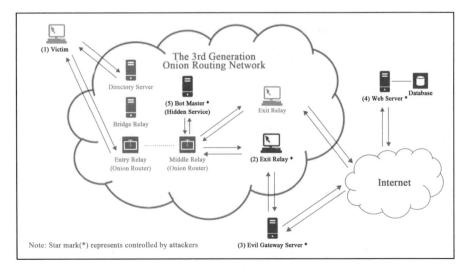

Fig. 2. System architecture

2.3 System Components

1. Victim: The client computers of using tor network
2. Exit Relay: Besides executing the original works in tor network, the exit relay that we setup and control executes to gather statistics of all connections from tor network.
3. Evil Gateway Server: It takes charge of filter and interception on all network packages from the exit relay, and executes to the assignment of attacking victims.
4. Web Server: It works in coordination with evil gateway server and records all information that the victims' computer is attacked by evil gateway server over the man-in-the-middle attack and connect back to web server.
5. Bot Master: It runs tor hidden service, hides in tor network and take control of the commands of botnet nodes.

2.4 System Functions and Features

1. MITM Attack is easy: As more and more exit relays controlled by the attackers, the probability of choosing them by victims will be big. So the attackers have certain probability to go on the offensive of MITM attack to any tor network users. In addition to that the MITM attack executed by the attackers is a passive form, the attackers do not need to take active attacks toward tor network system. Therefore, they can hide well, not be found and attack easily.
2. Connection Record and Statics: The web server in the system architecture can record the information of the tor clients of executing the short codes successfully, identify the operating system version, the browser version and connection time, and gather statistics of the source of IP addresses.
3. DNS Poisoning Attack: Because the attackers control the gateway server on the exit relay, they do not need to attack real domain name servers. They just easily modify the domain name resolution information on the gateway server, and the service requests from tor network are redirected to fake service servers. To tor network clients, the domain name does not change, so they cannot become aware of the relation of domain name and IP address are poisoned.
4. Session Hijacking: They can filter and intercept the content of HTTP protocol, get the sensitive information of clients browsing on internet, and go on session hijacking attacks to specific target website.
5. Content Tampering: Attackers can tamper all the content of HTTP protocol responded from service server at the gateway server on the exit relay. For example, tampering the content of the user login page, getting sensitive individual information of users, or fabricating the response content from service server.

3 Conclusion

We still do not have many methods for full-scale defense to our system architecture of this thesis. This section only proposes defense methods or software techniques for discussing now. We hope more full-scale methods to prevent or decrease the threat of this system architecture to the tor clients. (1) Using multi-threads architecture to decrease the affection of the buffer overflow attacking: Recent years, every browser developer continues to develop some protected mechanism to the threats of web malicious codes. For example, to Chrome browser, it separates and executes every module with multi-thread architecture. Consequently, only single thread will crash and others will not affect or crash when the browser be attacked by the malicious codes. It helps to promote the security of browser usage. (2) Prevention of attacking in advance by intercepting malicious website link: To safe the browser clients, all new version browsers have the build-in or plug-in function packages to intercept malicious web link. They are in accordance with a black or white list to compare and distinguish the malicious websites from the clients connecting to the unsafe websites. However, this mechanism of protection toward our system architecture, it cannot fully defense all attacks from network gateway. (3) Verifying SSL certification of the websites to defense the attacking of the fabricated websites: New version browser can verify the SSL certificate of the website then prompt to block, and the clients need to agree to progress the connection of the website. Although this way can defense the attacking of the fabricate websites or domain name service poisoning attacks effectively, it cannot defense the tampering data of http protocol, software buffer overflow attacks or zero-day vulnerability attacks from network gateway.

References

1. Huang, Y.-W.: The research of attacking TOR network users. Master Thesis of NCU. Accessed 28 Jan 2014. http://ir.lib.ncu.edu.tw/handle/987654321/63536#.XilWjjMzbcs
2. Smith, M.: No Conspiracy Theory Needed: Tor Created for U.S. Gov't Spying. Accessed 28 Mar 2011. http://www.networkworld.com/community/blog/no-conspiracy-theory-needed-tor-created-us-go
3. Smith, M.: 25 More Ridiculous FBI Lists: You Might Be a Terrorist If. Accessed 6 Feb 2012. http://www.networkworld.com/community/blog/25-more-ridiculous-fbi-lists-you-might-be-terrorist-if
4. Wheatley, M.: Japan's Cops Want Tor Network Banned After Cyber-Terror Cat Humiliation. Accessed 22 Apr 2013. http://siliconangle.com/blog/2013/04/22/japans-cops-want-tor-network-banned-after-cyber-terror-cat-humiliation/

Preserving Tenacious DDoS Vitality
via Resurrection Social Hybrid Botnet

Jung-San Lee, Chit-Jie Chew, Ying-Chin Chen, Chih-Lung Chen,
and Kuo-Yu Tsai[✉]

Department of Information Engineering and Computer Science,
Feng Chia University, No. 100 Wenhwa Road, Seatwen, Taichung 407, Taiwan
kytsai@fcu.edu.tw

Abstract. Botnet is always a serious threat to the network security. The hacker has ability to launch the distributed denial of service (DDoS) attack via mastering the vulnerable devices. The enterprise and government suffer from the tremendous loss when the application cannot normally supply the services. The cyber engineer shall be able to takedown botnet as soon as possible once a new one appears. Therefore, through counterintuitive thinking, we actively predict a possibility variant of hybrid botnet based on social network. This is for cyber researcher to formulate the corresponding resistance strategy. We employ the resurrection mechanism and reputation strategy for having indomitable vitality botnet. We address the proposed scheme as resurrection social hybrid botnet (RSHB). According to the experiments, the survivability of our method is better than that of the related work. Consequently, the novel botnet might be a coming sample for cyber engineer to explore the defense strategy.

Keywords: Botnet · DDoS · Reputation · Resurrection mechanism

1 Introduction

As technology advances explosively, people's daily lives are filled with Internet of Things (IoT), such as health wearable devices and vehicles [1]. If there is no comprehensive consideration of information security, it is an opportunity for hacker to abuse them. For instance, the botmaster can execute individual malicious behaviors through the Internet botnet [2]. The victim may suffer from the threat of information stealing, distributed denial of service (DDoS) attack, and cryptography mining.

As to the botnet, the hacker intends to cover himself/herself identity and increase the vitality of botnet. So far, there are three classical architectures: centralized botnet [3], peer-to-peer (P2P) botnet [4, 5], and hybrid peer-to-peer botnet (HP2P) [6, 7]. Accordingly, the researchers gradually propose defending against botnet tactics. The attention of most researches focuses on how to detect and disintegrate the existed botnet and DDoS attack [8–13]. In the method of Rossow et al. [7], crawler and sensor node are able to effective enumerate botnet, which camouflage as a fake bot in the botnet and investigate botnet data. Regretfully, it is too passive to defend DDoS attack. Once a novel botnet is available, the solutions are unable to timely resist.

© Springer Nature Switzerland AG 2020
L. C. Jain et al. (Eds.): SICBS 2019, AISC 1145, pp. 39–45, 2020.
https://doi.org/10.1007/978-3-030-46828-6_5

Through counterintuitive thinking, we actively predict a possibility variant of hybrid botnet based on social network. This is for cyber researcher to formulate the corresponding against strategy. We employ the resurrection mechanism and reputation strategy for having indomitable vitality botnet. We address the proposed scheme as resurrection social hybrid botnet (RSHB). In the experiment evaluation, we compare the enumeration rate through crawler and sensor node. According to the outcomes, our proposed method has a great survivability which threaten our daily lives.

2   Proposed Scheme

2.1   Construction of Communication Layer

Here, we present social network into hybrid botnet for being the medium of command announcement, such as Twitter, Facebook, and Weibo. A botmaster can release the command strings via social website. Therefore, the command is able to be easily broadcasted to the whole botnet. This process is called command publishing point of social network (CPP). Since the communication of CPP is executed using HTTPS, the used ports among hacker, bots, and social network do not be filtered by firewall. This can provide greater covertness of network transmission. The diagram of social hybrid botnet is shown in Fig. 1.

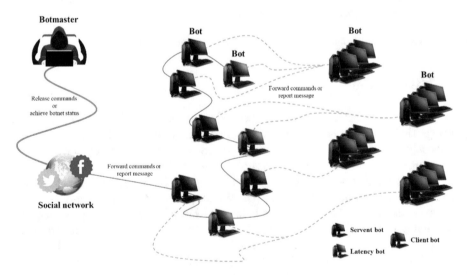

Fig. 1. The architecture of RSHB

2.2 Introduction of Reputation Strategy

Reputation strategy is for bot to determine whether peer servent bot can be trusted or not. Each bot has both peer and visitation lists, as defined in Eqs. (1) and (2).

$$PeerList = \{(IP_1, K_1, P_1, R_1), (IP_2, K_2, P_2, R_2), \ldots, (IP_m, K_m, P_m, R_m)\} \quad (1)$$

$$VisitList = \{(URL_1, R_1), (URL_2, R_2), \ldots, (URL_n, R_n)\} \quad (2)$$

where IP, URL, K, P, and R are IP address, uniform resource locator, key, connected port, and reputation value of neighbor bot, respectively. m is the total number of neighbor peers. n is the sum of the known social networks.

Assume that there exist two servent bots: Bot_A and Bot_B. If Bot_A successfully communicates with Bot_B, the reputation value of Bot_B in peer list of Bot_A is plus (+1), such as forwarding command and sharing peer. Otherwise, the reputation value is minus (−1). Furthermore, servent bot sends the trust message TM to its peer while the value of peer is reached R_{max}, where R_{max} is the maximum reputation in the botnet. After that, the trust counter TC of the peer is added.

Suppose that servent bot Bot_B requests sharing peer message to Bot_A. Bot_A only shares the information of peer Bot_C to neighbor, which has the highest reputation in its peer list. The reputation of new peer Bot_C is set to be default value R_b in Bot_B. On the other hand, the servent bot removes one bot once the sharing command is valid, which has the minimum reputation value. The servent bot has to execute requesting sharing peer when there exists a bot with the reputation value lower than R_{min} in peer list, where R_{min} indicates the minimum reputation in the botnet.

2.3 Design of Latency Bot

Latency bot is used to ensure that there exist a number of reliable servent bots in the botnet. It is also known as slumber bot. Servent bot is transformed into latency bot when it receives at least four times trust message TM from others. In other words, the trust threshold TT has reached to four. Latency bot has an individual pattern which is different from those of servent bot and client bot. It can only recognize and execute the resurrection command from botmaster.

2.4 Implementation of Resurrection Mechanism

Resurrection mechanism is launched for extending the operation ability of botnet when botmaster discovers that cyber engineers try to collapse the botnet, as shown in Fig. 2. The botmaster releases resurrection command to social websites. Latency bots are awakened to rectify the existing botnet. After reorganization, the resurrected botnet normally actives with a group of reliable bots. Here, we consider that the reputation value of bot which is higher than (Rb+2) is of high confidence. Otherwise, the bot is determined as low confidence. The servent bot which has low confidence is eliminated. The reborn botnet only reserves the high confidence servent bots. This can exclude the disguised bots which are simulated by cyber engineers.

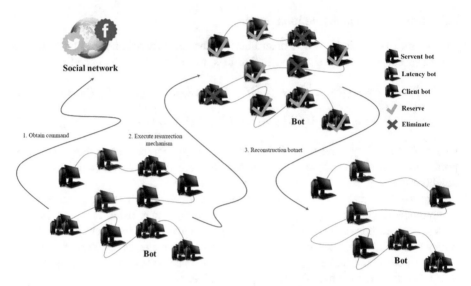

Fig. 2. Resurrection mechanism of RSHB

3   Performance Evaluation

3.1   Simulation Assumptions and Setup

A personal computer installed with Windows10 64bits is applied to comply the simulation environment, equipped with Intel-Core i7-4710HQ and 16 GB RAM. We employed C++ and C# to conduct the experiment. The simulator mimics the physical network, equipment, behavior of bot, and the topology change in botnet, as shown in Fig. 3. For reality, we consider the real world network peak and off-peak periods at 10 a.m. and 10 p.m., respectively. In our circumstance, there are 20% of devices which are always online [14]. The online rate of devices has reached to 80% during peak period, while that of the off-peak period has reduced to 20%. Furthermore, we set routable and non-routable devices each half [15]. The maximum number of bot in botnet is 20,000, and the number of vulnerable devices is 100,000. It means that the botnet will stop expanding once the botnet scale has achieved 20,000. In the part of peer list, the size of a bot is set to ten. The initial value of servent bot is set to 50, in which these servent bots are randomly assembled to each peer list of bot. We configured three social websites as publishing points for issuing command.

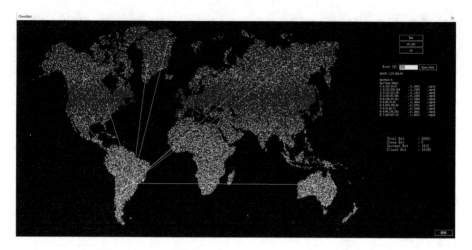

Fig. 3. Botnet simulator

3.2 Enumeration Results

Hereafter, we compare HP2P with our mechanism (RSHB) through enumeration. The experiment consistently runs for 24 h. The enumeration consequences of each botnet are illustrated in Table 1. We analyze the results according to different types of botnets, separately. The high rate represents that the bot is increasingly discovered. It is dangerous to botnet.

Table 1. Enumeration consequence of each botnet

Botnet	Enumeration			
	Crawling			Sensor
	$|V^C|$	$|V_r^C|$	$|V_r^C|/|V^C|$	$|V^S|$
HP2P	2602	2314	88.90%	21653
RSHB	218	106	48.62%	4909

The crawling state $|V^C|$ and $|V_r^C|$ of HP2P are 2602 and 2314. These can be determined as baseline results from crawling enumeration by cyber engineer. For $|V_r^C|/|V^C|$, it is used to determine the accuracy of crawling. The values of HP2P botnet is 88.90%. In the aspect of sensor outcomes, HP2P have been disclosed the whole bots in the botnet. The total number of bots is larger than 20,000 because there have added and eliminated bots during the process.

The values $|V^C|$, $|V_r^C|$, and $|V^S|$ of our architecture are declined to 218, 106, and 4904, according to the latency bot and resurrection mechanism. The values are decreased due to evolving addition defense strategy, such as instruction determination, reputation mechanism and resurrection mechanism. Under the instruction determination, a bot first judges whether received instruction is replicated or not when it gets a new command. This can disrupt the crawler operation. Even the crawler shortens the effective time of instruction, it is infeasible transmitted to destination in a brief time. Regards to reputation mechanism, crawler and sensor have low reputation R because the bot in the proposed architecture has great probability to have a session with high reputation bot based on the reputation strategy. Even a sensor is successfully injected into peer list of one true bot, the bot seldom talks to the sensor. Moreover, the sensor is hard to be shared out since the bot only distributes the peer of the highest R. Furthermore, the resurrection mechanism can eliminate the low confidence bots to reconstruct a robustness botnet. In the part of high confidence bot, it shall have stable peer list based on properly maintaining the operation of botnet. That means it is severe for security crew to position the bot after the procedure of resurrection. That is the reason why RSHB can wipe out a number of unreliable bots and remain high quality bots.

4   Conclusions

In this paper, we speculate a mutation botnet through counterintuitive thinking. We introduce the property of strain and resurrection mechanism in botnet. It can be considered as an active defense technique. From the experimental results, the proposed scheme has preferable resistance compared with current architectures when the cyber engineer enumerates the botnet. This study improves the abilities of concealment and vitality to a botnet, which is a future sample of botnet for security member to exploit. Hence, cyber staff can react the corresponding tactic at the first moment.

References

1. Khan, W.Z., Aalsalem, M.Y., Khan, M.K.: Communal acts of IoT consumers: a potential threat to security and privacy. IEEE Trans. Consum. Electron. **65**(1), 64–72 (2019)
2. Silvaa, S.S.C., Silvab, R.M.P., Pintob, R.C.G., Salles, R.M.: BotNets: a survey. Comput. Netw. **57**(2), 378–403 (2013)
3. Feily, M., Shahrestani, A., Ramadass, S.: A survey of botnet and botnet detection. In: The Third International Conference on Emerging Security Information, Systems and Technologies, Athens, Glyfada, Greece, pp. 268–273. IEEE (2009)
4. Tetarave, S.K., Tripathy, S., Kalaimannan, E., John, C: eBot: approach towards modeling an advanced P2P botnet. In: The 17th IEEE International Conference on Trust, Security and Privacy in Computing and Communications/12th IEEE International Conference on Big Data Science and Engineering (TrustCom/BigDataSE), pp. 391–396. IEEE, New York (2018)
5. Yin, J., Cui, X., Li, K.: A reputation-based resilient and recoverable P2P botnet. In: The IEEE Second International Conference on Data Science in Cyberspace (DSC), Shenzhen, China, pp. 275–282. IEEE (2017)
6. Han, Q.T., Yu, W.Q., Zhang, Y.Y., Zhao, Z.W.: Modeling and evaluating of typical advanced peer-to-peer botnet. Perform. Eval. **72**, 1–15 (2014)

7. Wang, P., Sparks, S., Zou, C.C.: An advanced hybrid peer-to-peer botnet. IEEE Trans. Dependable Secure Comput. **7**(2), 113–127 (2008)
8. Rossow, C., Andriesse, D., Werner, T., Stone-Gross, B., Plohmann, D., Dietrich, C.J., Bos, H.: SoK: P2PWNED - modeling and evaluating the resilience of peer-to-peer botnets. In: The IEEE Symposium on Security and Privacy, Berkeley, CA, USA, pp. 97–111. IEEE (2013)
9. Debashi, M., Vickers, P.: Sonification of network traffic for detecting and learning about botnet behavior. IEEE Access **6**, 33826–33838 (2018)
10. Albanese, M., Jajodia, S., Mason, G., Venkatesan, S.: Defending from stealthy botnets using moving target defenses. IEEE Secur. Priv. **16**(1), 92–97 (2018)
11. Meidan, Y., Bohadana, M., Mathov, Y., Mirsky, Y., Shabtai, A., Breitenbacher, D., Elovici, Y.: N-BaIoT—network-based detection of IoT botnet attacks using deep autoencoders. IEEE Pervasive Comput. **17**(3), 12–22 (2018)
12. Bhuyan, M.H., Kashyap, H.J., Bhattacharyya, D.K., Kalita, J.K.: Detecting distributed denial of service attacks: methods, tools and future directions. Comput. J. **57**(4), 537–556 (2014)
13. Lima, N.A.S., Fernandez, M.P.: Towards an efficient DDoS detection scheme for software-defined networks. IEEE Latin Am. Trans. **16**(8), 2296–2301 (2018)
14. Gkantsidis, N.C., Karagiannis, T., VojnoviC, M.: Planet scale software updates. In: The 2006 Conference on Applications, Technologies, Architectures, and Protocols for Computer Communications, Pisa, Italy, pp. 423–434. ACM (2006)
15. Bhagwan, R., Savage, S., Voelker, G.M.: Understanding availability. In: Peer-to-Peer Systems II, vol. 2735, pp. 256–267 (2003)

Internet of Things and Big Data

Impact of Big Data Analytics Capability and Strategic Alliances on Financial Performance

Shing-Mei Lee[✉]

Department of Finance, Chaoyang University of Technology, 168, Jifeng E. Rd.,
Wufeng District, Taichung 41349, Taiwan
sml@gm.cyut.edu.tw

Abstract. The study aims to explore how big data analytics capability and strategic alliances impact on financial performance. This study provides an integrated framework that conceptualizes multifaceted antecedents pertaining to financial performance of Taiwan's industry relation to big data analytics capability and strategic alliances. However, due to the limited amount of samples taken, those study results would be biased. Therefore, this study applied quantities research and develops multiple-item measures of multiple-dimensions amounted to 215 questionnaire in 2018, in a total sample of 206 with valid collection rates of 96%. The results indicate that (1) a high level of big data analytics management capabilities, big data analytics software and hardware infrastructure, and big data analytics personnel professional competence could enable firms to align their financial strategies to achieve high sales growth, market share growth, profitability, and return on investment. (2) Complementarity of partners' resource-based contributions will be positively related to strategic alliance performance. (3) Higher levels of cooperation in marketing-based alliance will be positively associated with higher levels of Taiwan high technology firms' financial performance. (4) The evidence indicate that higher levels of organizational learning-based will be positively related to financial performance.

Keywords: Big data analytics capability · Strategic alliances · Financial performance

1 Introduction

This study aims to investigate how big data analytics and strategic alliance practices related to the financial performance in the Taiwan manufacturing and service industries. Big data analytics capability is defined as the ability of a firm to effectively deploy technology and talent to capture, store, and analyze data, toward the generation of insight [1]. Although big data analytics capability dimensions differ in their terminology, the taxonomy schemes proposed by the literature are similar as they reflect big data analytics management capability, big data analytics infrastructure capability and big data analytics talent capability-related aspects. In order to enhance financial performance, Taiwan's firms adopt various solutions to enhance their big data analytics

© Springer Nature Switzerland AG 2020
L. C. Jain et al. (Eds.): SICBS 2019, AISC 1145, pp. 49–63, 2020.
https://doi.org/10.1007/978-3-030-46828-6_6

capability and strategic alliances, which can be defined as a firm's ability to capture, share, disseminate and apply alliance management knowledge. The Concept Model we test is illustrated in Fig. 1.

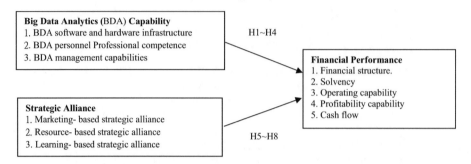

Fig. 1. The concept model

2 Literature Review

2.1 Big Data Analytics Capability and Financial Performance

Recent studies argue that effects of big data analytics capability on business performance are indirect and are mediated by changes in firm's organizational capabilities [2]. Big data analytics capability provides business insights using data management, infrastructure (technology) and talent (personnel) capability to transform business into a competitive force [3]. In addition, there is scarce research on how organizations should proceed to apply big data analytics into the organizational financial performance and little knowledge toward the strengthening of which organizational capabilities they should leverage their investments [4].

Hypothesis 1: *Organizational big data analytics capability will be positively related to financial performance.*

2.2 Big Data Analytics (BDA) Software and Hardware Infrastructure Capability and Financial Performance

The rapid adoption of big data frameworks, which benefit from maximum usage of software and hardware infrastructure in parallel, has amplified the need for a more unified approach to IT. [5] indicated that big data analytics infrastructure capability refers to the ability of the big data analytics capability infrastructure (e.g., applications, hardware, data, and networks) to enable the big data analytics capability staff to quickly develop, deploy, and support necessary system components for a firm. There are also challenges in combining multiple workloads onto a single infrastructure including: managing service level agreements (SLAs); harmonizing workload and resource managers; and supporting different hardware and file systems on-premises and in the cloud. [6] highlight that a growing number of firms are accelerating the deployment of their big data analytics initiatives with the aim of developing critical insight that can ultimately provide them with a competitive advantage. Thus, the following hypothesis was investigated:

Hypothesis 2: Organizational big data analytics software and hardware infrastructure capability will be positively related to financial performance.

2.3 Big Data Analytics (BDA) Personnel Professional Competence and Financial Performance

[5] indicated that Big data analytics personnel capability refers to the BDA staff's professional ability (e.g., skills or knowledge) to undertake assigned tasks, a firm's competence to change existing business processes better than competitors do in terms of coordination/integration, cost reduction, and business intelligence/learning. A big data analytics capability strengthens a firm's sensing, seizing, and transforming capabilities, which ultimately leads to stronger marketing and technological capabilities [7]. [8] proposes that analysts should be competent in four important skill sets: technical knowledge (e.g., database management); technology management knowledge (e.g., visualization tools, and techniques management and deployment); business knowledge (e.g., understanding of short-term and long-term goals). Thus, the following hypothesis was investigated:

Hypothesis 3: Organizational Big data analytics personnel Professional competence will be positively related to financial performance.

2.4 Big Data Analytics (BDA) Management Capabilities and Financial Performance

A big data analytics management capability is increasingly becoming a crucial component of decision-making processes in businesses [9]. [3] focus on management culture, data management infrastructure, and skills. [5] indicated that big data management capability refers to the big data analytics capability unit's ability to handle routines in a structured (rather than ad hoc) manner to manage IT resources in accordance with business needs and priorities. [10] identifies how the big data-based models can improve firm performance. Thus, the following hypothesis was investigated:

Hypothesis 4: Organizational big data analytics management capabilities will be positively related to financial performance.

2.5 Strategic Alliances and Financial Performance

Strategic alliances are now widespread. The dynamic aspects of strategic alliances have received increasing attention from both academics and practitioners in the past decade. A strategic alliance is defined as a formal relationship between two or more parties to pursue a set of agreed upon goals or to meet a critical business need while remaining independent organizations [11]. Some possible benefits of a strategic alliance includes: (1) Enter new markets with products and services [12]. (2) extend your market reach (Perry et al. 2002). (3) Increase the scale of your production output. (4) Get better prices through bulk purchasing [14]. (5) Get access to new technology [13]. (6) Accelerate research and development by sharing costs and resources [15]. The strategy of allying with other organizations has become increasingly prevalent with many organizations opting for strategic alliances in order to strengthen their market positions and improve on their financial performance [16]. Thus, the following hypothesis was investigated:

Hypothesis 5: *Organizational strategic alliances will be positively related to financial performance.*

2.6 Resources-Based Strategic Alliances and Financial Performance

Resources-based strategic alliances maximized firm value through gaining access to other firms' valuable resources [17]. Resource-based allows an organization to access complementary capabilities in a situation where there are resource constraints, which include financial, technological, production capability, sales channel. [18] indicated that resources-based strategic alliances attempt to find the optimal resource boundary through which the value of their resources is better realized than through other resource combinations. [19] indicated that resource based motives cited as key reasons for allying include (1) access to newly developed technologies, capabilities, or general research resources of the partner firm. Resource constraints may direct an organization towards collaboration in a situation where collaboration is not an efficient response to the exchange conditions [20]. (2) Improved market valuation of the technology, transferring tacit knowledge [21] and additional funding or financial resources [22]. Thus, the following hypothesis was investigated:

Hypothesis 6: *Organizational Resources-based strategic alliances will be positively related to financial performance.*

2.7 Marketing-Based Strategic Alliances and Financial Performance

Marketing-based alliance is marketing activity undertaken by more than one entity, which can include a mix of different levels of governments as well as private sector organizations, jointly to promote and sell a concept, product or service which has benefit to all the stakeholders [23]. [24] indicated that marketing alliances can boost opportunities for product success in local markets for global brand companies and entering an alliance partnership with different industry brands allows firms to reinforce their brands image, and improve chances of commercial success. A marketing alliance that is created by the joint naming and the technical co-operation of two brands of an equal they might perceive two product categories are well paired under co-marketing alliances, which aim at integrating different resources, customers, suppliers, and markets from each partner [25]. [26] indicated that Marketing-based alliance' complementarity and substitutability are mainly related to use and the marketing alliance is the strategically complementary alliance between two brands. For example, when planning its expansion into Mexico, Wal-Mart formed a joint venture with Mexico's Cifra, shortening Wal-Mart's learning curve and providing a firm base in the Mexican market. However, it is assumed that marketing alliances are focused on implementing alliances to conduct business activities such as product development, promotion, or distribution; and alliances put their resources on extending alliances to enlarge product lines, cross-selling, brand alliances, market share, and globalization. Co-marketing alliances are lateral relationships between firms at the same level in the value-added chain and represent a form of "symbiotic marketing" [27, 28]. Thus, the following hypothesis was investigated:

Hypothesis 7: *Organizational Marketing-based strategic alliances will be positively related to financial performance.*

2.8   Learning-Based Strategic Alliances and Financial Performance

[29] indicated that effective organizational learning will enhance firm's alliance capability. A firm with strong alliance capability can conduct organizational learning more effectively and at the same time deal with relationship with alliance partner better. This knowledge can be converted into new products, processes, and services, which contribute directly to the firms' final financial results [30]. Often alliances are formed for the purpose of learning and enhancing core competencies through a partner's complementary knowledge [31] and [32]. While the level of learning within an alliance depends upon the absorptive capacity of partners, the success of the collective learning effort is determined by the cooperative learning among partners [29]. [33] indicate that the increasing global harmonization of high technology research procedures and learning-based strategic alliances, Partners can learn together in the course of collaboration, and jointly develop new capabilities and skills, useful knowledge and information for discovery have become more codified, divisible and transferable. This provides greater opportunity for high technology MNEs to ally with, and learn from, technology start-up firms and even global competitors with complementary R&D strengths and strategies. Thus, the following hypothesis was investigated:

Hypothesis 8: *Organizational Learning-based strategic alliances will be positively related to financial performance.*

3   Research Methodology

This research tests the hypotheses using structural equation modeling(SEM) technique. Dependent variable: Financial performance could be measured through an examination of (1) financial structure, (2) solvency, (3) operating capability, (4) profitability capability, and (5) cash flow. Independent variable 1: Big data analytics could be measured through an examination of (1) big data analytics (BDA) soft and hardware infrastructure, (2) big data analytics (BDA) personnel Professional competence and (3) big data analytics (BDA) management capabilities. Independent variable 2: The strategic alliance could be measured through an examination of (1) marketing-based (2) resource-based (3) learning-based. However, SEM may arguably be most general data-analytic framework at present time [34].

3.1   Sample and Data Sources

For this study, all measurement items were taken from the existing literature and were adapted to fit the big data analytics context. The sample of firms for this study was drawn from the 2018 *CommonWealth Magazine "*Taiwan *Top 2000 Enterprises"* list of Taiwan companies. This study applied quantities research and develops multiple-item measures of multiple-dimensions amounted to 215 companies in 2018, in a total sample of 206 companies with valid collection rates of 96%. The industry involves large and well-known companies including 128 manufacturing companies, 78 Service and financial companies.

After that a large-scale questionnaires survey was administered to secure more information from top managers or key persons. Given that a comprehensive questionnaire such as this one requires much time to answer, most of the samples were approached through acquaintants. After removing some companies with some missing financial numbers, the remaining number of samples in the final data set for analysis is 206, as Table 1.

Table 1. Sample distribution of Taiwan's companies

Manufacture industry	Samples of companies	Service and financial industry	Samples of companies
Computer system	6	Financial industry	18
Petrochemical	3	Construction engineering service	10
Semiconductor	12	Department store, wholesale and retail	4
Electronics	26	International trade	15
PC/peripherals	24	Logistics	5
Optoelectronics	19	E-commerce	1
Metals	9	Tourism and catering	8
Food	3	Freight forwarder and customs broker	4
Metal products	18	Information technology	10
Plastic/rubber products	7	Telecommunications and cable TV	3
Total samples	**128**	**Total samples**	**78**

3.2   Variables and Measures

- *Dependent Variable: Measures of Financial Performance*
  In this research, the financial performance measures of financial performance include (1) financial structure, (2) solvency, (3) operating capability, (4) profitability capability, and (5) cash flow.
- *Independent Variables: Measures of Big data analytics (BDA) Capability*
  In this research, big data analytics capability provides business insights using (1) big data analytics (BDA) soft and hardware infrastructure, (2) big data analytics (BDA) personnel Professional competence and (3) big data analytics (BDA) management capabilities [3].
- *Independent Variables: Measures of Strategic Alliance*
  The strategic alliance factors view constructs use reflective scales developed by [34], having three indicates including (1) marketing-based (2) resource-based (3) learning-based. Marketing- based strategic alliance refers to share market operation risk, open new market, increase market share and develop new product.

Resource-based strategic alliance refers to build economic scale, ensure the resource of raw material, cost reduction and competitive edge improvement and investment portfolio and risk-avoidance orientation. Learning-based strategic alliance refers to reduce R&D cost, accelerate technology transfer and improve learning curve effect.

4 Results

4.1 *The Structural Equation Modeling (SEM) Results for* Strategic Alliance, Big Data Analytics *Financial Performance*

In Fig. 2, the Structural Equation Modeling (SEM) results for financial performance, big data analytics capability and strategic alliance, all of the path coefficients are significant. The following path model output reports the overall model coefficients as Fig. 2.

Fig. 2. Full SEM of strategic alliance, big data analytics and financial performance

(1) Figure 2 showed that the financial performance measures include (1) financial structure, (2) solvency, (3) operating capability, (4) profitability capability, and (5) cash flow. Financial performance was reflected by financial structure ($\beta = 0.84$), solvency ($\beta = 0.96$), operating capability ($\beta = 0.93$), profitability capability ($\beta = 0.92$) and cash flow ($\beta = 0.87$), although we identified these differences in measuring the importance of financial performance, we note that differences are very small, thus all the dimensions should be given equal importance in building financial performance.

(2) Figure 2 showed that big data analytics capability provides business insights using (1) big data analytics (BDA) soft and hardware infrastructure, (2) big data analytics (BDA) personnel professional competence and (3) big data analytics (BDA) management capabilities (Kiron et al. 2014). Big data analytics capability was reflected by big data analytics (BDA) software and hardware infrastructure ($\beta = 0.92$), big data analytics (BDA) personnel professional competence ($\beta = 0.93$) and big data analytics (BDA) management capabilities ($\beta = 0.92$), all the dimensions should be given equal importance in building big data analytics capability.

(3) Figure 2 showed that strategic alliance factors having three indicates including (1) marketing-based (2) resource-based (3) learning-based (Morgan and Hunt 1994). Strategic alliance was reflected by Marketing-based ($\beta = 0.90$), Resource-based ($\beta = 0.92$) and Learning- based ($\beta = 0.94$). Accordingly, variance of strategic alliance were calculated to reflect their corresponding components (Fig. 2), thus all the dimensions should be given equal importance in building strategic alliance.

4.2   The Influence of the Big Data Analytics Capability (BDAC) on Financial Performance

Table 2 and Fig. 2 showed that big data analytics capability enhanced financial performance, with a path coefficient of 0.734*** (p $\leq$ 0.000), explaining 82% (R^2) of the variance. Big data analytics capability provides business insights using data management, infrastructure (technology) and talent (personnel) capability to transform business into a competitive force [3]. In addition, there is scarce research on how organizations should proceed to apply big data analytics into the organizational financial performance and little knowledge toward the strengthening of which organizational capabilities they should leverage their investments [4]. Therefore, the result of SEM model supports Hypothesis 1 (*Organizational big data analytics capability will be positively related to financial performance.*).

4.3   The Influence of Strategic Alliance on Financial Performance

Table 2 and Fig. 2 showed that Strategic alliance enhanced financial performance, with a path coefficient of 0.651*** (p $\leq$ 0.000), explaining 82% (R^2) of the variance. The strategy of allying with other organizations has become increasingly prevalent with many organizations opting for strategic alliances in order to strengthen their market positions and improve on their financial performance [16]. Therefore, the result of SEM model supports Hypothesis 5 (*Organizational strategic alliances will be positively related to financial performance.*) was supported.

Table 2. Standardized regression weights and p-value of the big data analytics (BDA) capability and strategic alliance SEM model

Hypothesis	Path	Standardized regression weight	p
H1	Big data analytics capability → Financial performance	0.734***	0.000
H5	Strategic alliance → Financial performance	0.651***	0.000

4.4 The Influence of the Big Data Analytics (BDA) Software and Hardware Infrastructure Capability on Financial Performance

Table 3 and Fig. 3. showed that the Big data analytics (BDA) capability software and hardware infrastructure capability enhanced financial performance, with a path coefficient of 0.620** (p ≤ 0.005), explaining 68% (R^2) of the variance. Big data analytics infrastructure capability refers to the ability of the big data analytics capability infrastructure (e.g., applications, hardware, data, and networks) to enable the big data analytics capability staff to quickly develop, deploy, and support necessary system components for a firm [5]. [6] highlight that a growing number of firms are accelerating the deployment of their big data analytics initiatives with the aim of developing critical insight that can ultimately provide them with a competitive advantage. Therefore, the result of SEM model supports Hypothesis 2. (*Organizational big data analytics software and hardware infrastructure capability will be positively related to financial performance.*)

4.5 The Influence of the BDA Personnel Professional Competence on Financial Performance

Table 3 and Fig. 3 showed that BDA personnel professional competence enhanced financial performance, with a path coefficient of 0.369** (p ≤ 0.003), explaining 68% (R^2) of the variance. [5] indicated that big data analytics personnel capability refers to the BDA staff's professional ability (e.g., skills or knowledge) to undertake assigned tasks, a firm's competence to change existing business processes better than competitors do in terms of coordination/integration, cost reduction, and business intelligence/learning. Therefore, the result of SEM model supports Hypothesis 3 (*Organizational Big data analytics personnel professional competence will be positively related to financial performance.*)

4.6   The Influence of the BDA Management Capabilities on Financial Performance

Table 3 and Fig. 3 showed that the BDA management capabilities enhanced financial performance, with a path coefficient of 0.608** (p $\leq$ 0.007), explaining 68% (R^2) of the variance. It is evident that the higher levels of BDA management capabilities will be positively related to financial performance (Hypothesis 4). [5] indicated that big data management capability refers to the big data analytics capability unit's ability to handle routines in a structured (rather than ad hoc) manner to manage IT resources in accordance with business needs and priorities. Therefore, the result of SEM model supports Hypothesis 4. (*Organizational big data analytics management capabilities will be positively related to financial performance.*)

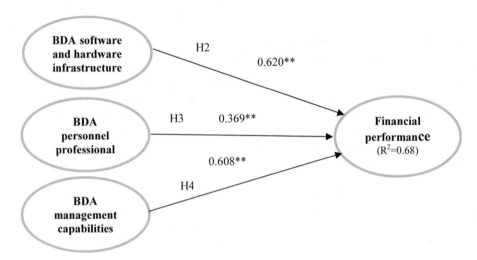

Fig. 3. The SEM of BDA

Table 3. Standardized regression weights and P-value of the big data analytics (BDA) capability SEM model

Hypothesis	Path	Standardized regression weight	p
H2	BDA software and hardware infrastructure → financial performance	0.620**	0.005
H3	BDA personnel professional competence → financial performance	0.369**	0.003
H4	BDA management capabilities → financial performance	0.608**	0.007

* p $\leq$ 0.05; ** p $\leq$ 0.01; *** p $\leq$ 0.001

4.7 The Influence of the Resource-Based Strategic Alliances on Financial Performance

Table 4 and Fig. 4 showed that the resource-based strategic alliance enhanced financial performance, with a path coefficient of 0.517^{**} ($p \leq 0.004$), explaining 74% (R^2) of the variance. Strategic alliances "is about creating the most value out of one's existing resources and by combining these with others' resources" [14]. The resource-based motivation explains that firms form an alliance that they prefer to combine their respective resources, such as manufacturing resources and distribution resources, to achieve certain mutually agreed business goals [35]. However, the degree of resource complementarity will be a critical factor in determining an alliance's future course and outcome. Therefore, the result of SEM model supports Hypothesis 6. (*Organizational resource-based strategic alliances will be positively related to financial performance.*)

4.8 The Influence of the Marketing-Based Strategic Alliances on Financial Performance

Table 4 and Fig. 4 showed that marketing-based strategic alliances enhanced financial performance, with a path coefficient of 0.308^{**} ($p \leq 0.002$), explaining 74% (R^2) of the variance. The alliance relationship management of the firm marketing alliances create a synergic effect which can amplify and build user awareness of benefits derived from these complementarities [36] and allow firms to reinforce their brands image, augment brand awareness, and improve chances of commercial success [24, 37]. Therefore, the result of SEM model supports Hypothesis 7. (*Organizational marketing-based strategic alliances will be positively related to financial performance.*)

4.9 The Influence of the Learning-Based Strategic Alliances on Financial Performance

Table 4 and Fig. 4 showed that the learning-based strategic alliances enhanced financial performance, with a path coefficient of 0.507^{**} ($p \leq 0.001$), explaining 74% (R^2) of the variance. [38] indicate that a firm which is continuously engaged in learning tends to stand a better chance of tracking and responding to customer needs, sensing and seizing on market opportunities, and offering appropriate and finely targeted products, results which lead to superior levels of profitability, sales growth, and customer retention Therefore, the result of SEM model supports Hypothesis 8. (*Organizational Learning-based strategic alliances will be positively related to financial performance.*)

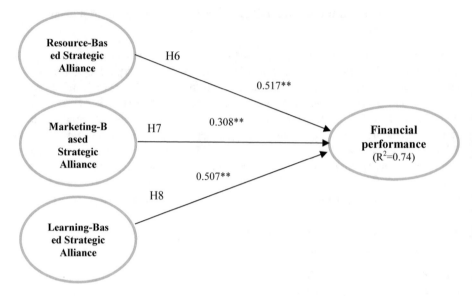

Fig. 4. The SEM of strategy alliance

Table 4. Standardized regression weights and P-value of the strategic alliance SEM model.

Hypothesis	Path	Standardized regression weight	p
H6	Resource-based strategic alliance → Financial performance	0.517**	0.004
H7	Marketing-based strategic alliance → Financial performance	0.308**	0.002
H8	Learning-based strategic alliance → Financial performance	0.507**	0.001

* $p \leq 0.05$; ** $p \leq 0.01$; *** $p \leq 0.001$

5   Conclusions

5.1   Contribution of This Research

This study demonstrates that (1) higher levels of cooperation in big data analytics capability will be positively associated with higher levels of financial performance. Big data analytics capability provides business insights using data management, infrastructure (technology) and talent (personnel) capability to transform business into a competitive force. These terms are adopted into big data analytics capability research, as prior research on IT capability and firm performance has already highlighted the important role of several types of dynamic capabilities. (2) big data analytics infrastructure capability refers to the ability of the big data analytics capability infrastructure (e.g., applications, hardware, data, and networks) to enable the big data analytics capability staff to quickly develop, deploy, and support necessary system components for a firm. (3) Complementarity of partners' resource-based contributions will be positively related to strategic alliance performance. Resource based motives cited as key reasons for Taiwan firms allying include access to newly developed technologies, capabilities, or general research resources of the partner firm, transferring tacit knowledge, improved market valuation of the biotechnology technology. (4) Higher levels of cooperation in marketing-based alliance will be positively associated with higher levels of Taiwan high technology firms' financial performance. Marketing alliances create a synergic effect which can amplify and build user awareness of benefits derived from these complementarities and allow Taiwan high technology firms to reinforce their brands image, augment brand awareness, and improve chances of commercial success. (5) The evidence indicate that higher levels of organizational learning-based will be positively related to financial performance. Taiwan high technology firms which are continuously engaged in learning tends to stand a better chance of tracking and responding to customer needs, sensing and seizing on market opportunities, and offering appropriate and finely targeted products, results which lead to superior levels of profitability, sales growth, and customer retention.

5.2   Suggestion for Future Research

The alternative explanations discussed in the previous section also provide potential avenues for future research. However, this study highlighted alliance type and alliance experience in impacting alliance management capability in the entrepreneurial context. (1) The further study should investigate what other factors, besides alliance type, alliance experience and the establishment of a dedicated alliance functions are critical in developing and enhancing a firm's alliance strategies. (2) In future research, if it is possible to obtain data from the other country sample, a better insight about understanding the significant relationships between strategic alliance success factors and financial performances. (3) A limitation of this study, of course, is that it deals only with Taiwan high technology firms perceptions about their alliance strategies and financial performance. Future research that focuses on manufacturers in other countries of origin might produce results different from the findings of this study. (4) Future research should investigate how different alliance types interact with one another, and thus this study was not able to discuss the differential impact of portfolios made up of different mixtures of upstream, horizontal and downstream alliances on a firm's alliance management capability.

5.3   Limitations of This Research

This study was largely motivated by the great interest of scholars and practitioners on the phenomenon of big data. This study has several potential limitations. First, it is difficult to follow a strictly randomized sampling procedure. Samples are collected on a convenient basis in order to increase the response rate and ensure the quality of data. It is recommended that a larger and more randomized sample be taken for a more comprehensive future study. However, the results of the model should be considered as limited to the variables specified within the model. It is possible that other variables outside the model may also affect the relationships.

References

1. Gupta, M., George, J.F.: Toward the development of a big data analytics capability. Inf. Manag. **53**, 1049–1064 (2016)
2. Côrte-Real, N., Oliveira, T., Ruivo, P.: Assessing business value of big data analytics in European firms. J. Bus. Res. **70**, 379–390 (2017)
3. Kiron, D., Prentice, P.K., Ferguson, R.B.: The analytics mandate. MIT Sloan Manag. Rev. **55**, 1–25 (2014)
4. Mikalef, P., Pappas, I.O., Krogstie, J., Giannakos, M.: Big data analytics capabilities: a systematic literature review and research agenda. Inf. Syst. E-Bus. Manag. **16**, 1–32 (2018)
5. Kim, G., Shin, B., Kwon, O.: Investigating the value of sociomaterialism in conceptualizing IT capability of a firm. J. Manag. Inf. Syst. **29**, 327–362 (2012)
6. Constantiou, I.D., Kallinikos, J.: New games, new rules: big data and the changing context of strategy. J. Inf. Technol. **30**, 44–57 (2015)
7. Protogerou, A., Caloghirou, Y., Lioukas, S.: Dynamic capabilities and their indirect impact on firm performance. Ind. Corp. Chang. **21**, 615–647 (2011)
8. Davenport, T.H., Patil, D.: Data scientist: the sexiest job of the 21st century. Harvard Bus. Rev. **90**, 70–77 (2012)
9. Hagel, J.: Bringing analytics to life. J. Accountancy **219**, 24–25 (2015)
10. Barton, D., Court, D.: Making advanced analytics work for you. Harvard Bus. Rev. **90**, 78 (2012)
11. Mohr, J., Spekman, R.: Characteristics of partnership success: partnership attributes, communication behavior, and conflict resolution techniques. Strategic Manag. J. **15**, 135–152 (1994)
12. Perks, H., Easton, G.: Strategic alliances: partner as customer. Ind. Market. Manag. **29**, 327–338 (2000)
13. Perry, C., Cavaye, A., Coote, L.: Substance in business-to-business relationships. J. Bus. Ind. Market. **17**, 75–89 (2002)
14. Das, T.K., Teng, B.S.: A resource-based theory of strategic alliances. J. Manag. **26**, 31–61 (2000)
15. Harrigan, K.R.: Strategic Alliances and Partner Asymmetries, pp. 205–226. Lexington Books, Lexington (1988)
16. Morgan, R.M., Hunt, S.D.: The commitment-trust theory of relationship marketing. J. Market. **58**, 20–39 (1994)
17. Ramanathan, K., Seth, A., Thomas, H.: Explaining joint ventures: alternative theoretical perspectives. In: Beamish, P.W., Killing, J.P. (eds.) Cooperative Strategies: North American Perspectives, pp. 51–85. New Lexington Press, San Francisco (1997)
18. Lambe, C.J., Spekman, R.E., Hunt, S.D.: Alliance competence, resources, and alliance success: conceptualization, measurement, and initial test. J. Acad. Market. Sci. **30**, 141–158 (2002)
19. Pfeffer, J., Salancik, G.: The External Control of Organizations: A Resource Dependence Perspective. Harper & Row, New York (1978)
20. Kale, P., Dyer, J., Singh, H.: Alliance capability, stock market response, and long-term alliance success: the role of the alliance function. Strateg. Manag. J. **23**, 747 (2002)

21. Collins, J.D., Hitt, M.A.: Leveraging tacit knowledge in alliances: the importance of using relational capabilities to build and leverage relational capital. J. Eng. Technol. Manag. **23**, 147–167 (2006)
22. Coombs, J.E., Deeds, D.L.: International alliances as sources of capital: evidence from the biotechnology industry. J. High Technol. Manag. Res. **11**, 235 (2000)
23. Abratt, R., Motlana, P.: Business horizons: capitalizing on collaboration. Market Leader (19) (2002)
24. Geylani, T., Inman, J.J., Hofstede, F.T.: Image reinforcement or impairment: the effects of co-branding on attribute uncertainty. Market. Sci. **27**, 730–744 (2008)
25. Dussauge, P., Garrette, B., Mitchell, W.: Asymmetric performance: the market share impact of scale and link alliances in the global auto industry. Strateg. Manag. J. **25**, 701–711 (2004)
26. Thoumrungroje, A., Tansuhaj, P.: Globalization effects, co-marketing alliances, and performance. J. Am. Acad. Bus. **5**, 495–502 (2004)
27. Adler, L.: Symbiotic marketing. Harvard Bus. Rev. **44**(November-December), 59–71 (1966)
28. Vardarajan, P.R., Rajaratnam, D.: Symbiotic marketing revisited. J. Market. **50**(January), 7–17 (1986)
29. Muthusamy, S.K., White, M.A.: Learning and knowledge transfer in strategic alliances: a social exchange view. Organ. Stud. **26**, 415–441 (2005)
30. Cohen, W.M., Levinthal, D.A.: Absorptive capacity: a new perceptive on learning and innovation. Adm. Sci. Q. **35**, 128–152 (1990)
31. Anand, B.N., Khanna, T.: Do firms learn to create value: the case of alliances. Strateg. Manag. J. **21**, 295–315 (2000)
32. Grant, R.M., Baden-Fuller, C.: A knowledge accessing theory of strategic alliances. J. Manag. Stud. **41**, 61–84 (2004)
33. Inkpen, A.C., Tsang, E.W.K.: Social capital, networks, and knowledge transfer. Acad. Manag. J. **30**, 146–165 (2005)
34. Tomarken, A.J., Waller, N.G.: Structural equation modeling: strengths, limitations, and misconceptions. Ann. Rev. Clin. Psychol. **1**, 31–65 (2005)
35. Bulter, R., Sohod, S.: Joint-venture autonomy: resource dependence and transaction costs perspectives. Scand. J. Manag. **11**(2), 159–176 (1995)
36. Bucklin, L.P., Sengupta, S.: Organizing successful co-marketing alliances. J. Market. **57**, 32–46 (1993)
37. Gammoh, B.S., Voss, K.E., Chakraborty, G.: Consumer evaluation of brand alliance signals. Psychol. Market. **23**, 465–486 (2006)
38. Slater, S.F., Narver, J.C.: Market orientation and the learning organization. J. Market. **59**, 63–74 (1995)

A Progressive Secret Image Sharing Scheme Based on GEMD Data Hiding

Wen-Chung Kuo[1], Yi-Jiun Chen[1], Chun-Cheng Wang[2]([✉]),
and Yu-Chih Huang[3]

[1] Department of Computer Science and Information Engineering,
National Yunlin University of Science and Technology, Yunlin, Taiwan, R.O.C.
{simonkuo,m10617004}@yuntech.edu.tw
[2] National Center for High-Performance Computing,
National Applied Research Laboratories, Tainan 744, Taiwan, R.O.C.
1703158@narlabs.org.tw
[3] Department of Information Management, Tainan University of Technology,
Tainan 71002, Taiwan, R.O.C.
t00232@mai.tutl.edu.tw

Abstract. With the development of the Internet, people spread digital information through the Internet. However, the Internet is a public channel. Avoiding digital information be tampered with to keep the security of data is the most important thing today. The main method of data hiding is to embed secret information into multimedia and then transmit it to the recipient. The stego multimedia would not attract someone and is protected from tampering during transmission. In addition to data hiding technology, this paper is combined with secret image sharing. We use Progressive Secret Image Sharing (PSIS) that the secret image is divided into multiple regions and hidden in N stego images. If the recipient wants to extract the secret image, the recipient must have a part of or all stego images to extract a similar secret image or a complete secret image. In this paper, we propose the method of secret image sharing based on GEMD data hiding. The secret image is not in sequence, but can be overlapped and divided. The experimental results present that the stego images are PSNR $\geq$ 50dB. The recipient can extract be similar or complete with stego images and judge whether stego images is tampered or not.

Keywords: Data hiding · GEMD (General exploiting modification directions) · Secret image sharing

1 Introduction

With the development of the Internet, we always spread digital information through the Internet. However, the Internet is a public channel. Therefore, how to avoid that information is tampered with to protect the security of data is the most important issue today.

The most common way to protect data is cryptography. The data is encrypted so that even if the data is stolen, the decrypted data cannot be obtained without the key.

© Springer Nature Switzerland AG 2020
L. C. Jain et al. (Eds.): SICBS 2019, AISC 1145, pp. 64–75, 2020.
https://doi.org/10.1007/978-3-030-46828-6_7

However, the encrypted information meaningless, which may cause suspicion to others and may be intercepted and tampered.

The scholars have proposed the concept of data hiding in order to solve the above problems [2, 6, 7, 9, 10, 13, 16–18, 20, 24]. Data hiding is to embed the most primitive and complete information into other files, and transmit them to the recipients without being detected. The preliminary classifications of the data hiding are covert channels, anonymity, steganography, and copyright marking. The types of methods proposed in this paper are steganography. The secret image is embedded in the original image to generate a stego image. Since the stego media does not generate garbled characters like encryption technology, it will not cause suspicion to others during the transmission. Finally, the receiver will obtain the secret image through the data hiding method as shown in Fig. 1.

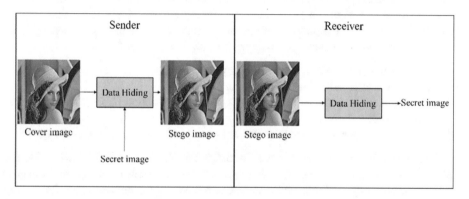

Fig. 1. Data hiding process

In general, favorable data hiding techniques must the following requirements:

- Imperceptibility: In order to avoid being easily suspected or even falsified by others, the quality of the stego image must be above a certain level.
- Payload: The secret information embedded in the stego image can reach as large amount as possible.
- Security: Any illegal receiver can not extract secret message from the stego image.

As the amount of information load increases, the quality of the stego image decreases. Therefore, the balance between the imperceptibility and payload is also an important issue.

Data hiding is mainly aimed at the safe transmission of messages without suspicion and has often been used for military purposes in the past. Most people send messages in a way that is not a letter, but is transmitted through a communication software. It may also be stolen by someone who is in the process of transmission. Generally, secret messages are encrypted before transmission to prevent confidential messages from being known to the any illegal receiver. However, the encrypted messages are unnaturally garbled and are easily intercepted by any illegal receiver, and can not be delivered to the receiver. At this time, data hiding technology is used, even if the stego media is stolen by others, it will not cause suspicion of interception, and the message can be transmitted safely.

Secret Image Sharing (SIS) is extended by Secret Sharing. In the past encryption technology, the key was used to protect the secret data, but there are many ways to protect the key. One of them is to divide the main key into multiple sub keys, when the main key is needed, all the sub keys must be assembled to obtain the main key. That is, suppose the owner of a design company encrypts the secret data in order to protect an important secret information, and then divides the key into several sub key for a number of trusted people to keep. When the boss needs the secret information, the sub key to all trusted people must be assembled in order to obtain the complete key. Although this method can improve security, when the secret data is to be extracted, it must have the complete sub key. If any sub golden key is damaged and the correct key can not be extracted, the secret information can not be obtained successfully. In order to solve the above problems, in 1979, the secret sharing method proposed by Shamir [19] and Blakey [1] solved the problem of the management of the main key, indicating that when the conditions are met, the main key can be obtained, as follows:

(1) The main key is divided into n sub keys by the operation, and the sub keys are distributed to n individuals for storage.
(2) Among n individuals, as long as any t personal collection $(k \leq n)$, the main key can be jointly derived.
(3) If only $k - 1$ golden keys, they can not know any message of the main key.

The secret sharing concept is extended to SIS. The SIS is to embed the secret image into two or more stego images and distribute them to different receivers for management. However, only a single stego image can not extract a complete secret image. In the past, there have been many studies [11, 12, 14, 22, 23, 25] that hides secret image segments into multiple stego images and finally transmit them to the receivers, that is, data hiding technology combined with secret information and SIS.

In addition to data hiding technology, this paper is combined with secret image sharing. The pixels of the secret image are divided into n block of secret image in a progressive manner, and are hidden in n stego images. If the receiver wants to extract the secret image, he must have a certain number of stego images to extract a similar secret image or a complete secret image.

This paper proposes secret image sharing scheme based on general exploiting modification direction (GEMD) of data hiding method. To convert pixel of secret images to binary and divide it into overlapping n blocks. By using the least significant bit (LSB) of the pixel in the secret image, it is possible to judge whether or not the stego image has been tampered with, and the purpose is to practice the integrity of the secret image hiding.

The structure of this paper: The second section is to review the data hiding technology and related research. The third section is the description of the method of this paper, which is enhance integrity hiding. The fourth section is the experimental results and analysis. The fifth section is the conclusion and future prospects.

2 Related Works

This section will introduce EMD and GEMD of data hiding in sequence, and secret image sharing.

2.1 Exploiting Modification Direction (EMD) [26]

A data hiding method using the extraction function is proposed by Zhang and Wang in 2006. According to the characteristics of the EMD modulus extraction function, n adjacent pixels are grouped. During the hiding process, just only one pixel is modified ± 1in each group. EMD maintains the very good stego image quality and hides the secret message with the smallest change in n pixels. To use the extraction function (1) in EMD to calculate f_{EMD}:

$$f_{EMD}(x_1, x_2, \ldots, x_n) = \left[\sum_{i=1}^{n} x_i \times i \right] \bmod (2n+1) \tag{1}$$

Where x_i is a group pixel, i is a weight value, and all pixels in the process are integers. Let us explain the steps of EMD embedding and extraction, which is shown in following scheme 1:

Embedding phase:
Input: cover image I, secret message S
Output: stego image I'

(1) Extract n pixels into a group x, nest into formula (1), and calculate f_{EMD}.
(2) To select secret message conversion modulus is $2n+1$ such as $(S)_{2n+1}$.
(3) Calculate the D_{EMD} between the secret message S and f_{EMD}.

$$D_{EMD} = (S - f_{EMD}) \bmod (2n+1) \tag{2}$$

(4) If $D_{EMD} > n$, $x_{(2n+1-D_{EMD})} - 1$; else if $D_{EMD} \leq n$, $x_{(D_{EMD})} + 1$; else $D_{EMD} = 0$, $x_i = x_i$.

Output stego image I'
Extracting phase:
Input: stego image I'
Output: secret message S

(1) After obtaining n pixels, the value obtained by the formula (1) is converted into binary, which is the secret message S.
(2) Repeat step 1 until all pixel group extractions of the image are completed.

2.2   General Exploiting Modification Direction (GEMD) [6]

In order to increase the secret information payload, by adjusting the weight value of the function, this method increases the modulus of group pixels. The amount of payload can be maintained at 1bpp(bits per pixel) or more. Select the same n pixels as a group, and modify the modulus value $(2n+1)$ of the EMD extraction function (1) to (2^{n+1}), and the weight value i to $(2^i - 1)$. Before hiding the secret message, must first convert the secret message to binary, and then use the extraction function (3) in GEMD to calculate f_{GEMD}:

$$f_{GEMD}(x_1, x_2, \ldots, x_n) = \left[\sum_{i=1}^{n} x_i \times (2^i - 1) \right] \bmod (2^{n+1}) \qquad (3)$$

where x_i is a group pixel, $(2^i - 1)$ is a weight value.

In order to increase the amount of GMED data hiding method, the (2^{n+1}) modulus must be increased. It can be found that the difference between the modified pixel and the original pixel of GMED still remains at most ± 1, and Maintain good stego images. Next, the steps of embedding and extraction, which is shown in following scheme 2:

Embedding phase:
Input: cover image I, secret image S
Output: stego image I'

(1) Extract n pixels into a group g, nest into formula (3), and calculate f_{GEMD}.
(2) To convert s to binary and select $n+1$ bit secret message s.
(3) Calculate the D_{GEMD} between the secret message S and f_{GEMD}.

$$D_{GEMD} = (s - f_{GEMD}) \bmod (2^{n+1}) \qquad (4)$$

(4) If $D_{GEMD} > 2^n$, $D'_{GEMD} = 2^{n+1} - D_{GEMD}$, and transform binary $D'_{GEMD} = (b_n, b_{n-1}, \ldots, b_1, b_0)_2$.

> *For* $(i = n; i \geq 1; i - -)$
> $\{if\ (b_i == 0\ and\ b_{i-1} == 1), x_i - 1;$
> $else\ if\ (b_i == 1\ and\ b_{i-1} == 0), x_i + 1\}$

Else if $0 < D_{GEMD} < 2^n$, D_{GEMD} transform binary $D_{GEMD} = (b_n, b_{n-1}, \ldots, b_1, b_0)_2$.

> *For* $(i = n; i \geq 1; i - -)$
> $\{if\ (b_i == 0\ and\ b_{i-1} == 1), x_i + 1;$
> $else\ if\ (b_i == 1\ and\ b_{i-1} == 0), x_i - 1\}$

Else $D_{GEMD} = 2^n$, $x_1 = x_1 + 1$ and $x_n = x_n + 1$

Output stego image I'
Extracting phase:
Input: stego image I'
Output: secret message S

(1) After obtaining n pixels, the value obtained by the formula (3) is converted into binary, which is the secret message S.
(2) Repeat step 1 until all pixel group extractions of the image are completed.

2.3 Progressive Secret Image Sharing (PSIS)

The first section mentions that SIS is extended by secret sharing. All number of stego images can be used to extract a complete secret image. The concept of progressive secret image is that even if it has partial stego images, it can extract part of the secret image.

The secret image is divided into non-overlapping k blocks, which are respectively hidden in N images, and N stego images are generated. When the receiver has a number of stego images, part of the secret image can be restored. Progressive secret image sharing is more secure than simple secret image sharing.

Figure 2 illustrates the embedding process of SIS. When the secret image is divided into three equal parts, the secret image 1 is partially embedding in the stego images 1–3; the secret image 2 is partially embedding in the stego image 2–4; and the secret image 3 is partially embedding in the stego images 3–5. A total of three images are embedded in five cover images to form five stego images.

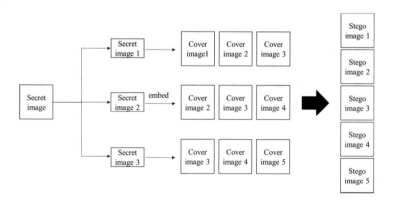

Fig. 2. Embedding of PSIS

If there are three stego images given to the receivers, a small part of the secret image can be extracted. If there is a stego images 1–3, the secret image 1 can be obtained. when there are four stego images, two blocks of the secret image can be extracted, such as If there is a stego image 1–4, the secret image 1–2 can be obtained. Conversely, when all the stego images are possessed, the secret image can be completely extracted, as shown in Fig. 3.

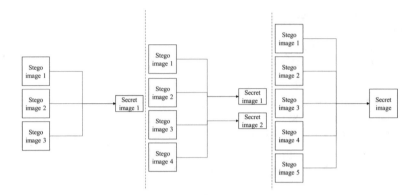

Fig. 3. Extraction of PSIS

3 Proposed Method

This paper uses the literature in the second section to explore the GEMD and the extension of the PSIS concept in related research. The main goal is to effectively practice the secret image sharing of the data hiding method. When the secret image is extracted, the LSB can be used for the purpose of determining whether the stego image has been tampered with. The secret image is divided into two parts or more after converting the secret image pixels into binary. Wherein some of the bits overlap, and finally the GEMD method is embedded, and all the N stego images can be generated. When the receiving end obtains the stego image, the secret is obtained. In the case of images, the overlapping bits can be used to determine the correctness of the secret image.

The pixels of the secret image are converted into binary, and then k blocks are divided. The secret image bit length to be hidden for each k block is l_k and the last 1 bit of s_i. Therefore, the secret image pixel is divided into k blocks, wherein the last one bit is repeatedly hidden, as shown in Fig. 4 is a flow architecture diagram of the method, and experiments are performed using the secret image.

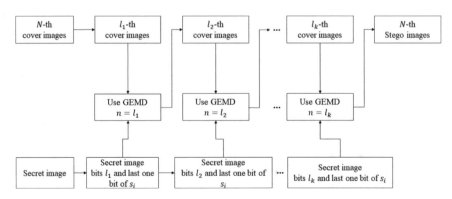

Fig. 4. Flow architecture diagram of the Proposed method

3.1   Segmentation of Secret Image

The pixel of secret image divided block k before the process of embedding, the block k represents the binary length of the secret image pixel. The block k also affects the number of stego images. k is the number of divided blocks of the secret image. l_k is a shared image added by the k-th level. According to the number of grayscale image bits, it must satisfy $l_1 + \cdots + l_k = N$ stego images.

The bits of the repeated embedding of this paper method do not need to be embedding in order. The bit can be selected at intervals, combined with the LSB concept, the purpose is to convert the value extracted by the stego image into a binary, and then compare whether the last bit is the same. When the receiver extracts the secret image pixels, the GEMD method can be used to calculate whether the duplicated bits are consistent with each other, that is, whether the masquerading image of the receiving end is falsified, as shown in the example: $162 = (10100010)_2$, it is divided into two parts $l_1 = 3$ and $l_2 = 4$, $(1010)_2$, $(00010)_2$, indicating that the bit position of the repeated hiding is the 8th bit, that is $(0)_2$.

3.2   The Method of Embedding and Extracting

The experimental method of this paper, N pictures of 512×512 size are used to create N images by hiding a secret image of 512×512 size. The pixels $x_{(l_1,i)}$ at the same position are respectively taken out from the l_1 cover images, and the cover image pixels are modified by the GEMD method. Secondly, select the pixels $x_{(l_2,i)}$ of the same position in l_2 images, and a total of l_2 images are embedding in the GMED mode. The note is that the length of the secret image bits is including l_2 and the last one bit of s_i, the pixels of stego image modification is still modified according to the GEMD.

Embedding phase:
Input: $H \times W$ cover images N, secret images S, l_1, ..., l_k
Output: stego images N

(1) To sequentially read N pixels of images $x_{(1,i)}, x_{(2,i)}, \ldots, x_{(N,i)}$ and pixels of secret images s_i.
(2) To select l_1 images to be embedding from the N images.
(3) The secret image pixel s_i is converted into binary, the bit of l_1 bit and the last one bit of s_i are taken, the GEMD mode is calculated, and l_1 image pixels are modified.
(4) To select l_2 images to be embedding from the N images.
(5) Converting the secret image pixel s_i into binary, taking l_2 bit and s_i last one bit from the not yet embedded, performing GEMD and modifying l_2 image pixels.
(6) To select l_k images to be embedding from the N images.
(7) Converting the secret image pixel s_i into binary, taking l_k bit and s_i last one bit from the not yet embedded, performing GEMD and modifying l_k image pixels.
(8) The secret image pixels s_i are all embedding in stego images N.

In the extraction process, as long as the l_{k-1} and l_k stego images are substituted into the formula (3) and then converted into binary, the bits can be obtained, and then the last calculated by l_{k-1} is compared with each other. If the bit is different with the last bit of l_k, the stego image has been tampered with by others.

Extracting phase:
Input: $(l_1 + l_2 + \cdots + l_k)$ stego image, $l_1, \ldots, l_k$
Output: a similar or complete secret image

(1) The obtained l_i stego image is sequentially taken out of the pixel $x_{(1,i)}, \ldots, x_{(k,i)}$.

(2) The secret image pixels $s'_{l_1+1}, s'_{l_2+1}, \ldots, s'_{l_k+1}$ are calculated by the GEMD method.

(3) To convert s'_{l_i+1} and $s'_{l_{i+1}}$ to binary, and discriminate the last one bit of s'_{l_i+1} and $s'_{l_{i+1}}$ whether is the same. If they are the same, the stego image is not destroyed and extract a complete secret image; else it means that the stego image has been destroyed and only extract a similar secret image.

4   Experimental Results

The first section introduces the requirements for quality of stego image. The results of this experiment are based on image quality requirements, using Mean Squared Error (MSE) and Peak Signal to Noise Ratio (PSNR), can determine whether the performance is superior, the formula is as follows:

$$\text{MSE} = \frac{1}{M \times N} \sum_{i=1}^{M} \sum_{j=1}^{N} \left(p'_i - p_i\right)^2 \tag{5}$$

$$\text{PSNR} = 10 \times log_{10}\left(\frac{255^2}{MSE}\right) \tag{6}$$

The grayscale image size of the experimental result is M × N, p_i is the cover pixel, and p'_i is the stego pixel. However, the MSE is the difference between the cover image and the stego image, the PSNR of the stego image is calculated by using Eq. (6). The larger the value, the more similar to the cover image. The unit of PSNR is dB(Decibels). In general, PSNR $\geq$ 30 dB, which means that the human vision can not recognize the difference of the image after disguise, it is also that the related data hiding technology referred to.

The experimental environment is Python 2.7, using seven 512 × 512 grayscale cover images and a 512 × 512 secret image, resulting in seven stego images. The method of this paper can also use the same size, but different cover images were embedding. The extracted secret image pixels have repeat one bit, but do not need to be embedding in order, in order to improve security and ensure that the message has not been tampered and destroyed by others. 0 shows the original image and seven stego images shown in this mothed, and 0 shows the secret image. When $N = 7$, divided two blocks $(l_1, l_2) = (3, 4)$, the first $n = l_1$, the modified stego image is Fig. 5(b)–(d); the

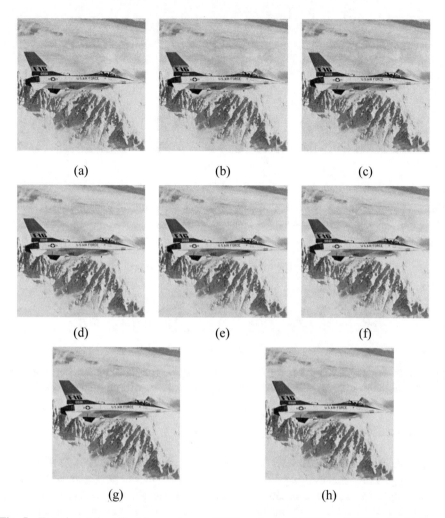

Fig. 5. Experimental results (a) cover image; (b) Stego image 1 PSNR = 50.7991 dB; (c) Stego image 2 PSNR = 50.7937 dB; (d) Stego image 3 PSNR = 50.8021 dB; (e) Stego image 4 PSNR = 51.1403 dB; (f) Stego image 5 PSNR = 51.1446 dB; (g) Stego image 6 PSNR = 50.8853 dB; (h) Stego image 7 PSNR = 50.8672 dB

second $n = l_2$, the modified stego image is Fig. 5(e)–(h), PSNR $= \infty$ represents the same as the secret image.

The average value calculated by 2521 grayscale images is shown in Fig. 6. When $l_1 = 3$, the pixels of stego images 1–3 are modified; and $l_2 = 4$, and the pixels of stego images 4–7 are modified. Since this method is for enhanced security, the calculated PSNR will be similar to the value embedding by the GEMD. However, since this paper

(a)   (b)   (c)

Fig. 6. (a) secret image; (b) $l_1 = 4$, PSNR = 29.2071 dB, extracted a similar secret images; (c) $l_1 + l_2 = 7$, PSNR $= \infty$, extracted a complete secret images

Table 1. The average PSNR of 2521 strgo images

	Stego images						
	Stego1	*Stego2*	*Stego3*	*Stego 4*	*Stego 5*	*Stego 6*	*Stego7*
PSNR	50.8	50.8	50.8	51.0	51.0	51.0	51.0

uses secret images for hiding, if use secret messages to embed, then according to the number of stego images, different secret message content will be displayed (Table 1).

5   Conclusions

This paper proposes a secret image sharing method combined with GEMD data hiding technology, which makes it possible to discriminate whether or not the stego image has been tampered with, and to ensure that the extracted secret image can be correct. The proposed method not only retains the original features of GEMD, but also maintains an image quality PSNR > 50 dB. However, this paper uses grayscale images. If use color images as the carrier and the embedding secret messages are grayscale images, then the payload will increase.

Acknowledgment. This work was supported in part by the Ministry of Science and Technology of the Republic of China under Contract No. MOST 107-2221-E-224 -008 -MY2 and MOST 108-2218-E-492 -006 -.

References

1. Blakley, G.R.: Safeguarding cryptographic keys. In: International Workshop on Managing Requirements Knowledge (AFIPS), vol. 48, pp. 313–317 (1979)
2. Chan, C.S., Liu, C.L., Tsai, Y.: Dual image reversible data hiding by modifying least significant-bit matching revisited rules. J. Electron. Sci. Technol. **15**, 217–221 (2017)

3. Chan, C.K., Cheng, L.M.: Hiding data in images by simple LSB substitution. Pattern Recogn. **34**, 469–474 (2004)
4. Fridrich, J., Goljan, M., Du, R.: Detecting LSB steganography in color and gray-scale images. IEEE Multimedia **8**, 22–28 (2001)
5. Kieu, T.D., Chang, C.C.: A steganographic scheme by fully exploiting modification directions. Expert Syst. Appl. **38**, 10648–10657 (2011)
6. Kuo, W.C., Wang, C.C.: Data hiding based on generalized exploiting modification direction method. Imaging Sci. J. **61**, 484–490 (2013)
7. Kuo, W.C., Wang, C.C., Hou, H.: Signed digit data hiding scheme. Inf. Process. Lett. **116**, 183–191 (2011)
8. Lai, I.J., Tsai, W.H.: Secret-fragment-visible mosaic image–a new computer art and its application to information hiding. IEEE Trans. Inf. Forensics Secur. **6**, 936–945 (2011)
9. Lee, C.F., Weng, C.Y., Chen, K.C.: An efficient reversible data hiding with reduplicated exploiting modification direction using image interpolation and edge detection. Multimedia Tools Appl. **76**, 9993–10016 (2017)
10. Li, P., Yang, C.N., Zhou, Z.: An efficient reversible data hiding algorithm using two steganographic images. Digit. Sig. Process. **50**, 98–109 (2016)
11. Li, P., Yang, C.N., Wu, C.C., Kong, Q., Ma, Y.: Essential secret image sharing scheme with different importance of shadows. J. Vis. Commun. Image Represent. **24**, 1106–1114 (2013)
12. Li, P., Yang, C.N., Zhou, Z.: Essential secret image sharing scheme with the same size of shadows. Digit. Sig. Process. **50**, 51–60 (2016)
13. Liu, Y.X., Yang, C.N., Sun, Q.D.: Enhance embedding capacity of generalized exploiting modification directions in data hiding. IEEE Access **6**, 5374–5378 (2017)
14. Liu, Y.X., Yang, C.N., Chou, Y.S., Wu, S.Y., Sun, Q.D.: Progressive (k, n) secret image sharing Scheme with meaningful shadow images by GEMD and RGEMD. J. Vis. Commun. Image Represent. **55**, 766–777 (2018)
15. Lu, T.C., Tseng, C.Y., Wu, J.H.: Dual imaging-based reversible hiding technique using LSB matching. Sig. Process. **108**, 77–89 (2015)
16. Lu, T.C., Wu, J.H., Huang, C.C.: Dual-image-based reversible data hiding method using center folding strategy. Sig. Process. **115**, 195–213 (2015)
17. Puteaux, P., Puech, W.: An efficient MSB prediction-based method for high-capacity reversible data hiding in encrypted images. IEEE Trans. Inf. Forensics Secur. **13**, 1670–1681 (2018)
18. Qu, X., Kim, H.J.: Pixel-based pixel value ordering predictor for high-fidelity reversible data hiding. Sig. Process. **111**, 249–260 (2015)
19. Shamir, A.: How to share a secret. Commun. ACM **22**, 612–613 (1979)
20. Wang, C.C., Kuo, W.C., Huang, Y.C., Wuu, L.C.: A high capacity data hiding scheme based on re-adjusted GEMD. Multimedia Tools Appl. **77**, 6327–6341 (2017)
21. Wang, R.Z., Lin, C.F., Lin, J.C.: Image hiding by optimal LSB substitution and geneticalgorithm. Pattern Recogn. **34**, 671–683 (2001)
22. Wang, Z.H., Di, Y.F., Li, J., Chang, C.C., Liu, H.: Progressive secret image sharing scheme using meaningful shadows. Secur. Commun. Netw. **9**, 4075–4088 (2016)
23. Yang, C.N., Chu, Y.Y.: A general (k, n) scalable secret image sharing scheme with the smooth scalability. J. Syst. Softw. **84**, 1726–1733 (2011)
24. Yang, C.N., Hsu, S.C., Kim, C.: Improving stego image quality in image interpolation based data hiding. Comput. Stand. Interfaces **50**, 209–215 (2017)
25. Yang, C.N., Li, P., Wu, C.C., Cai, S.R.: Reducing shadow size in essential secret image sharing by conjunctive hierarchical approach. Sig. Process. Image Commun. **31**, 1–9 (2015)
26. Zhang, X., Wang, S.: Efficient steganographic embedding by exploiting modification direction. IEEE Commun. Lett. **11**, 1–3 (2006)

Device Discovery Techniques
for Industrial Internet of Things
Through Predictive Analytic Mechanism

Santosh Pattar[1]([✉]), Ritika[1], Darshil Vala[1], Rajkumar Buyya[2],
K. R. Venugopal[3], S. S. Iyengar[4], and L. M. Patnaik[5]

[1] IoT Lab, University Visvesvaraya College of Engineering, Bengaluru, India
santoshpattar01@gmail.com
[2] University of Melbourne, Melbourne, Australia
[3] Bangalore University, Bengaluru, India
[4] Florida International University, Miami, USA
[5] National Institute of Advanced Studies, Bengaluru, India

Abstract. Maintenance and management of a manufacturing process involve the collection and processing of the machine data. Integral data from a running machine is to be gathered first to accomplish the effective management of the machine in an industry. This collected data provides the current status of the machine and eventually helps in predicting the machine failures beforehand. These sequences of events are called predictive maintenance of a machine. The approach of device discovery helps to predict the machine failures in a better way by transmitting the machine data to edge devices. In the proposed work, machine failures are predicted with the help of aptly designed detection engines. These engines confirm the machine failures by comparing the current status of the machine data with the set of pre-defined rules. Therefore, a machine is serviced right before any unexpected failure and thereby stops the unusual crash-down of a device.

Keywords: Device discovery · Failure detection · Ontology · Predictive maintenance · Industrial Internet of Things

1 Introduction

Industrial Internet of Things (IIoT) is an applied domain of Internet of Things (IoT) whose primary purpose is to provide an efficient management system for Industrial applications. It works on the principle of building systems by adopting a network of devices that are embedded with sensors to retrieve necessary data and use the same to monitor the industrial equipment for effective and efficient management. For example an IIoT application designed and implemented for the textile industry, monitors the spindles used for weaving and advice when to change or service them before they breakdown [11].

© Springer Nature Switzerland AG 2020
L. C. Jain et al. (Eds.): SICBS 2019, AISC 1145, pp. 76–89, 2020.
https://doi.org/10.1007/978-3-030-46828-6_8

Predictive maintenance system has to deal with risks and significant challenges (that are explained in subsequent sections) that prevent it from the effective operation. Thus, questions like "What if the service done to any equipment is carried out way too earlier than it is required?" have to be addressed. It is a demanding task for a predictive maintenance system to determine the exact time for replacement or assistance of the devices [22].

Device discovery aims to search for an object that closely matches a given query to complete a required function. Many parameters like the location of the device, network identity document (ID), protocol, *etc*, are required to identify a device [14]. The implementation process of device discovery strategy in industries improves reliability of the devices, communication, time complexity, *etc*, leading to a better outcome in the production process [2]. It is complex task to locate a device due to the presence of a large number of devices and heterogeneity among them. The issue of scalability also arises when a large number of devices are available to perform the same task in different ways and still produce the same output (continuous monitoring of devices and performance analysis is to be cross-checked for every such device in order to select the optimal one). This process consumes a huge amount of time due to the large size of the network and complex relationship among them [21].

1.1   Motivations

Predictive Maintenance is the most recently researched topic which explains how to increase the lifespan of equipment used in any industry. The study of existing work reveals that fault monitoring and condition of the machinery is supervised so that unpredicted breakdown is averted. Device discovery approach enables the detection of physical connections around so that the devices can communicate with each other to build a social relationship. They have a set of rules according to which the devices are standardized [13]. There are multiple drawbacks in the current system of device discovery some of which includes increased power consumption. Moreover, these prevailing criteria of device discovery are not applicable for ultra-dense networks [8].

The proposed predictive maintenance model is based on the set of rules that are already defined and stored in the database. When the sensor gives the surrounding readings, it is mapped with the rules present in the database. These rules define the failure conditions and are defined in simple if-else clauses.

1.2   Contributions

The contributions of this paper are as follows.

– *Ontology Model for PdM*: We have developed an ontological model for PdM. This model is designed based on the generic features of the machine and its failure and helps to monitor the system through semantic technologies. We also demonstrate a use case example and its application.

- *Reference Architecture*: A reference architecture is laid out that depicts the predictive maintenance of each device and these devices are regulated using edgent components.

1.3 Organization

The rest of paper is systematically organised as follows. The background work and literature are surveyed in Sect. 2. Use case scenario description is specified in the subsequent Sect. 3. Nextly, in Sect. 4 with the help of system architecture, our approach of resource discovery for predictive maintenance is explained. Sequentially, the experimental setup is shown along with dataset description and implementations in Sect. 5. This also includes evaluations, results and discussions. Lastly, we conclude in Sect. 6.

2   Literature Survey

In this section, we describe some of the most recent works in the IIoT for predictive maintenance and device discovery techniques, in their respective subsections.

2.1   Related Works on IIoT

Rapid advancements in the manufacturing techniques and development have led to the increased use of computational methods to overcome challenges such as productivity, management, and effective resource utilization. A recent paper [1] briefed about the use of IIoT healthcare applications for context-sensitive access to the information. Similarly Liao *et al.* [10] systematically reviewed the insights and literature of IIoT which finds the root cause of product failure along with inclusion-exclusion criteria. Jeschke *et al.* [7] describes the use of IIoT in the manufacturing of cyber systems and other applications. They concluded that an increase in adaptability and robustness plays a major role in the cyber-physical system for smart factories. In the following paragraphs, we review a few recent publications that address the predictive maintenance issue in IIoT.

Huynh *et al.* [5] proposed a parametric predictive maintenance decision-making framework that involves no risks for maintenance. It provides generic and flexible maintenance along with improved performance model. However, the methodology is applicable to a single system only causing sub optimal results with limited resources (that promote in high inspection cost). Likewise, Wang *et al.* [20] implemented a predictive maintenance system based on event-log analysis. The most prominent feature of the selection method used for model construction that can be customized and optimized for any kind of equipment. False alarm rate of the system is not handled efficiently in this work which leads to wastage of human labor and replacement cost without the display of system error log.

Vianna and Yoneyama [19] worked on optimization techniques for redundant systems in aircraft subjected to numerous wear profiles. The operation cost estimates and identification of future degradation (is the most favourable gain whereas the unfavourable opinion is that) the technique does not incorporate troubleshooting tasks while planning.

2.2   Device Discovery Techniques

A device discovery technique aims to locate an appropriate device matching the requirements based on various properties of the devices, relationship with other devices, *etc* [15,16]. In the following paragraphs, we present a survey on such recent techniques for device discovery in IoT.

Suntholap *et al.* [18] demonstrated the intelligent device discovery in IoT for the domain of robot society. The system is fast, scalable and makes use of a set of criteria for device ranking such that, it measures the device's degree of social relationship, clustering coefficient and betweenness among them. However, concerns like how to standardize the expression of computing requirements are not considered.

Ngu *et al.* [12] surveyed on middle-ware available for the IoT. Their work is broadly focused on enabling technologies using middleware and its related issues. The advantage is that it supports heterogeneity among the IoT devices and is a lightweight platform. But, the system is dependent on the context and forces the users to create their IoT applications according to that context only. Ishino *et al.* [6] discovered relay mobile device with proximity services for user-provided IoT networks. These services are feasible with reduced traffic and improve the existing crowdsourcing based application which can be reused. However, these services are more than the number of user equipment along with their deployment environment and thus the system is not scalable.

Device discovery system proposed by Epstein *et al.* [4] includes a data storage medium that is used for storing clustering data structure. But, the security of this data is not addressed by the proposed work. Although the system includes a processor for device identification helps with decision making, there is an issue of complexity over-heading in hardware system architecture.

Lakshmanan *et al.* [9] worked on the concept of methods and systems for device detection and authorization in IoT framework. The proposed methodology builds a time schedule of proximity events and ranks them according to the assigned weighting factor of every device. However, interrelated proximity events do not consider the dynamic factors where a device's attributes can change abruptly for several devices.

The concept of energy efficient device discovery was proposed by Sharma *et al.* [17] for reliable communication in IoT which is based on 5G. The system provides energy, offloading and fault tolerance models. Due to the extra amount of energy that is spent to evaluate percentage packet loss energy expenses increase. Device-to-Device (D2D) communication technologies is explained by Bello *et al.* [3] with a major focus on network layer interoperation in the IoT. Scalable integration and interoperability in D2D technology is an added advanced feature but the TCP/IP protocol stack is limited for future implementations of D2D communication.

3   Use-Case Scenario and Predictive Maintenance Ontology

In this section, we outline the research challenges in predictive maintenance of devices in IIoT in detail with the help of relevant use cases and examples. These expected results benefit the stakeholders by improvising the life expectancy of the equipment. For each such failure, we construct a rule and design a Predictive failure Detection Engine (PDE) to analyze the present condition of the machine and thus, aid in its predictive maintenance. The data obtained from the sensors is fed to a processing unit PDE, that identifies the failure condition based on a certain set of rules. The output of the same is further processed by the Predictive maintenance Detection Engine (PmDE) to decide on the failure of the machine.

3.1   Boiler

A boiler is a most commonly used machine to generate steam in industries that manufacture automobiles, locomotives *etc.* These boiler machines play a crucial role in the functioning of the industries and their sudden unexpected failure may lead to heavy financial loss and also pose a threat to the safety of the workers. Figure 1 depicts the boiler's PmDE.

Some of the root causes for failure of a boiler tank are high-temperature creeps, substantial tube well thinning or graphitization of matrix probe. There can be one or more than one such conditions that cause failure. Every condition is sensed and detected by using and processing data gathered from various sensors that are connected to the boiler tank.

We have designed rules for boiler tank failure based on these conditions as follows. We formulate two rules for temperature creep and wall-thinning of the boiler to detect an early failure. Firstly, the boiler failure Rule 1 (R_1) says,

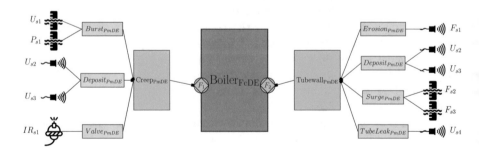

Fig. 1. Boiler predictive maintenance engine

```
IF ((waterLevel > maxWaterLevel) AND (waterPressure < minPressureLevel))
   OR ((sediment > maxSedimentLevel) AND (limescale > maxLimescaleLevel))
   OR (fillValve == broken)
THEN
  tempretureCreep == true
```

When the water level and the water pressure inside the boiler tank exceeds a given threshold value then a boiler can burst. Similarly, another condition like sediment/limescale deposit can be monitored with the help of deposit-sensors. Also if the boiler tank has a broken fill-valve, the possibility of temperature creep is also present. Secondly, the boiler failure Rule 2 (R_2) is expressed as:

```
IF (waterVelocity > maxWaterVelocity)
   OR ((sediment > maxSedimentLevel) AND (limescale > maxLimescaleLevel))
   OR ((steamFlow <= minRate) AND (firingRate >= max ))
   OR (tubeLeak == true)
THEN
  substantialTubeWallThininng == true
```

Here, the conditions are erosion, deposit, surge and tube leak that detect the substantial tube wall thinning and boiler tube blocking state. If the water velocity is exceeding the maximum velocity inside the tank, then erosion of the tank occurs. The condition for failure due to deposit is similar to that of previous rule condition *i.e.,* limescale/sediment deposit. Surge refers to the water flow rate inside the boiler tank. When a boiler is started and water begins to rush inside the tank, that flow rate of water is referred to as firing rate. Whereas the tube leak condition can alone determine the wall thinning of a boiler tank. Hence, we can derive to the conclusion that these conditions lead to the boiler tank failure scenario. To detect the malfunctioning or failure of the boiler, we construct a PmDE to process the data obtained from various sensors that are installed at/on the boiler machine.

```
IF (tempCreep == true)
   OR (tubewallThinning == true)
THEN
  boilerFailure == true
```

3.2   Predictive Maintenance Ontology

Predictive Maintenance Ontology (PMO) gives a backbone architecture for the entire model by defining structures. These defined structures assist the PMO to detect machine failures. The conditions that cause a machine to fail are predefined and protocols are set such that the failure can be predicted. PMO ontology is extremely necessary because it builds the system model by taking semantic knowledge as its basic foundation. Semantic knowledge is a domain-oriented language which takes the conceptual ideas and frames a semantic model upon which an ontology can be constructed. PMO not only defines the failure conditions of a machine but also predicts the failure and instructs how to avoid such conditions.

The components of PMO includes machines, failures, failureConditions, predictive Maintenance Detection Engine (PmDE) and Failure Condition Detection Engine (FcDE), as shown in Fig. 2. PMO is constructed with respect to Industrial domain, specifically considering the use cases of elevators, turbines, and boilers. The detection engines (DE) utilize the knowledge from other components to derive to a conclusion and thereby deciding the machine failure. For example, the boiler machine FcDE detects the conditions where there are possibilities for a boiler to fail. To detect these failure conditions, the FcDE uses data from failureConditions which is another component of PMO.

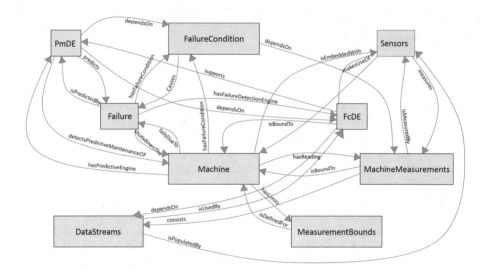

Fig. 2. PMO architecture

Multiple classes, sub-classes, properties *etc* are defined in the PMO. Classes are related to one another by defining relationship properties between them. Machine class defines the three use cases considered, failure class includes the types of failure for every instance of machine class. Nextly, the failureCondition class holds the scenarios which can cause a breakdown. A sensor class is implemented which has several subclasses. These comprise all the sensors required in PMO to monitor the current condition of a machine. There are two DEs enforced as classes and they're FcDE and PmDE. FcDE detects the failure of a machine due to some context or component failure whereas PmDE predicts the maintenance required by a machine through FcDE. Machine measurements are characterized in another class and their bounds specify the range of machine attribute measurements. Set of data items are captured and are transferred to the users sequentially and continuously at certain intervals of time. These set of data items are illustrated in the data stream class.

Object property characteristics show the attributes of a particular object. There are several functions available that give a better framework for the defined object properties like functional, transitive, reflexive, etc. The object properties denote a connotation between two classes. For example, failure conditions cause failure. Here, failure conditions and Failure are two different classes and causes is the object property defined with failure condition as domain and failure as a range. There can be one/more object properties between any two classes, hence to distinguish uniquely, the functional characteristics are provided. Every object property in the PMO is defined with a specific domain and range (which happens to be class again).

Data properties are the attributes that define a class. For example, sensor attributes in a data property that defines the sensor class and the information like sensitivity, linearity, accuracy, range *etc* can be derived for a particular sensor instance. These are listed in Table 1.

Table 1. PMO ontology details

Domain	Object property	Range	Description
– PmDE – FcDE	dependsOn	– FcDE – Failure condition	Failure depends on failure condition
Machine failure	hasFC	Failure condition	Machine fails on meeting one of its failure condition
Failure	hasFDE	– Failure condition – FcDE	Failures are detected using detecting engines
Measurement bounds	isDefinedFor	– Machine measurements – Machine	Limiting bounds are defined for measuring every parameter of machine
Machine measurements	isUsedBy	FcDE	Failure condition detection engine uses machine measurements

4   PdM Architecture

In this section, we discuss the architecture designed for the discovery techniques of devices for industrial internet using predictive analytic mechanism.

4.1   Overview of Architecture

The PM architecture can be explicitly explained based on a layered architecture as shown in Fig. 3. This architecture comprises four layers, each with specific functionality. Here, the layered architecture focuses on the processing of edgent components. The PM architecture includes four layers: Sensor Layer (SL), Topology Layer (TL), Provider Layer (PL), Application Layer (AL). The sensor layer lies at the bottom part of the architecture. It holds the collection of sensors that are embedded on the machine to monitor its working condition. These sensors collect the information about its surrounding environment and a set of data is taken for consideration. These data collected from the sensor layer is given to the second layer in architecture *i.e.*, topology layer that creates a specific data stream out of the data collected from the sensor layer based on the failure conditions that are to be monitored by the experts. These data streams are fed above by the provider layer which is responsible for handling the execution of failure prediction and failure detection of the machine.

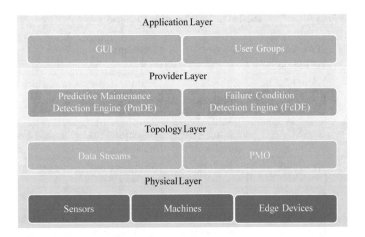

Fig. 3. PdM component architecture

Firstly, the *physical layer* consists of a hub, a machine and a group of edge devices. These three components are embedded and interconnected to one another in the physical layer itself. The machine components are further embedded with respective sensors to read the machine environment conditions. Secondly, the *topology layer* includes a set of databases that stores the data stream values and a PMO database that backlogs the populated datastreams. Thirdly, the *provider layer* frames four different data containers namely FcDE database (FcDE-DB), PmDE database (PmDE-DB), Rule-Set database (RS-DB) and Processed Data Stream (PDS-DB) database. The FcDE-DB holds the prediction results, RS-DB comprises of well-defined rules that determine machine failure, the PDS-DB incorporates the processed data stream from the previous layer. Finally, the components of *application layer* are the more relevant real world. The three main components of the application layer are users group, enterprises that use our application and lastly the repair consultant/handyman.

4.2 Workflow

The real-time sensors planted within the machine environment, reads the values to fetch its surrounding area conditions. Every stream is stored, pre-processed and populated into the PMO-DB. It encompasses all of the processed data stream values and sends the same to the provider layer. The FcDE-DB excavates the failure condition of a machine and compares the data with the input data streams whereas the PmDE-DB detects a fault in a machine by correlating the processed input data stream with pre-defined set of rules from RS-DB. The result of PmDE is given to the enterprise via the user interface in the application layer.

4.3 Example

Fault Detection in Boiler Machine is describes as example. There are two main conditions for a boiler machine to fail. It can fail either if there is creep in temperature of boiler machine or due to the substantial wall thinning. The circumstances that lead to temperature creep of a boiler machine are tank-burst, tank-deposit or broken-fill-valve. On the other hand, we have the boiler machine failure due to substantial wall thinning condition. There are four cases that margin the wall thinning conditions *i.e.,* tank-erosion, tank-deposit, tank-surge, and tank-tube-leak. Correspondingly, the erosion-FcDE, deposit-FcDE, surge-FcDE and leak-FcDE are devoted failure condition detection engines.

As mentioned earlier the sensors nested within a boiler machine are ultrasonic sensors, pressure sensors, infra-red sensors, and flow sensors. These dedicated FcDEs are fed with the corresponding data streams like the tempCreepDS (temperature creep data stream) and thinWallDS (wall thinning data stream). The tempCreepDS is further populated with burstDS, depositDS, and valveDS. Likewise, the thinWallDS is colonized with erosionDS, depositDS, surgeDS, and leakDS. The respective sensors for these, collect the datasets and send the data to PMO-DB. In turn, the PMO-DB compiles the rules set and analyzes the machine condition measurements, measurement bounds and hence PmDE gives the output in terms of fault detection.

5 Experiments and Results

5.1 Dataset Description

For the use-case boiler machine, we have taken nine sensors into consideration which measures required environmental parameters. We collected the sensor data from internet. These data sets are fed into the PdM machines in the form of input data.

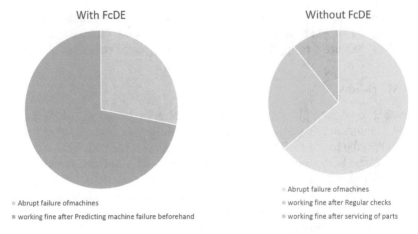

Fig. 4. Result evaluations

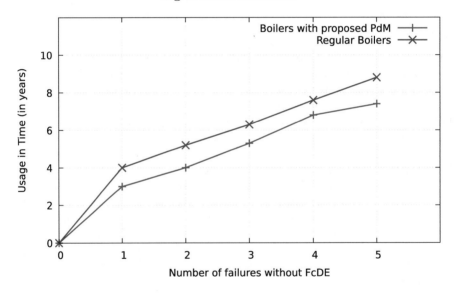

Fig. 5. Without FcDE

5.2 Implementation and Experimental Setup

We analyzed the proposed failure detection engines and compared the observations with old result statistics. Firstly, we created detection engines for every use case as shown in Fig. 4 and made use of reference PMO architecture. In order to detect the failure we require the sensor readings that are fit into the use case machines. Using edgent technology we get specific data streams from every sensor for *e.g.*, Temperature sensor gives temperature reading of boiler every 1 to 2 ms in the form of data streams. Similarly all the sensors that are involved

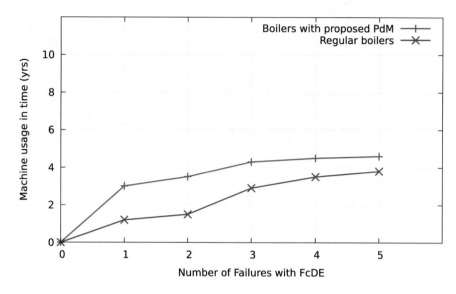

Fig. 6. With FcDE

in the process of failure detection of boiler, are activated and data streams are collected and then fed into detection engines.

As shown in Fig. 5 the number of failures grows when the machine is used for a longer period of time. Here, the proposed model of predictive maintenance is not being implemented. Therefore, the boiler machine failures are increasing linearly. In similar conditions when the proposed model of PdM is implemented, the results are shown in Fig. 6. Here, the number of failures are relatively less when compared.

6   Conclusions

This paper proposes a novel and efficient yet simple way of predicting a machine's failure beforehand. The set of rules are to standardize the conditions of machine parts. If the machine parts performance does not match the rule standards, the probability of machine failing is discovered. Apparently, these predictive maintenance engines are set up exclusive to a machine and hence is effective as there's no redundancy of data.

References

1. Al-Turjman, F., Alturjman, S.: Context-sensitive access in Industrial Internet of Things (IIoT) healthcare applications. IEEE Trans. Ind. Inf. **14**(6), 2736–2744 (2018)
2. Anjum, N., Yang, Z., Saki, H., Kiran, M., Shikh-Bahaei, M.: Device-to-Device (D2D) communication as a bootstrapping system in a wireless cellular network. IEEE Access **7**, 6661–6678 (2019)

3. Bello, O., Zeadally, S., Badra, M.: Network layer inter-operation of device-to-device communication technologies in Internet of Things (IoT). Ad Hoc Netw. **57**, 52–62 (2017)
4. Epstein, S., Darshan, E., Cain, H., Shali, M.: Device discovery system. US Patent App. 15/068,754, 14 September 2017
5. Huynh, K.T., Grall, A., Bérenguer, C.: A parametric predictive maintenance decision-making framework considering improved system health prognosis precision. IEEE Trans. Reliab. **22**(99), 1–22 (2018)
6. Ishino, M., Koizumi, Y., Hasegawa, T.: Relay mobile device discovery with proximity services for user-provided IoT networks. IEICE Trans. Commun. **100**(11), 2038–2048 (2017)
7. Jeschke, S., Brecher, C., Meisen, T., Özdemir, D., Eschert, T.: Industrial Internet of Things and cyber manufacturing systems. In: Proceedings of the Industrial Internet of Things, pp. 3–19. Springer (2017)
8. Kadechkar, A., Moreno-Eguilaz, M., Riba, J.R., Capelli, F.: Low-cost online contact resistance measurement of power connectors to ease predictive maintenance. IEEE Trans. Instrum. Measur. **68**, 4825–4833 (2019)
9. Lakshmanan, A., Osterweil, E., Fregly, A.: Systems and methods for device detection and authorization in a IoT framework. US Patent 9,633,197, 25 April 2017
10. Liao, Y., Loures, E.D.F.R., Deschamps, F.: Industrial Internet of Things: a systematic literature review and insights. IEEE IoT J. **5**(6), 4515–4525 (2018)
11. Liu, M., Yu, R., Teng, Y., Leung, V., Song, M.: Performance optimization for blockchain-enabled Industrial Internet of Things (IIoT) systems: a deep reinforcement learning approach. IEEE Trans. Ind. Inf. **15**(6), 3559–3570 (2019)
12. Ngu, A.H., Gutierrez, M., Metsis, V., Nepal, S., Sheng, Q.Z.: IoT middleware: a survey on issues and enabling technologies. IEEE IoT J. **4**(1), 1–20 (2017)
13. Pattar, S., Buyya, R., Venugopal, K.R., Iyengar, S.S., Patnaik, L.M.: Searching for the IoT resources: fundamentals, requirements, comprehensive review, and future directions. IEEE Commun. Surv. Tutorials **20**(3), 2101–2132 (2018)
14. Pattar, S., Dwaraka, S.K., Darshill, V., Buyya, R., Venugopal, K.R., Iyengar, S.S., Patnaik, L.M.: Progressive search algorithm for service discovery in an IoT ecosystem. In: Proceedings of the 2019 International Conference on Internet of Things (iThings) and IEEE Green Computing and Communications (GreenCom) and IEEE Cyber, Physical and Social Computing (CPSCom) and IEEE Smart Data (SmartData), pp. 1041–1048 (2019)
15. Pattar, S., Sandhya, C.R., Darshill, V., Buyya, R., Venugopal, K.R., Iyengar, S.S., Patnaik, L.M.: SoCo-ITS: service oriented context ontology for intelligent transport system. In: Proceedings of the 2019 The 7th International Conference on Information Technology: IoT and Smart City (ICIT-2019), pp. 1–6 (2019)
16. Pattar, S., Sandhya, C.R., Darshill, V., Chouhan, D., Buyya, R., Venugopal, K.R., Iyengar, S.S., Patnaik, L.M.: Context-oriented user-centric search system for the IoT based on fuzzy clustering. In: Proceedings of the International Conference on Computational Intelligence, Security & IoT (ICCISIoT), pp. 1–14 (2019)
17. Sharma, V., Song, F., You, I., Atiquzzaman, M.: Energy efficient device discovery for reliable communication in 5G-based IoT and BSNs using unmanned aerial vehicles. J. Netw. Comput. Appl. **97**, 79–95 (2017)
18. Sunthonlap, J., Nguyen, P., Ye, Z.: Intelligent device discovery in the Internet of Things-enabling the robot society. arXiv preprint arXiv:1712.08296 (2017)
19. Vianna, W.O.L., Yoneyama, T.: Predictive maintenance optimization for aircraft redundant systems subjected to multiple wear profiles. IEEE Syst. J. **12**(2), 1170–1181 (2018)

20. Wang, J., Li, C., Han, S., Sarkar, S., Zhou, X.: Predictive maintenance based on event-log analysis: a case study. IBM J. Res. Dev. **61**(1), 11–121 (2017)
21. Xiaowei, W., Jie, G., Xiangxiang, W., Guobing, S., Lei, W., Jingwei, L., Zhihui, Z., Kheshti, M.: High impedance fault detection method based on variational mode decomposition and Teager-Kaiser energy operators for distribution network. IEEE Trans. Smart Grid **10**, 6041–6054 (2019)
22. Yu, W., Dillon, T.S., Mostafa, F., Rahayu, W., Liu, Y.: A global manufacturing big data ecosystem for fault detection in predictive maintenance. IEEE Trans. Ind. Inf. **16**, 183–192 (2019)

SmileJob: A Lightweight Personalized Accompanying System for Home Security

Pin-Yu Lin[1], Yow-Shin Liou[1], Jian-Xing Li[1], Joon-Kui Liew[1], Po-Wen Chi[2], and Ming-Hung Wang[1]([✉])

[1] Department of Information Engineering and Computer Science,
Feng Chia University, Taichung, Taiwan
`qaz1232009@gmail.com`, `a757777@gmail.com`, `zz0978628963@gmail.com`,
`ahkui@outlook.com`, `mhwang@mail.fcu.edu.tw`
[2] Department of Computer Science and Information Engineering,
National Taiwan Normal University, Taipei, Taiwan
`neokent@gapps.ntnu.edu.tw`

Abstract. With the rapid development of deep learning techniques in computer vision, emotion detection has been an emerging topic for different industries, including healthcare, mobile entertainment, and even securities. These organizations leverage emotion detection and apply it to both real-time cameras and offline videos to provide more interactivity between users and devices. In this study, we provide a lightweight personalized accompanying system for home security using emotion detection and IoT devices. We implement our design and demonstrate a practical scenario using Nvidia Jetson Nano. According to our design, we hope the system can benefit the health monitoring and interactivity for those elderly living alone.

Keywords: Home security · Accompanying system · Deep learning · IoT applications

1 Introduction

With the popularity of deep learning techniques in computer vision, many applications related to the security monitoring at home and health care with the elderly are emerging. Among these applications, emotion detection has been an emerging topic for different industries, including healthcare, mobile entertainment, and even securities. People can use emotion detection techniques to acknowledge suspected people for further investigation. These organizations leverage emotion detection and apply it to both real-time cameras and offline videos to provide more services.

According to the above context, we observe that many of the healthcare applications are simply collecting and reporting statistics, lacking companionship and interaction which are important and essential in this digital age. With the rapid development of artificial intelligence and IoT (Internet of Things) toolkit, in this

L. C. Jain et al. (Eds.): SICBS 2019, AISC 1145, pp. 90–101, 2020.
https://doi.org/10.1007/978-3-030-46828-6_9

study, we propose to construct a lightweight personal accompanying system. The main objective of this work is to provide a conceptual prototype as a ubiquitous accompanying for both adolescents and the elderly. Our contributions in this work are three-fold:

- We leverage the power of deep learning in computer vision to monitor and identify the emotions of users.
- We also implement our design on a lightweight development board Jetson Nano to validate our proposal.
- In addition to emotion detection, we further provide the interactivity between users and our system by using LEDs. The interface is composed of a small number of LEDs to feedback users using graphics according to the user's emotions and actions.

The remainder of this paper is organized as follows. We discuss the related studies of our work in Sect. 2. In Sect. 3, we present the design of our framework. In Sect. 4, we present the implementation detail of our proposal. In Sect. 5, we provide a series of demonstration and evaluation of our proposal. The conclusion and future issues of this study are described in Sect. 6.

2   Related Works

In the modern age, loneliness remains a significant issue in our society [2]. For people who are not familiar with social activities, the lonesome feeling may result in depression and negative emotions [9]. However, with the popularity of the Internet and mobile devices, instead of making friends in the real world, an increasing number of users may turn to the Internet, dating applications, and the virtual world to discover another way to escape from loneliness [5,7,10]. On the other hand, in Taiwan, the aging society results in the increasing population of Alzheimer's disease patients. According to the Ministry of Health and Welfare, the Taiwan Alzheimer's Disease Association [6] conducted a random sample of the population over the age of 65 and conducted a visit to the field by specially trained interviewers. From the results, 1 in 20 people aged over 65 in Taiwan suffers from dementia.

With the increasing popularity of the lightweight development board such as *Raspberry Pi*, a number of smart healthcare devices are proposed [4,8,11]. Such devices not only help to monitor health conditions but also contribute to the lightweight camera system for home security [1,3]. Though the proposals are practical, one of the major weaknesses is the restricted computing power of the board. Thus, applications with intensive computing and real-time interactivity are not easy to be deployed on a single board. To conquer this, in this study, we combine the development kit with small size and reasonable computer power with the remote server and a set of LED displays to propose a personalized accompanying system.

3   Proposed Design

Our proposed design is composed of the following three components.

1. Nvidia Jetson Nano with Pi camera
2. Led Matrix Output System
3. Prediction Model Building

The currently proposed method is mainly based on Openpose and Nvidia Jetson Nano's photographic lens to realize the integration of human body detection, light display, and image capture. Nvidia Jetson Nano stores the image in the local Redis database and sends the image to the server for identification. Then the server sends the result back to the Nvidia Jetson Nano, and the web server that executes the image locally and controls the LED bulb through the GPIO. In the following sections, we will describe our proposal and implementation in detail.

3.1   Proposed System Architecture Building

To build the framework, SmileJob, we initially use the Jetson Nano with pi-camera as the basic components. We build a learning model to recognize the face emotions and skeletons on a remote server with strong computing power. According to the above components, we can identify user emotions and use a series of LEDs to present the corresponding graphics to the user. According to the result, we also build a DBMS system to save the result of the previous activities for further enhancement. The diagram of our architecture is shown in Fig. 1.

Nvidia Jetson Nano with Pi Camera. The Jetson family has supported MIPS-CSI cameras, which stands for Camera Serial Interface. This protocol is for high-speed transmission between cameras and host devices. Currently, one of the popular CSI-2 cameras is the *Raspberry Pi* Camera Module V2. The cameras have a ribbon connector that connects to the board using a simple connector. It connects directly to the GPU, not CPU, and is capable of 1080 p video encoding, 5 MP stills is pretty decent quality. Because it's attached to the GPU, the CPU will not be affected and retain its computing power for other processing.

Led Matrix Output System. The image data from the camera is stored in the Redis database. The system retrieves data from Redis for further visualization. We connect 8×8 led matrix and 74HC595 (Shift Register) to build a 16×16 output led matrix system as a prototype which is easy to prove our concept. We also use a web led editor which can simulate a led matrix to enable users to design the animation on the output system.

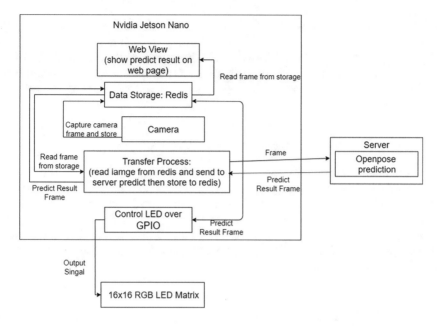

Fig. 1. Architecture diagram

Prediction Model Building. The models for processing image data from data storage are composed of face recognition model and skeleton prediction model. The face recognition model makes SmileJob be able to determine user emotions so that SmileJob can respond to the user according to their current mood. To analyze user emotions, face recognition could detect 68 facial landmarks in real-time. We use key facial structures to recognize the user's mood. On the other hand, the skeleton recognition model makes SmileJob be able to realize body actions. As a result, SmileJob can interact with users by understanding their poses.

3.2   Visualization

In order to present a suitable real-time visualization on LED, we design a web page that can control the LEDs to display the user drawing on the web page. This interface can help them to train their hand muscles and concentration. For patients with mild dementia, they cannot control their muscles very well, so we add a game similar to the white flag game. The machine uses the LED panel to display the action, and if the elder makes the same action on the machine, it gives feedback. This helps the muscles of the patient to slow down the speed of dementia. In addition, we use the photographic lens on the Nvidia Jetson Nano to let users interact with the machine with different experiences, not just plain input and ordinary screen. We assembled 4 LED panels and self-connected circuit to make a lightening output, and use pose detection to interact with the body movements.

4   Implementation

Our concept implementation is based on Nvidia Jetson Nano and an Nvidia GTX 1050 graphics card that can support Nvidia CUDA technology. The computing server is installed Linux OS and based on i5-8500 and 8 GB RAM for emotion and pose recognition.

4.1   LED Display, Wiring, and Wafer Control

We use 4 pieces of RGB 8*8 2388 LEDs matrix to display. To make the display different from the traditional LCD screen, we use several LEDs to construct an analog monitor. Therefore, we also need the breadboard and the wafer to control them. We use the 74hc595 wafer because it can use 3 pins to control 8 pins. However, a 74hc595 wafer outputs only 8 bits signal. Thus, we use 8 74hc595 wafers to control 4 pieces of RGB color and the row columns. The circuit diagram is shown in Fig. 2.

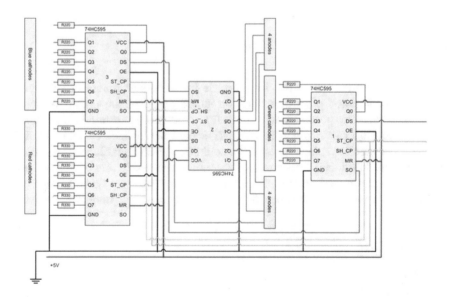

Fig. 2. Circuit diagram for signal control

4.2   Nvidia Jetson Nano Environment and Settings

(A) Nvidia Jetson Nano environment:
Our system input is composed of web pages and camera input; thus, we installed OpenCV and Flask on the Nvidia Jetson Nano. The Nvidia Jetson Nano is based on the ARM architecture, then we use the command "make"

to compile OpenCV and command "pip" to install Flask. When we finish the installation, we connect the camera supporting Nvidia Jetson Nano and import program to control the camera.

(B) Nvidia Jetson Nano image storage and processing:

For camera input, we use the *Raspberry Pi* camera toolkit, which Nano can support by mounting the lens on the Jetson Nano camera connector. The part of the program uses Opencv to process the captured image of the camera. When receiving the image captured by the camera, the size of the photo can be set by parameters. The processed image is encoded by base64 and stored in Data Storage. The other program will get the data in the Data Storage and transfer it to the server for processing.

(C) The Nvidia Jetson Nano display builds using GPIO:

We set up the server to receive the picture which is captured by Jetson Nano. When transferring the picture, it encodes utf-8 on the client, and then transfer to the server. After the server receives the data, it uses ByteIO to restore the picture, recognize and transfer to the web on the client. The web page shows two kinds of pictures, one is a recognized picture and the other is the original picture. Figure 3 shows the scenario diagram. When the camera takes a picture, it will send the picture to the server. After receiving the picture, the server will capture the feature points on the face of the person, save as a list type and input it to the model for expression. Then, it will return the result to the client and it can infer the data of expression to control LEDs to show corresponding graphics to interact with the user. For instance, when the machine detects disappointed or sad expressions, it takes the initiative to ask what the user needs help with. When detecting other expressions, it also shows another corresponding mode.

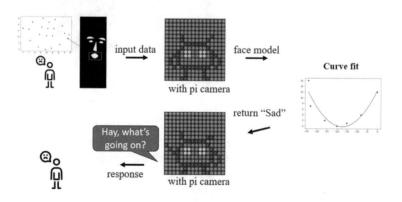

Fig. 3. The flow chart of our design

4.3 Server Environment and Settings

(A) Server environment:
For the reason that limb recognition requires using OpenPose that needs higher the computational and Jetson Nano can't process independently, we use the extra server to compute. Therefore, the server is required to install GPU which supports CUDA to maintain stable and optimize the OpenPose procession. However, installing OpenPose also needs a "make" command. We choice Nvidia GTX 1050 GPU to install on our server to process OpenPose.

(B) Server limb and expression recognition:
First, the server received the picture for decoding. Afterward, they use OpenPose to recognize. Incidentally, the OpenPose recognization is two kinds, one is COCO kind and the other is BODY_25 kind. Resulted from using GPU, we select BODY_25 kind to process faster than the other. On the opposite, if we use CPU to process, we will select COCO kind. The BODY_25 kind can recognize 25 points of the body, like knee and elbow etc., and these points in the list are fixed, so we can access the specific position from the list to get the point we need. For this reason, we can access it and the current point to recognize the user action. Lastly, the server will send the user action and recognize the user's body points to the client.

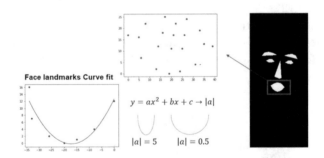

Fig. 4. The face landmarks model

Figure 4 explains the way about face recognition. Currently, the most common facial landmark model is 5 points and 68 points models which are provided by Dlib. Generally, it usually uses 68 points of landmark to correct. After we get facial features, we can use it to deal with facial correction. Then we carry on the step of recognition to increase the definition after making the facial correction. Therefore, Face align is a pretty important step of facial recognition. Besides, it will get on facial description after detecting 68 points. Then it calculates the arc degree of the edge and makes a description by using the parabolic formula. After that, it will describe those points into a face.

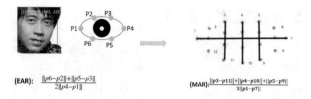

(EAR): $\dfrac{\|p6-p2\|+\|p5-p3\|}{2\|p4-p1\|}$ (MAR): $\dfrac{\|p3-p11\|+\|p4-p10\|+\|p5-p9\|}{3\|p1-p7\|}$

Fig. 5. The EAR and MAR calculation

Figure 5 depict the eye aspect ratio (EAR) and mouth aspect ratio (MAR) calculation. The EAR represents 6 coordinate points of the eye. Starting from the left corner of the eye, and then EAR will mark remaining points clockwise on the area around the eye. As there is a relationship between the width and height of these coordinates, we use this formula to avoid image processing. We can depend on it to confirm the user's current eye situation. However, we call Fig. 5 MAR formula is "L" and the distance between the first point and seventh point is "D". When the mouth is closed, the distance between the first point and the seventh point will increase, and the distance between top and bottom will decrease, causing L to decrease and D to increase as calculating MAR. Not only that, the program not only detects the smile but also depicts the circular arc of the lips edge using the parabola. We utilize the simple way to detect the user the current mood and if the user smile currently.

5   Results

Figures 6 and 7 show the implementation of the LED panel through the web page. In this way, users can draw their designed frame by clicking the light bulb component on the web, and the frame on the web would be displayed on the LEDs panel immediately. Furthermore, the number of frames created by users could be displayed as animations and also can be displayed on the LEDs panel at once. The drawn content provides a combination of three colors such as RGB.

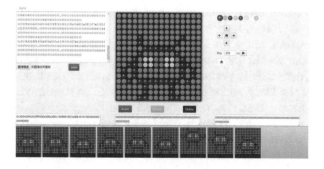

Fig. 6. Web page

Fig. 7. Web page manipulation LED panel

Fig. 8. OpenPose identification

A total of 7 colors can be used, and the drawing is completed. After that, the drawn content can be stored and can be used by reading the file when needed.

Figure 8 shows the image recognize by OpenPose. This image shows on the web page. When the camera continues to take a photo, the Jetson Nano also sends these to the server persistently. Next, the server will receive these images to detect the human body, hand, etc., and return the result to the client. The client will display the corresponding way to present. Since, we mainly identify the limb movements, judge the limb movements of the characters in the current image, and make the LEDs matrix display the respond differently, Therefore, no hand joint detection is performed, and the nodes of the fingers will not be displayed, only the body nodes of the character will be displayed.

Figure 9 presents that the system detected different expressions to display corresponding expressions. Then, it can be judged by the curvature of the lips and the MAR to determine the user's expression. After that, the way of comforts and corresponding machine expressions will be placed in this judgment method.

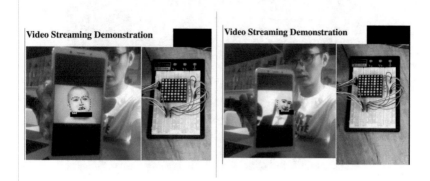

Fig. 9. Expression detection

Table 1. Expression and judgment value

Value	Expression
$MAR > 0.6$	Surprise
$Mbot_curse < 0.021$	Sad
$Mtop_curse > -0.007$	Smile

Table 1 shows that when the different values like MAR to detected the user's expressions. The mbot_curse represent the lower lip curvature, otherwise mtop_curse represent the upper lip curvature. For example, if the MAR is more than 0.6, the system will decided to surprise and show corresponding expressions on the LEDs.

Table 2. Reaction time

Start	End	Time(s)
Start: 2019-11-21 19:32:14.937	End: 2019-11-21 19:32:15.213543	Time: 0.276
Start: 2019-11-21 19:32:15.213666	End: 2019-11-21 19:32:15.470991	Time: 0.2573
Start: 2019-11-21 19:32:15.470991	End: 2019-11-21 19:32:15.7172	Time: 0.2413
Average: 0.2582		

Tabel 2 shows the start time after the photo was taken and the end time after the server returned the result of the recognition, and the execution time and the average time of the reaction are displayed.

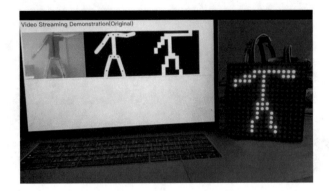

Fig. 10. Pixelated movements

Figure 10 shows that the system detects the body movements. The body movements will be converted into the pixel pictures by the program, and the pixel pictures will be transformed into a 16 × 16 matrix and converted into one-dimensional vectors, which are similar to the given posture comparison. When the comparison is greater than a certain value, it will identify the same action and a successful animation will be displayed on the matrix panel. Then it switches to the next round.

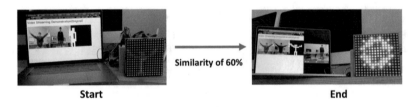

Fig. 11. Games: stretching exercises

Figure 11 shows that the system will give a pattern on the LED matrix. The user needs to make this action in front of the camera lens. When the system compares the two and the similarity is higher than 0.6, a firework animation will be played to represents success. Incidentally, the difficulty can be adjusted. The higher the difficulty is, the higher the similarity will be required. There are five rounds in a game, and the game ends when all five rounds are successful.

6 Conclusion

The research enables image recognition technology to be approachable for people as an IoT device for home security. According to our design, the proposed scheme can increase human interactions between those with loneliness or disability. When the lonely user is in a depressed mood, our proposal can play as a friend for the user. Also, when you are in a good mood, it would be someone who you can share your joy with. In addition, it can provide home entertainment to increase parent-child relationships.

In the future works, we aim to enhance the devices for long-term care and accompanying users. It can be used for home security. For example, to elderly people living alone or single people, our system can obtain images through the lens to identify the user's expression and body movements. Then it can monitor the user's security status or health condition from the lens of the device. If the user encounters an accident, the user's family can be notified immediately and conduct further actions.

References

1. Abaya, W.F., Basa, J., Sy, M., Abad, A.C., Dadios, E.P.: Low cost smart security camera with night vision capability using raspberry Pi and OpenCV. In: 2014 International Conference on Humanoid, Nanotechnology, Information Technology, Communication and Control, Environment and Management (HNICEM), pp. 1–6. IEEE (2014)
2. Cacioppo, J.T., Cacioppo, S.: Loneliness in the modern age: an evolutionary theory of loneliness (ETL). In: Advances in Experimental Social Psychology, vol. 58, pp. 127–197. Elsevier (2018)
3. Chowdhury, M.N., Nooman, M.S., Sarker, S.: Access control of door and home security by raspberry Pi through Internet. Int. J. Sci. Eng. Res 4(1), 550–558 (2013)
4. Gupta, M.S.D., Patchava, V., Menezes, V.: Healthcare based on IoT using raspberry Pi. In: 2015 International Conference on Green Computing and Internet of Things (ICGCIoT), pp. 796–799. IEEE (2015)
5. Hardie, E., Tee, M.Y.: Excessive Internet use: the role of personality, loneliness and social support networks in Internet addiction. Aust. J. Emerg. Technol. Soc. 5(1), 34–47 (2007)
6. Ministry of Health Welfare: Taiwan alzheimer disease rate survey, Taiwan elderly over 65 years old was 4.97%. (2013). https://www.mohw.gov.tw/cp-3211-23536-1.html. Accessed 12 Apr 2013
7. Karapetsas, A.V., Karapetsas, V.A., Zygouris, N.C., Fotis, A.I.: Internet addiction and loneliness. Encephalos 52, 4–9 (2015)
8. Kumar, R., Rajasekaran, M.P.: An IoT based patient monitoring system using raspberry Pi. In: 2016 International Conference on Computing Technologies and Intelligent Data Engineering (ICCTIDE 2016), pp. 1–4. IEEE (2016)
9. Matthews, T., Danese, A., Caspi, A., Fisher, H.L., Goldman-Mellor, S., Kepa, A., Moffitt, T.E., Odgers, C.L., Arseneault, L.: Lonely young adults in modern britain: findings from an epidemiological cohort study. Psychol. Med. 49(2), 268–277 (2019)
10. Nowland, R., Necka, E.A., Cacioppo, J.T.: Loneliness and social Internet use: pathways to reconnection in a digital world? Perspect. Psychol. Sci. 13(1), 70–87 (2018)
11. Pardeshi, V., Sagar, S., Murmurwar, S., Hage, P.: Health monitoring systems using IoT and raspberry Pi—a review. In: 2017 International Conference on Innovative Mechanisms for Industry Applications (ICIMIA), pp. 134–137. IEEE (2017)

Secure Authentication Key Agreement Protocol with eCK Model in Heterogeneous IoT Environment

Chien-Ming Wang and Chih-Hung Wang[✉]

Department of Computer Science and Information Engineering,
National Chiayi University, Chiayi, Taiwan
altis1119@gmail.com, wangch@mail.ncyu.edu.tw

Abstract. Internet of Things (IoT) has been developed increasingly and is a next wave in the era of computing. IoT is the concept in which many of the objects are getting networked and connected anytime and anyplace. It uses any network and any service in a heterogeneous circumstance. According to recent research, it is estimated nearly 30 billion devices on IoT; therefore, the security issues on IoT turn into a vital fundamental in present tendency. Although IoT brought convenience life, most existing IoT devices were controlled by multiple units (or parties), including owners, employees, and maintenance vendors. Distinct units may join different Certificate Authorities (CAs) for the public key management. How to share a common session key between IoT devices certified by distinct CAs is a practical issue worth to be studied. Therefore, we propose an authentication key agreement protocol using elliptic curve algorithm with extended Canetti-Krawczyk (eCK) security that can solve heterogeneous IoT session key problem for distinct CAs and satisfies all security requirements of AKA protocols.

Keywords: Internet of Things (IoT) · Heterogeneous IoT communication · Elliptic Curve Cryptography (ECC) · Authentication Key Agreement (AKA)

1 Introduction

Internet of Things (IoT) [7, 20, 24] is the emerging technology in the field of computer science and information technology. The internetworking of IoT physical devices embedded with sensors, actuators, software, electronics, web connectivity that enable these devices to exchange and collect data. This brings to greater value and service for networking. Each object is uniquely identifiable through its embedded control system. IoT transforms the real-world objects into intelligent virtual objects. Moreover, IoT is useful in gaining large amounts of data from far off location.

Due to IoT devices being uniquely identified, they can be controlled from anywhere and anytime. They are located with the help of IPv6 addressing protocol that offers large address space. IoT devices are controlled over the internet through sensorial and actuation capabilities of the devices that are monitored by the control system. In real world, IoT is implemented by combining many technologies such as Wireless Sensor Network (WSM), Radio Frequency Identification (RFID), IPv6 address scheme,

L. C. Jain et al. (Eds.): SICBS 2019, AISC 1145, pp. 102–112, 2020.
https://doi.org/10.1007/978-3-030-46828-6_10

Machine to Machine (M2M). The IoT operation allows a device to access a website via legal authentication; therefore, the IoT front-end secure authentication key agreement mechanism problem should be paid much attention.

IoT applications including smart buildings, machines, vehicles, etc. has made human life more comfortable. Protecting IoT privacy network is a challenge issue due to the heterogeneous environment of the IoT components. For the security assurance, IoT devices acquire their public keys from certificate authorities (CAs). One single CA can issue public keys to the users of a small size group. However, it is impractical to assume that one single CA can certify public keys for the users belonging to large amount of different groups, not to mention the entire country or the entire world. Further, it is also impractical to assume that distinct CAs provide the same system parameters and/or security capabilities.

There is a critical problem needed to be solved when two or more heterogeneous IoT devices communicate with each other. For example, when receiving an alert, the security guard with smart device/card wants to enter a smart building to check the security status. In this situation, the security guard needs a permission from the IoT smart lock. How do these two heterogeneous IoT devices, smart device/card and smart lock (as shown in Fig. 1), securely communicate with each other? It is assuming that the smart device/card and smart lock are made by different manufactures which own different system parameters. Therefore, the communication problem between the heterogeneous IoT devices that employ distinct system parameters is a crucial challenge needed to be solved.

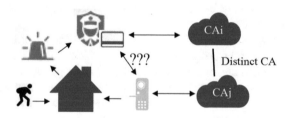

Fig. 1. Problem of key exchange between heterogeneous IoT devices.

Since IoT includes small objects that are considered to be limited capability, elliptic curve cryptography (ECC) [15, 21] can be an appropriate choice for the implementation in the public key cryptography. ECC is used to ensure high-level protection against the security risks such as privacy, integrity and confidentiality, and can be more efficient than RSA system. Elliptic curve discrete logarithm problem (ECDLP) is not easy to find the discrete logarithm of a point on an elliptical curve. ECC is widely used in many fields and has the benefits of smaller key size, less storage memory, and low computing power.

ECC was proposed by Miller [22] and independently presented by Koblitz [14] in the mid-1980s. As stated above, ECC takes smaller key size than RSA or other cryptographic algorithm, and provides the equal or higher level security that is why here we focus on ECC. Most of the IoT manufacture companies produce IoT devices that make use of ECC algorithm. ECC is widely used in IoT devices such as electronic tickets, bank cards, smart cards, etc. with the advantages of less memory storage. The system uses smaller size of keys that means it requires less memory and can reduce the computational power in operation. With a little bit of hardware assistance to speed up the operation, ECC provides better performance than the conventional cryptographic algorithms.

The first Key Agreement (KA) protocol in public channel was proposed by Diffie and Hellman [8], but their protocol did not give an authentication proof of the communication entities. The Authentication Key Agreement (AKA) [3, 16, 23, 25] uses symmetric or asymmetric cryptography to solve MITM problem and the advantage of AKA protocol is that it allows each entity has security communication to deliver a conversation or message exchange with an authentication.

The first security model with authentication security for AKA protocols was proposed by Bellare and Rogaway [1, 2]. Afterwards, numerous related researches in recent years have been proposed such as Canetti and Krawczyk model [6, 13]. The both Bellare-Rogaway and Canetti-Krawczyk models were developed on strong assumption that the adversary cannot reveal the long-term secret key or ephemeral secret key in session communication. LaMacchia, Lauter and Mityagin [18] developed the extension to the Canetti and Krawczyk (eCK) model which can resist all possible attacks including the leakage of long-term secret key and ephemeral secret key. Therefore, many scholars [5, 12, 26] proposed AKA protocols under eCK security model. The eCK model is secure if the adversary only knows either long-term secret key or ephemeral secret key of one entity. If the adversary can learn both long-term secret key and ephemeral secret key, he can easily perform an impersonation attack and the AKA protocol would be destroyed.

As mentioned above, the IoT front-end secure authentication key agreement mechanism problem and heterogeneous IoT devices with distinct system parameters communication problem are worthy to be studied. In this paper, we propose a secure IoT authentication key agreement with distinct CAs using ECC algorithm and eCK security protocol. Our protocol combines the long-term private key and the ephemeral private key of both IoT communication devices so that the security can satisfy the eCK model. Moreover, our protocol uses elliptic curve algorithm to provide higher memory storage and computation efficiency.

The remainder of this paper is organized as follows. Section 2 presents the IoT security phases. In Sect. 3, we review the assumptions and preliminaries of the protocol. The details of the proposed protocol are described in Sect. 4. The security analysis of the proposed protocol and the performance comparison are shown in Sect. 5. Finally, Sect. 6 concludes this paper.

2   Authentication Protocol for Heterogeneous IoT

The immense number of devices that connect through heterogeneous infrastructures increases the risk of attacking threats. Therefore, cryptography is the practice of encapsulating information that is the critical technology to keep information secure. In this section, three security phases that can be deployed in IoT are shown in the following [27].

Phase 1. Device authentication and communication security
Phase 2. Data procession security
Phase 3. Data storage security

Although these three IoT security phases focus on different IoT operating orientations, the device authentication and communication security phase is the most important one because it can protect IoT process in the front-end. Our contribution is to provide highly security IoT front-end authentication key agreement protocol in heterogeneous IoT devices communication using ECC algorithm, which also can satisfy AKA protocol security requirements.

3 Preliminaries

In this section, we briefly describe the mathematical techniques and related works that is used in our proposed protocol including elliptic curve cryptography, ECC assumptions, extension Canetti-Krawczyk assumptions and the NAXOS security model protocol.

3.1 Elliptic Curve Cryptography

Some public key cryptosystems like RSA, Diffie-Hellman key agreement are very suitable for the device with high power computation capability. However, for IoT, it is possible that low power computing devices cannot support such types of computations. Elliptic Curve Cryptography (ECC) is an emerging popular mathematical techniques because it requires less memory, communication bandwidth, and computational power when compared to other cryptosystems.

ECC works on the elliptic curve group with finite field. An elliptic curve E is given by an equation for cubic or quadratic polynomials. An equation will be of the form

$$y^2 = x^3 + ax + b,$$

where $a, b \in$ finite field F_p, and $4a^3 + 27b^2 \neq 0$.

O is a point at infinity. The set of points $E(F_p)$ forms an abelian group. $G = \{(x, y) : x, y \in F_p; (x, y) \in E(F_p)\} \cup \{O\}$ with the following rules:

1. **Identity**: $\forall P \in E(F_p)$, $P + O = O + P = P$.
2. **Point addition**: if $P_1, P_2 \in E(F_p)$, we can find $P_3 \in E(F_p)$ where $P_3 = P_1 + P_2$.

ECC involves key generation (public key and private key), encryption and decryption. User encrypts the message using receiver's public key. Receiver decrypts the ciphertext using his own private key.

3.2 eCK Security Assumption

In extension to the Canetti and Krawczyk (eCK) [18] security model, there are n honest parties $\{P_1, P_2, \ldots P_n\}$, each of whom is a PPT turing machine model. Their communications are controlled by the *Adversary M* which is also a PPT turing machine model. Each party neither knows whether the other parties exist or not, nor knows whether the message is sent from the *Adversary M* or the other parties. We called an AKA session which means that a particular instantiation of the AKA protocol executed by one entity. The matching session means that there are two entities P_i and P_j whose lawful execution of an AKA protocol consists of two AKA sessions. The AKA protocol is distinct between the executor being the initiator and the responder.

3.3 NAXOS Authentication Key Agreement Security Model

The NAXOS protocol [18] has n entities $\{P_1, P_2, \ldots P_n\}$ and each of them is modeled as a PPT turing machine. Furthermore, the NAXOS is modeled as a two-entity AKA protocol and each party has a private-public key pair to construct a session. Both entities A and B execute two AKA protocol sessions which are called matching session.

4   Proposed Protocol

The IoT device may need to communicate with other heterogeneous IoT devices. These heterogeneous IoT devices were made by distinct manufactures which provide different system parameters. In this section, we present a heterogeneous IoT authentication key agreement protocol using elliptic curve algorithm with eCK security. Our protocol is composed of three phases including system initialization phase, long-term private/public key generation phase and key agreement phase.

4.1 System Initialization Phase

In our protocol, assume that there are n distinct CAs which announce different system parameters. Each CA runs an elliptic curve AKA protocol in the initialization phase to provide the public system parameters to IoT devices certified by it. The steps involved in this phase are listed as follows:

Step 1.   Each CA (denoted by CA_i) chooses a nonsingular elliptic curve $E_i(F_{q_i})$ over a finite field F_{q_i}. The set of points $E_i(F_{q_i})$ forms an abelian cyclic group G_i. Then chooses a generator point P_{ipub} on the curve $E_i(F_{q_i})$ and a prime factor n_i where $n_i P_{ipub} = O$ and $n_i < q_i$.

Step 2. CA_i announces the public system parameters $\{F_{q_i}, E_i(F_{q_i}), G_i, q_i, P_{ipub}\}$ on the authenticated public domain

Only the authorized CA_i can update the public system parameters in the public domain. That means the system parameters are protected against tampering and anyone can read these parameters from the public domain but cannot change or rewrite these public parameters unless receiving the authorization from CA_i.

4.2 Long-Term Private/Public Key Generation Phase

Each IoT entity has a unique ID, and generates its own long-term private key and public key (authenticated and certified by CA). These keys will be used in securing the communication between heterogeneous IoT devices in the network. The extraction algorithm is described that each IoT entity (denoted by U) is assumed to have a long-term private/public key pair x_U/P_{KUi}, where $x_U \in Z^*_{q_i}$, and a certificate that binds the entity's long-term public key $P_{KUi} = x_U P_{ipub}$. The certificate can be verified with the public key of CA.

4.3 Key Agreement Phase

Two IoT entities A and B of two distinct CAs want to communicate with each other (entity A is certified by CA_i and entity B is certified by CA_j, respectively). They need to establish an authenticated common session key. The key agreement steps are shown as follows:

Step 1. As stated in Sect. 4.1, CA_i announces the public system parameters $\{F_{q_i}, E_i(F_{q_i}), G_i, q_i, P_{ipub}\}$ and CA_j announces the public system parameters $\{F_{q_j}, E_j(F_{q_j}), G_j, q_j, P_{jpub}\}$.

Step 2. As stated in Sect. 4.2, the entity A chooses long-term private key $x_A \in Z^*_{q_i}$, computes long-term public key $P_{KAi} = x_A P_{ipub}$, and then announces P_{KAi}. The entity B chooses long-term private key $x_B \in Z^*_{q_j}$, computes long-term public key $P_{KBj} = x_B P_{jpub}$, and then announces P_{KBj}.

Step 3. The entity A chooses an ephemeral private key $a \in Z^*_{q_i}$, and calculate $\tilde{a}_i = H_{1i}(a, x_A) \in Z^*_{q_i}$ and $\tilde{a}_j = H_{1j}(a, x_A) \in Z^*_{q_j}$, where both $H_{1i} : \{0,1\}^* \to Z^*_{q_i}$ and $H_{1j} : \{0,1\}^* \to Z^*_{q_j}$ are collision resistant hash functions. Then entity A computes $D_1 = \tilde{a}_j P_{jpub}$ and $D_2 = \tilde{a}_i P_{ipub}$, and sends $\{D_1, D_2\}$ to B.

Step 4. The entity B chooses an ephemeral private key $b \in Z^*_{q_j}$, and calculate $\tilde{b}_i = H_{1i}(b, x_B) \in Z^*_{q_i}$ and $\tilde{b}_j = H_{1j}(b, x_B) \in Z^*_{q_j}$. Then entity B computes $E_1 = \tilde{b}_i P_{ipub}$ and $E_2 = \tilde{b}_j P_{jpub}$, and sends $\{E_1, E_2\}$ to B.

Step 5. After receiving $\{E_1, E_2\}$ from B, A computes the keys $K_{A1} = x_A E_1$, $K_{A2} = \tilde{a}_j P_{KBj}$, $K_{A3} = \tilde{a}_i E_1$, and $K_{A4} = \tilde{a}_j E_2$. A computes the common session key with an ordinary one-way hash function H such as SHA-2:

$$K_{AB} = H(K_{A1}, K_{A2}, K_{A3}, K_{A4}, A\|B).$$

Step 6. After receiving $\{D_1, D_2\}$ from A, B computes the keys $K_{B1} = \tilde{b}_i P_{KAi}$, $K_{B2} = x_B D_1$, $K_{B3} = \tilde{b}_i D_2$, and $K_{B4} = \tilde{b}_j D_1$. B computes the common session key with an ordinary one-way hash function as:

$$K_{BA} = H(K_{B1}, K_{B2}, K_{B3}, K_{B4}, A\|B).$$

5 Security Analysis and Performance Comparison

5.1 Security Analysis

In this section, we analyze the security of our protocol, and also give a proof of the security for our protocol under ECGDHP assumption. Each entity only can receive the common attributes sent by other entities and cannot reveal any additional private information. Furthermore, it can be proved that if an adversary intends to break in our protocol, he cannot reveal any useful private messages and also cannot derive the common session key. We describe the security proofs of our protocol as follows.

Theorem 1 (Correctness): If two entities complete matching session, they obtain the same session key.

Proof: If two entities, assumed to be entities A and B, complete the matching session, they both obtain the same following session key.

$$
\begin{aligned}
K_{AB} &= H(K_{A1}, K_{A2}, K_{A3}, K_{A4}, A\|B) \\
&= H(x_A E_1, \tilde{a}_j P_{KBj}, \tilde{a}_i E_1, \tilde{a}_j E_2, A\|B \\
&= H(x_A \tilde{b}_i P_{ipub}, x_B \tilde{a}_j P_{jpub}, \tilde{a}_i \tilde{b}_i P_{ipub}, \tilde{a}_j \tilde{b}_j P_{jpub}, A\|B) \\
&= H(\tilde{b}_i P_{KAi}, x_B D_1, \tilde{b}_i D_2, \tilde{b}_j D_1, A\|B) \\
&= H(K_{B1}, K_{B2}, K_{B3}, K_{B4}, A\|B) \\
&= K_{BA}
\end{aligned}
$$

Therefore, session key $K = K_{AB} = K_{BA}$. We here complete the proof of Theorem 1.

Theorem 2: Similar to Hu et al.'s proof in [12], if G_1 is a finite cyclic group, H is a random oracle and the ECGDHP assumption holds, our protocol is a secure authentication key exchange protocol under the eCK security model.

Proof: Let M be an adversary against our protocol with the security parameter k who runs at most $n(k)$ honest entities and $s(k)$ sessions. Our protocol in this mechanism with the common session key K includes 5-tuple, where $H(\cdot)$ is a random oracle. If M reveals our common session key K from a random key string successfully, M can distinguish the session key in the following two ways after M queries the test session:

(1) **Key replication attack.**
   The adversary compels the establishment of another session successfully which is neither the test session nor the matching session of the test session to reveal the same session key. Our protocol uses one-way hash function with 5-tuple in common session key under eCK security model, and the collisions happen at most with negligible probability $O(s(k)/2^k)$.
(2) **Forging attack.**
   The adversary queries random oracle hash function H on the 5-tuple in the test session. According to distinct information, the adversary may reveal the long-term private key during the attack. We distinguish the following cases.

Case 1. M reveals the long-term private key of both entities of the test session.

Case 2. M makes Ephemeral Key Query to both the test session and its matching session.

Case 3. M reveals the long-term private key of one entity of the test session and makes Ephemeral Key Query to its matching session.

Case 4. M makes Ephemeral Key Query to the test session and learns the long-term private key of the other entity of the test session.

The probability of one of the four cases is 1/4. If M can make a forging attack successfully, than a simulator S can be constructed to solve the ECGDHP by using M as a subroutine. S takes a ECGDHP instance $\{aP, bP\}$ as input and is asked to compute $ECCDHP(aP, bP) = abP$. After an execution of an eCK experiment with M, S learns the solution efficiently with non-negligible probability.

Proof of Case 1: As mentioned above, there are 4 cases, and the probability of case 1 happens is 1/4. Given ECGDHP challenge instance $U, V \in G$, simulator S prepares n entities, and randomly allots long-term key pairs for these entities. In the beginning, S randomly chooses two sessions named s_1 and s_2 independently and uniformly. S guesses that M will choose two entities denoted by A as the test session and B as the peer. S allots A's public key to be U, B's to be V. All queries asked by M will be answered by S faithfully except s_1 or s_2. M activates s_1 or s_2, and the output messages are replaced by U and V, respectively.

M chooses the session s_1 as the test session and s_2 as its matching session with probability at least $1/s(k)^2$. M queries random oracle H with $\varphi = (D_1, D_2, D_3, D_4, sid)$, and randomly allots D_4 (or D_3) such that $D_4 = ECCDHP(U, V)$. S checks whether $D_4 = ECCDHP(U, V)$ or not by means of ECDDHP oracle. Unless the four possible cases happen or a random answer occurs, S just needs checking queries made by M and then gives the right answer. In this case, S simulates the random oracle H as usual by maintaining an initially empty list H_{list} with entries of the form (φ, h_{sid}). For a query $H(\varphi)$, if there exists an entry (φ, h_{sid}) in H_{list}, S responds with the stored value h_{sid}. If it does not, S selects a value $h_{sid} \in \{0.1\}^k$ at random, sends it to M and stores the new (φ, h_{sid}) in H_{list}. As above, S simulates M's circumstance perfectly. If M wins in forging attack, he must query H with φ that contains $ECCDHP(U, V)$ as a substring. That means S can solve ECGDHP. Therefore, the advantage of S in this case is estimated as

$$Adv^{ECGDHP}(S) \geq \frac{1}{4} \cdot \frac{1}{s(k)^2} Adv^{AKA}(M)$$

Proof of Case 2, 3 and 4: Similar to the proof process of case 1, proceeding cases 2, 3 and 4, we obtain the advantage of S estimated as

$$Adv^{ECGDHP}(S) \geq \min\{\frac{1}{4} \cdot \frac{1}{n(k)^2}, \frac{1}{4} \cdot \frac{1}{s(k)^2}, \frac{1}{4} \cdot \frac{1}{s(k)n(k)}\} Adv^{AKA}(M)$$

Here complete the proof of Theorem 2.

5.2   Security and Performance Comparison

In the following, as shown in Table 1, we will compare the performance of our heterogeneous IoT AKA protocol with other well-known AKA protocols. First, in security comparison, only the proposed protocol is proven to be secure in the eCK security model. Second, since the pairing computation is more time consuming than other cryptographic computations, our protocol is more efficient than other approaches due to without pairing operations. In order to compare the computational costs of these seven protocols listed in Table 1, we take three operations required in each protocol into account, including scalar multiplication, addition and hash function. Our protocol only requires one scalar multiplication operation for each CA, four scalar multiplication operations and two hash function operations for each user, and does not need pairing operations. Therefore, our protocol is more suitable for IoT environment than other protocols because of low computational overheads. Moreover, the key escrow problem was unsolved in other protocols since they applied the identity-based mechanism; however our protocol has no this kind of problem as the certificate-based model is used. In Table 1, we also show the number of key-items used in each protocol. It can be concluded that our protocol achieves security requirements and lower computational overheads in AKA for secure IoT communication.

To explain the computational cost, we use some notation as "P" for time complexity of pairing operation, "M" for scalar multiplication in ECC, "H" for hash function and "A" for addition in ECC.

Table 1. Security and performance comparison

Protocol	eCK security	Pairing		Computational cost		ECC	Key-items	Non-key escrow problem
		Each	All	Each CA	Each user			
Chen and Kudla [4]	✗	2P	4P	2M1H	2M2H	✗	4	✗
Kim et al.'s [17]	✗	2P	4P	2M1H	5M2H	✗	6	✗
Lee et al.'s [19]	✗	2P	4P	2M1H	4M3H	✗	4	✗
Farash and Attari [9]	✗	0	0	3M1H1A	8M1H1A	✓	8	✗
Farash and Attari [10]	✗	0	0	3M1H1A	8M1H1A	✓	8	✗
Farash and Attari [11]	✗	0	0	3M1H1A	8M1H1A	✓	8	✗
Ours	✓	0	0	1 M	4M2H	✓	4	✓

6 Conclusion

It is clear that IoT vision has a remarkable impact to the world. Compared with RSA or other key agreement schemes using large number size for its keys, ECC only needs very small key size and low computational power that is why we focus on ECC. For the secure communications among large amounts of IoT devices, it is necessary to use more efficient method which contains small-size keys. Secure IoT front-end authentication key agreement phase is the most important issue in heterogeneous IoT environment. This paper proposed a highly secure authentication key agreement protocol based on ECC between two heterogeneous IoT devices authorized by distinct CAs which provides different public system parameters. We prove the security of our protocol with eCK security model using random oracle. Our protocol solves the heterogeneous IoT session key agreement problem under distinct CAs and satisfies most security requirements of the AKA protocol.

Acknowledgements. This work was supported by Ministry of Science and Technology of Taiwan Grants MOST 108-2221-E-415-010 and Taiwan Information Security Center at National Sun Yat-sen University (TWISC@NSYSU).

References

1. Bellare, M., Rogaway, P.: Entity authentication and key distribution. In: Advances in Cryptology CRYPTO 1993, pp. 110–125. Springer (1993)
2. Bellare, M., Rogaway, P.: Random oracles are practical: a paradigm for designing efficient protocols. In: ACM Conference on Computer and Communications Security, pp. 62–73 (1993)
3. Chen, L., Lim, H.W., Yang, G.: Cross-domain password-based authenticated key exchange revisited. In: Proceedings of the 32nd IEEE International Conference on Computer Communications (INFOCOM), pp. 1052–1060 (2013)
4. Chen, L., Kudla, C.: Identity-based authenticated key agreement protocols from pairings. In: Proceeding of the 16th IEEE Computer Security Foundations Workshop, pp. 219–233. IEEE Computer Society Press (2003)
5. Cheng, Q.F., Han, G.G., Ma, C.G.: A new efficient and strongly secure authenticated key exchange protocol. In: Fifth International Conference on Information Assurance and Security, pp. 499–502 (2009)
6. Canetti, R., Krawczyk, H.: Analysis of key-exchange protocols and their use for building secure channels. LNCS, vol. 2045, pp. 453–474. Springer (2001)
7. Da, B., Esnault, P.P., Hu, S.H., Wang, C.: Identity/identifier-enabled networks (IDEAS) for Internet of Things (IoT). In: IEEE 4th World Forum on Internet of Things (WF-IoT), pp. 412–415 (2018)
8. Diffie, W., Hellman, M.E.: New direction in cryptography. IEEE Trans. Inf. Theory **22**(6), 644–654 (1976)
9. Farash, M.S., Attari, M.A.: An ID-based key agreement protocol based on ECC among users of separate networks. In: 9th International ISC Conference on Information Security and Cryptology, pp. 32–37 (2012)
10. Farash, M.S., Attari, M.A.: Provably secure and efficient identity-based key agreement protocol for independent PKGs using ECC. ISC Int. J. Inf. Secur. **5**(1), 1–16 (2013)

11. Farash, M.S., Attari, M.A.: A pairing-free ID-based key agreement protocol with different PKGs. Int. J. Netw. Secur. **16**(3), 168–173 (2014)
12. Hu, X.X., Cheng, Q.F., Liu, W.F., Zhang, Q.H.: A new AKE protocol with stronger security in the eCK model. In: WASE International Conference on Information Engineering, vol. 4, pp. 10–13 (2010)
13. Krawczyk, H.: A high-performance secure Diffie-Hellman protocol. In: Advances in Cryptology - CRYPTO 2005, LNCS, vol. 3621, pp. 546–566. Springer (2005)
14. Koblitz, N.: Elliptic curve cryptosystem. Math. Comput. **48**(177), 203–209 (1987)
15. Kushwaha, P.: Towards the equivalence of Diffie-Hellman problem and discrete logarithm problem for important elliptic curves used in practice. ISEA Asia Security and Privacy (ISEASP), pp. 1–4 (2017)
16. Kumari, P.L.S., Damodaram, A.: An alternative methodology for authentication and confidentiality based on zero knowledge protocols using Diffie-Hellman key exchange. In: 2014 International Conference on Information Technology (ICIT), pp. 368–373 (2014)
17. Kim, S., Lee, H., Oh, H.: Enhanced ID-based authenticated key agreement protocols for a multiple independent PKG environment. In: Proceedings of ICICS, pp. 323–335. Springer (2005)
18. LaMacchia, B., Lauter, K., Mityagin, A.: Stronger security of authenticated key exchange. LNCS, vol. 4784, pp. 1–16. Springer, Heidelberg (2007)
19. Lee, H., Kim, D., Kim, S., Oh, H.: Identity-based key agreement protocols in a multiple PKG environment. In: Proceedings of the International Conference on Computer Science and its Applications, ICCSA. LNCS, vol. 3483, pp. 877–886. Springer (2005)
20. Miladinovic, I., Schefer-Wenzl, S.: NFV enabled IoT architecture for an operating room environment. In: IEEE 4th World Forum on Internet of Things (WF-IoT), pp. 98–102 (2018)
21. Mehibel, N., Hamadouche, M.: A new approach of elliptic curve Diffie-Hellman key exchange. In: 5th International Conference on Electrical Engineering - Boumerdes (ICEE-B), pp. 1–6 (2017)
22. Miller, V.S.: Uses of elliptic curve in cryptography. In: Advances in Cryptology - CRYPTO '05, LNCS, vol. 218, pp. 417–428, Springer (1986)
23. Singh, A., Rishi, R.: An improved two-factor authenticated key exchange protocol in public wireless LANs. In: Third International Conference on Advanced Computing and Communication Technologies (ACCT), pp. 226–230 (2013)
24. Wang, S.L., Hou, Y.B., Gao, F., Ji, X.R.: A novel IoT access architecture for vehicle monitoring system. In: IEEE 3rd World Forum on Internet of Things (WF-IoT), pp. 639–642 (2016)
25. Yao, A.C.C., Zhao, Y.: Privacy-preserving authenticated key-exchange over Internet. IEEE Trans. Inf. Forensics Secur. **9**(1), 125–140 (2014)
26. Zhong, Y.T., Ma, J.F.: A highly secure identity-based authenticated key-exchange protocol for satellite communication. J. Commun. Netw. **12**(6), 592–599 (2010)
27. IoT Security Solutions, White Paper. https://www.insidesecure.com/Media/Files/Whitepapers/IoT-Security-Solutions

Multimedia Security

Robust Syndrome-Trellis Codes for Fault-Tolerant Steganography

Bingwen Feng[1,2,3], Zhiquan Liu[1], Xiaotian Wu[1], and Yuchun Lin[4(✉)]

[1] College of Information Science and Technology, Jinan University,
Guangzhou 510632, China
bingwfeng@gmail.com,
{zqliu,wxiaotian}@jnu.edu.cn
[2] Guangdong Provincial Key Laboratory of Information Security Technology,
Guangzhou 520006, China
[3] State Key Laboratory of Information Security, Institute of Information
Engineering, Chinese Academy of Sciences, Beijing 100093, China
[4] Faculty of Information Technology, Guangzhou International Economics College,
Guangzhou 510540, China
ychlin@gmail.com

Abstract. Syndrome-Trellis Codes (STCs) have been widely used in adaptive steganography due to their high embedding efficiency. However, this type of steganographic codes is sensitive to stego damage, which may be incurred by compression, channel noise, active attacks, and so on, in practical covert communication. In this paper, a construction of robust STCs is proposed to achieve a good balance between robustness and embedding efficiency. The encoder of the proposed scheme performs a specified Error Correction Code (ECC) on the STC's intermediate outputs. Further, a Viterbi algorithm is suggested to effectively find the optimum stego vector. At the extraction phase, the received stego vector is first decoded by the decoder of the employed ECC and then sent to the STC decoder, which gives the extracted message bits. Experimental results show the proposed scheme presents a good robustness performance, meanwhile the stego remains an acceptable detection resistance against steganalyzers.

Keywords: Steganography · Robustness · Syndrome-Trellis Codes · Error correction codes

1 Introduction

Steganography is a technique about covert communication. It hides secret messages into covers without drawing suspicion. The detection resistance against steganalyzers is an important criterion for designing steganographic schemes. For this purpose, many steganographic schemes try to minimize the distortion between the cover and the corresponded stego objects. The most successful approaches among them may be the ones based on the Syndrome-Trellis Code

© Springer Nature Switzerland AG 2020
L. C. Jain et al. (Eds.): SICBS 2019, AISC 1145, pp. 115–127, 2020.
https://doi.org/10.1007/978-3-030-46828-6_11

(STC) framework [1], which can embed near the payload-distortion bound. Benefited from STCs, steganographic schemes can focus on the designment of distortion measurements, such as HILL [2], MiPOD [3] in the spatial domain, and J-UNIWARD [4], GUED [5] in the DCT domain.

The published stego data may suffer from information loss in practice. Videos and images are usually compressed before being transmitted through mobile devices or social applications, such as Facebook, Twitter, and WeChat, due to the limitation of memory, network traffic, bandwidth, and so on [6,7]. Furthermore, the adversary may also want to destroy the covert communication channel by active attacks [8]. As a result, a secure steganographic system should be capable of both undetectability and robustness. However, as a side effect of high embedding efficiency, STCs are quite sensitive to channel disturbance. A minor damage on the received stego will spread over a wide range of message bits it carries, which is referred to as the damage diffusion of STCs [9]. One can prevent the potential stego damage by employing invariant embedding domain, for example, embedding message bits into DCT blocks to gain the resistance against JPEG compression [4,5]. However, this strategy provides rather limited robustness, and the stego damage cannot be eliminated perfectly.

Introducing Error Correcting Codes (ECCs) is a popular way to combat channel disturbance. In [10], Zhang et al. integrated ECCs with the stego encoder, which are both in the form of cyclic coding, to gain the error correction capability. However, this pioneer approach only enhances the reliability of a specific matrix embedding scheme, which is not suitable for state-of-the-art adaptive steganography. Liu et al. liu 2015 damage suggested a general damage-resistance matrix embedding framework, where secret messages were precoded with ECCs before the matrix embedding. Schemes in [6] and [7] encoded the message with Reed solomon codes and then sent it into STC. Besides, JPEG compression resistant embedding domain was also extracted in these two schemes to maintain good compression resistant ability. These schemes all conduct STCs followed by ECCs to take fully advantage of the embedding efficiency provided by the original STC framework. However, due to the damage diffusion of STCs, the encoder has to provide great error correction capability to correct the errors caused by even one damage bit in the stego. Moreover, it is sometimes difficult for the communication parties to distinguish the damage area in the stego. In [11], a type of dual-STCs is proposed for the noisy channel, where LSBs of state labels are used as templates to correct the disturbed stego. However, it still requires encoding the message with forward ECCs to achieve considerable robustness.

In this paper, a type of robust STCs is proposed by combining ECCs and STCs. It corrects the errors in the stego vector rather than those in the decoded message. Therefore, a stronger error correction capability could be available given the designed free distance of ECCs. The ensemble encoder is obtained by performing an additional ECC encoder on every STC's codeword, e.g., a combination of columns in STC's parity-check matrix. With regard to the syndrome trellis construction, it means that the weight of each edge depends on the outputs of ECCs rather than the outputs of STCs. We employ the binary

Feedforward Convolutional Codes (FCCs) to implement robust STCs due to their advantage on processing short inputs, and develop the corresponded Viterbi algorithm to find the optimum stego vector effectively. In the experiments, we analyze the robustness performance of the proposed scheme and show its advantage compared with those approaches based on STCs followed by ECCs. Due to the similar form with the original STCs, the proposed scheme can replace the coder in the STC framework to enhance the robustness of existed steganographic schemes.

2   Review of Syndrome-Trellis Codes

The proposed scheme extends the Syndrome-Trellis Codes (STCs) [1] so that they can resist stego damage. Therefore, despite being widely studied, we briefly review the construction of STCs. They are essentially the syndrome coding form of convolutional codes. Supposing the cover and message vectors are $\mathbf{x} \in \{0,1\}^{l_n \times 1}$ and $\mathbf{m} \in \{0,1\}^{l_m \times 1}$. STCs try to solve

$$\mathbf{y} = \underset{H_s \mathbf{y}' = \mathbf{m}}{\arg\min} D(\mathbf{x}, \mathbf{y}') \tag{1}$$

$$s.t. \quad H_s \mathbf{y} = \mathbf{m} \tag{2}$$

where $\mathbf{y} \in \{0,1\}^{l_n \times 1}$ denotes the stego vector, and $D(\mathbf{x}, \mathbf{y})$ denotes the distortion function, defined as

$$D(\mathbf{x}, \mathbf{y}) = \sum_{i=1}^{l_n} \mathbf{w}[i] \, |\mathbf{x}[i] - \mathbf{y}[i]| \tag{3}$$

where $\mathbf{w}[i]$ denotes the cost function associated with i-th cover element. The parity-check matrix $H_s \in \{0,1\}^{l_m \times l_n}$ is obtained by repeating a small submatrix $\hat{H}_s \in \{0,1\}^{l_k \times l_h}$, where $l_h = l_n/l_m$ represents the embedding rate of STC, along its main diagonal. A syndrome trellis can then be organized with H and $\mathbf{m}$. Each stego vector $\mathbf{y} \in \{0,1\}^{l_n \times 1}$ satisfying $H_s \mathbf{y} = \mathbf{m}$ represents a path through the trellis.

STCs use the Viterbi algorithm to find the shortest path. This algorithm consists of two parts. The forward part includes l_m rounds of state grid construction. In each round a $2^{l_k} \times (l_h + 1)$ sized state grid is constructed according to $\hat{H}_s$. Half of the states in the rightmost column are then shifted and connected to the next grid (called states prune) according to the embedded message bit. During the backward part, the shortest path is traced back to recover the closest stego vector $\mathbf{y}$.

Despite achieving high embedding efficiency, it has been shown that the robustness of STCs appears very poor due to the damage diffusion [9]. Each bit error in the stego vector will incur multi-bit errors in the decoded message. The ratio of message error rate to stego error rate, or referred to as the damage gain, grows exponentially with the height of submatrix $\hat{H}$.

3   Robust Syndrome-Trellis Codes

We combine the binary Error Correcting Codes (ECCs) with the syndrome coding to enhance its robustness. Since the damage gain grows exponentially, correcting stego errors would be more effective than correcting extracted message errors. As a result, the stego vector is pose-coded by ECCs. Meanwhile, the embedding impact should be minimized for the detection resistance against steganalyzers. This section gives a complete description of the proposed robust STCs.

3.1   Message Encoding Algorithm

The proposed scheme preserves the structure of the parity-check matrix of STCs so that the new syndrome trellis can still represent every possible stego vector. Each partial syndrome constructed by the parity-check matrix of STC is then coded by ECC, resulting in a segment of a possible stego vector. The encoding algorithm should find the closest stego vector, formally formed as

$$\mathbf{y} = \underset{H_s \cdot H_c \mathbf{y}' = \mathbf{m}}{\arg\min} \ D(\mathbf{x}, \mathbf{y}') \tag{4}$$

$$s.t. \ H_s \cdot H_c \mathbf{y} = \mathbf{m} \tag{5}$$

where $H_c \in \{0,1\}^{l_n \times l_q}$ is the parity-check matrix of ECC, and the distortion function $D(\mathbf{x}, \mathbf{y})$ has been defined in Eq. (3).

Equation (4) can still be solved by finding the shortest path through the trellis. However, due to the post ECC processing, weights cannot be directly assigned to each trellis edge. Assuming that the employed ECC encodes l_a bits message to output l_b bits codeword each time, and we are in the first column (the column labeled 1) in the trellis, the weights of paths from the column labeled 1 to the one labeled l_a depend on the first l_b cover elements.

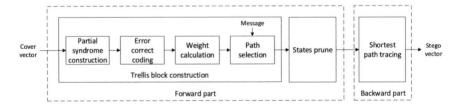

Fig. 1. The framework of the Viterbi algorithm in robust syndrome-trellis codes.

We still use the Viterbi algorithm to organize the syndrome trellis. The general framework of the algorithm is shown in Fig. 1. The number of paths entering a node equals to the cardinality of the message set of the employed ECC, 2^{l_a}, given the established paths to the previously processed slice. The one with the smallest total weight will be chosen as the final incoming path for that node. Therefore, the complexity of the Viterbi algorithm grows exponentially with the length of the ECC inputs. In view of this, we choose the Feedforward Convolutional Codes (FCCs) as the correction codes because of their efficiently addressing short inputs [12]. Some other superiorities of convolutional codes have been discussed in [1,9].

Herein we only simply write the encoding function of a rate l_a/l_b FCC as

$$(code, state') = \text{CCEnc}(info, G_c, state) \tag{6}$$

where $info \in \{0, \ldots, 2^{l_a} - 1\}$ represents the information bits and $code \in \{0, \ldots, 2^{l_b} - 1\}$ represents the codeword sequence. $state$ and $state'$ are the register states before and after the encoding, respectively. G_c stands for the code generator and is defined as $G_c \triangleq [G_c[:, 0], \ldots, G_c[:, l_b - 1]]$, where $G_c[:, i] \in \{0, \ldots, 2^{l_{g,i}}\}^{l_a \times 1}$, $l_{g,i}$ specifies the delay for the encoder's i-th output bits. Equation (6) implies that the Viterbi algorithm has to store the internal register states of FCC for each node in the syndrome trellis. Further, the width of STC submatrix l_h should be a multiple of l_a.

The Viterbi algorithm is detailed in Algorithms 1 and 2, where $\mathbf{x}[i]$ denotes the i-th element in vector $\mathbf{x}$, $X[i, j]$ denotes the (i, j)-th element in matrix X, and $\mathbf{x}[i : j]$ denotes a subvector containing the i-th element to the j-th element in $\mathbf{x}$. Function $\mathcal{B}(x)$ does the binary to decimal conversion and $\mathcal{D}(x)$ the inversion. The following is a simple explanation.

The forward part of the proposed Viterbi algorithm detailed in Algorithm 1 is used to construct the trellis composed of l_m blocks. This trellis describes all possible paths through STC states. In one block, which two states are connected is up to the columns of $\hat{H}_s$. Given an edge connecting two STC states, i.e., an earlier state and a later state, its weight is depended on the difference between the FCC output and the corresponding cover elements, where the FCC output is calculated by using the later state and the register state of FCC.

The backward part of the proposed algorithm, as detailed in Algorithm 2, is used to trace the shortest path. In order to reduce the tracing cost, we record the intermediate FCC output and STC state histogram associated with each STC state.

3.2 Example of Message Encoding

The proposed encoding algorithm is further explained by an example. Let the submatrix of STC and the code generator of FCC be

$$\hat{H} = \begin{bmatrix} 1 & 0 & 1 & 1 \\ 0 & 1 & 1 & 0 \end{bmatrix}, \quad G_c = \begin{bmatrix} 11 & 5 & 0 \\ 0 & 2 & 7 \end{bmatrix} \tag{7}$$

We now embed binary message sequence $\mathbf{m} = [0, 1]^T$ into cover vector $\mathbf{x} = [1, 1, 1, 0, 1, 0, 1, 0, 0, 0, 0, 1]^T$ with weight profile $\mathbf{w} = [12, 11, 10, 9, 8, 7, 6, 5, 4, 3, 2, 1]^T$. The encoding algorithm starts with STC state 00 and FCC state 00. Figure 2 shows the constructed syndrome trellis.

Algorithm 1. Viterbi Algorithm of Message Encoding (Forward Part)

1: $\mathbf{c}[0] = 0$; $\mathbf{c}[1 : 2^{l_k} - 1] = \inf$; /* cost vector */
2: $\mathbf{s}[0 : 2^{l_k} - 1] = 0$; /* FCC state vector */
3: $P_s[0 : l_n/l_b - 1, 0 : 2^{l_k} - 1] = 0$; /* STC state history */
4: $P_o[0 : l_n/l_b - 1, 0 : 2^{l_k} - 1] = 0$; /* output history */
5: $i_{col} = 0$; $i_{cov} = 0$; $i_{path} = 0$; $i_{msg} = 0$;
6: **while** $i_{msg} < l_m$ **do**
7:   /* 1.a) Trellis block construction */
8:   $\mathbf{c}' = \mathbf{c}$; $\mathbf{s}' = \mathbf{s}$;
9:   **for** $st = 0$ to $2^{l_k} - 1$ **do**
10:    **for** $sym = 0$ to $2^{l_a} - 1$ **do**
11:      $pst = st$;
12:      **for** $i_{bit} = 0$ to $l_a - 1$ **do**
13:        **if** $(l_k - i_{bit})$-th bit of sym equals to 1 **then**
14:          $pst = \mathcal{D}\left(\mathcal{B}(pst) \text{ XOR } \hat{H}_s[:, i_{col} + i_{bit}]\right)$;
15:        **end if**
16:      **end for**
17:      $(out, ccst) = \text{CCEnc}(sym, G_c, \mathbf{s}'(pst))$;
18:      $cost = \mathbf{c}'(pst) + \mathbf{w}[i_{cov} + i]^T \cdot (\mathcal{B}(out) \text{ XOR } \mathbf{x}[i_{cov} : i_{cov} + l_b - 1])$;
19:      **if** $sym == 0$ OR $cost_{min} > cost$ **then**
20:        $cost_{min} = cost$; $out_{min} = out$;
21:        $ccst_{min} = ccst$; $pst_{min} = pst$;
22:      **end if**
23:    **end for**
24:    $\mathbf{c}[st] = cost_{min}$; $\mathbf{s}[st] = ccst_{min}$;
25:    $P_s[i_{path}, st] = pst_{min}$; $P_o[i_{path}, st] = out_{min}$;
26: **end for**
27: $i_{col} = i_{col} + l_a$;
28: /* 1.b) State prune */
29: **if** $i_{col} > l_h$ **then**
30:    **for** $st = 0$ to $2^{l_k - 1} - 1$ **do**
31:      $\mathbf{c}[st] = \mathbf{c}[2 \times st + \mathbf{b}[i_{msg}]]$;
32:      $\mathbf{e}[st] = \mathbf{e}[2 \times st + \mathbf{b}[i_{msg}]]$;
33:    **end for**
34:    $\mathbf{c}[2^{l_k - 1} : 2^{l_k} - 1] = \inf$; $\mathbf{e}[2^{l_k - 1} : 2^{l_k} - 1] = 0$;
35:    $i_{col} = 0$; $i_{msg} + +$;
36: **end if**
37: $i_{cov} = i_{cov} + l_b$; $i_{path} + +$;
38: **end while**

Algorithm 2. Viterbi Algorithm of Message Encoding (Backward Part)

1: find the smallest value in $\mathbf{c}$, and set c and st with its value and index;
2: $i_{cov}--$; $i_{msg}--$; $i_{col} = -1$;
3: **for** $i_{path} = l_n/l_b - 1$ to 0 **do**
4:     **if** $i_{col} < 0$ **then**
5:         $i_{col} = l_h - 1$; $st = st \times 2 + \mathbf{m}[i_{msg}--]$;
6:     **end if**
7:     $out = P_o[i_{path}, st]$; $st = P_s[i_{path}, st]$;
8:     $\mathbf{y}[i_{cov} - l_b + 1 : i_{cov}] = \mathcal{B}(out)$;
9:     $i_{cov} = i_{cov} - l_b$;
10: **end for**
11: **return** stego vector $\mathbf{y}$, embedding distortion c.

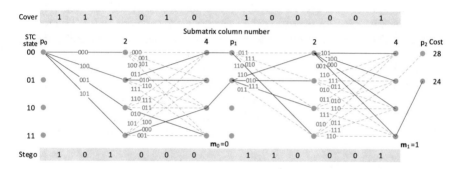

Fig. 2. Example of the syndrome trellis constructed by the proposed algorithm with $\hat{H}$ and G_c given in Eq. (7).

Since the employed FCC receives 2-bits inputs, there is no FCC output when STC processes the first or third submatrix column. As a result, the trellis only contains columns labeled 2 and 4. In Fig. 2, the number assigned to each edge represents the FCC output, which is also a segment of a possible stego vector. Take the STC state 01 in the column labeled 4 as an example. There are four edges entering this node, which correspond to partial STC codewords 00, 01, 10, and 11. These edges are from preceding STC states 01, 00, 10, and 11 in the column labeled 2, whose FCC states stored are $4, 0, 16$, and 20, respectively. Each STC codeword and associated FCC state are then input into FCC, resulting in four stego segments as 010, 001, 111, and 100, as marked out in Fig. 2. Changing the original cover segment 000 to these stego segments incurs embedding distortions as $8, 7, 24$, and 9. By adding these distortion values with the weights of paths reaching corresponded previous STC states, we get the total weights of each path as $21, 48, 39$, and 28. Therefore, the edge labeled 010 possesses the smallest weight and is thus retained in the trellis. It should be noted that the FCC outputs assigned to different trellis block do not duplicate each other due to the different FCC states initialized.

3.3   Message Decoding Algorithm

As indicated in Eq. (4), the message bits can be extracted from the received stego vector $\hat{\mathbf{y}}$ by

$$\mathbf{m} = H_{\mathrm{s}} \cdot H_{\mathrm{c}} \hat{\mathbf{y}} \tag{8}$$

The Viterbi decoding can efficiently reveal the damaged bits in $\hat{\mathbf{y}}$. We know that STC supports real time decoding thanks to its parity-check matrix structure. In view of this, the continuous operation mode is employed in the FCC decoder for the compatibility. The FCC decoder with continuous operation mode stores internal state information for successive decoder invocations, which can be simply written as

$$(info_i, tre_{i+1}) = \mathrm{CCDec}(code_i, G_{\mathrm{c}}, tre_i) \tag{9}$$

where tre_i specify the starting trellis information, such as the state metrics, traceback states, etc., for decoding the i-th codeword segment $code_i$. The message decoding procedure of the proposed robust STC consists of the following steps.

1. Divide the received stego vector $\hat{\mathbf{y}}$ into stego segments of length l_h.
2. Use Eq. (9) to decode the i-th stego segments, yielding the internal trellis information tre_{i+1}, as well as the temporary segment $info_i$.
3. Decode the i-th message bit $\mathbf{m}[i]$ by using $info_i$ and STC submatrix $\hat{H}_{\mathrm{s}}$.
4. Repeat Steps 2 and 3 until all the l_m message bits have been decoded.

4   Experimental Result

4.1   Performance Analysis

This section analyzes the robustness of the proposed robust STCs on artificial signals over the binary symmetric channel. When communicating over the channel with crossover probability ϵ_{c}, the exact bit error probability for the Viterbi decoding of FCCs can be estimated by [13]

$$\mathrm{Pe}_{\mathrm{c}}(\epsilon_{\mathrm{c}}) = \lim_{t \to \infty} \frac{E[W_t(state = 0|\epsilon_{\mathrm{c}})]}{t} \tag{10}$$

where $W_t(state = 0|\epsilon_{\mathrm{c}})$ denotes the weight of the information sequence corresponding to the code sequence decided by the Viterbi decoder at state $state$ and time t. Equation (10) can be determined by finding the stationary probability distribution of the cumulative normalized metrics, constructing the nonnegative matrix representing the linear part of the affine transformation of the weights, and computing this matrix's normalized eigenvector. The detail can be found in [13] and is omitted here.

The damage gain of STCs has been discussed in [9]. Herein we have a closer look at the bit error probability. The message bits satisfying Eq. (2) can be extracted by $m_i = \sum_{j=0}^{l_n - 1} H[i, j] \mathbf{y}[j]$. It indicates that a stego element $\mathbf{y}[j]$ associates to a message bit m_i if and only if $H[i, j] = 1$. Further, m_i will fail to be extracted if an odd number of associated stego elements happen to be damaged. As described in Sect. 2, except the beginning and truncation parts, each row of STC's parity-check matrix consists of $l_k \times l_h$ elements coming from all the l_k rows in $\hat{H}_s$. As a result, given the stego damage rate ϵ_s, the bit error probability of STCs can be evaluated by

$$\text{Pe}_s(\epsilon_s) = \sum_{i=1}^{\lceil \gamma/2 \rceil} \binom{\gamma}{2i-1} \epsilon_s^{2i-1}(1-\epsilon_s)^{\gamma+1-2i} \tag{11}$$

where γ represents the Hamming weight of $\hat{H}_s$.

As a combination of FCCs and STCs, the error probability of the proposed robust STCs can be derived as

$$\text{Pe} = \text{Pe}_s(\text{Pe}_c(\epsilon_c)) \tag{12}$$

We also experimentally evaluate the proposed scheme on pseudorandom binary cover signals of 1000-bit length at embedding rate $\rho \in \{0.125, 0.25\}$. The cost functions associated with each cover element are supposed to be linear ($\mathbf{w}[i] = i$) and square ($\mathbf{w}[i] = i^2$) to test the embedding efficiency of the proposed scheme. We employ the STC submatrices suggested in STC open-source codes [1], and the FCC generator matrices suggested in [12]. Their decimal representations are listed in Table 1. Note that Setting 1 in this table corresponds to the original STCs without error correcting. Table 2 reports their averaged embedding distortion.

The error probabilities of the proposed scheme with the above parameter settings are then tested in the cases of crossover probability $\epsilon_c \in [0, 0.3]$. Figure 3 compares the theoretical error probabilities with the experimental cases. It indicates that the proposed scheme effectively enhances the robustness compared with the original STCs. When $\epsilon_c \leq 0.1$, the proposed scheme with parameter setting 4 can well resist the stego damage at the cost of 15 times the embedding distortion. Comparison results on Setting 2 and Setting 3 also illustrate that the robustness declines with increasing the size of STC submatrix. This is because Eq. (11) is concave with respect to γ.

Table 1. Parameter Settings for Robust STC.

	STC submatrix (H_s)	FCC generator matrix (G_c)	Embedding rate
Setting 1	[17, 23, 29, 21, 19, 31, 27, 25]	[1]	$\rho = 0.125$
Setting 2	[17, 23]	[7, 5]	$\rho = 0.25$
Setting 3	[17, 23, 29, 21]	[7, 5]	$\rho = 0.125$
Setting 4	[17, 23]	[11, 13, 13, 15]	$\rho = 0.125$

Table 2. Embedding Distortion of Robust STCs with Different Parameter Settings.

	Setting I	Setting II	Setting III	Setting IV
Linear profile	$1.04e + 04$	$1.25e + 05$	$9.11e + 04$	$1.62e + 05$
Square profile	$6.74e + 06$	$8.22e + 07$	$6.01e + 07$	$1.06e + 08$

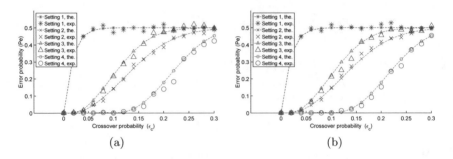

(a)   (b)

Fig. 3. Comparisons between theoretical and experimental error probabilities for (a) linear profile and (b) square profile. The employed parameter settings are listed in Table 1.

Some schemes suggest correcting the errors in the messages extracted by STCs from the damaged stego vectors [6,7,9]. Scheme in [9] (denoted as DR-STC) encodes the message with FCCs before the STC embedding while schemes in [6] and [7] (both denoted as RS-STC) encodes the message with Reed Solomon codes (RSs) before the embedding. We compare these embedding strategies with ours on 1000-bit length pseudorandom binary cover signals associated with linear cost function. The parameters in these schemes are set to keep the message length and embedding distortion similar, as listed in Table 3. The robustness comparison is illustrated in Fig. 4. It can be observed that the proposed outperforms the others when $\epsilon_c < 0.15$. In fact, the proposed scheme can provide much higher robustness than the compared ones under the same embedding rate. However, it usually incurs more serious embedding distortion due to the partial loss of STC's adaptivity.

Table 3. Parameter Settings for Different Schemes.

	Parameter setting	Embedding rate	Embedding distortion
The proposed	Setting 3	0.125	$9.11e + 04$
DR-STC	$G_c = [7, 7, 7, 7, 5]$	0.128	$9.49e + 04$
	STC's embedding rate $= 0.65$		
RS-STC	$m_{RS} = 4, k_{RS} = 3$	0.120	$8.83e + 04$
	STC's embedding rate $= 0.6$		

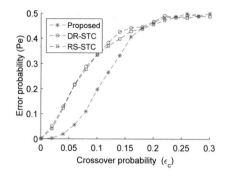

Fig. 4. Robustness comparisons among DR-STC, RS-STC, and the proposed. The employed parameter settings are listed in Table 3.

4.2 Comparison on Natural Images

At last, we test the proposed scheme on 10000 JPEG images of size 512×512. These test images are converted from the grayscale images in the BOSSbase database [14] by MATLAB function *imwrite* with quality factor 80. All the compared schemes are combined with the practical cost function J-UNIWARD [4] to construct steganographic methods. The parameters of these schemes are set as same as listed in Table 3. Further, to discard zero coefficients and enhance robustness, all the schemes only use coefficients randomly selected above the minor diagonal of quantized DCT blocks to embed message bits at 0.06 bpnc (bits per nonzero quantized DCT coefficients) payload. Binary sequences are pseudorandomly generated as secret messages.

The undetectability of these methods is evaluated by steganalyzers using 548-dimensional feature set CC-PEV [15] and SVM classifier [16]. In the evaluation, 5000 images are randomly selected for training, and the rest are used for testing. Tabel 4 reports the decision error probability (P_E) by 10 times of randomly splitting the training and the testing images. It demonstrates that these schemes present similar undetectability. This is because their parameters have been adjusted to keep similar embedding distortions.

100 images are randomly selected from the 10000 JPEG images above for the robustness comparison. Their stego versions are obtained by the three schemes with the same parameter settings used above. The Bit Error Rate (BER) results on these images in the present of JPEG compression and AWGN attacks are illustrated in Fig. 5. It can be observed that the proposed scheme outperforms the compared ones. This suggests that, given the excepted undetectability, the proposed scheme can better increase the message extraction correctness if the stego images are suffered from image processing and active attacks. DR-STC is designed for the binary erasure channel. However, identifying the damage area may be difficult in the compressed or noisy images (Table 4).

Table 4. Undetectability Comparison among Different Schemes.

	The proposed	DR-STC	RS-STC
P_E	0.1472	0.1510	0.1418

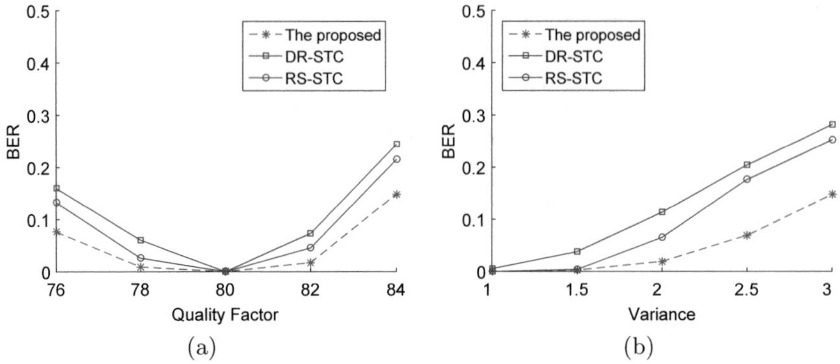

Fig. 5. Robustness comparison under (a) JPEG compression with different quality factors, and (d) AWGN with different variances.

5   Conclusion

In spite of high embedding efficiency, STCs present poor robustness. This paper proposes a construction of robust STCs by integrating ECCs into the framework of STCs. In the syndrome trellis of the proposed scheme, the weight of each edge depends on the outputs of ECCs, and each node has to store necessary information of both ECCs and STCs. A Viterbi algorithm is then suggested to effectively find the optimal path through the syndrome trellis. Different from existed robust steganographic codes where ECCs are carried out on the inputs of STCs, the proposed scheme performs ECCs on the outputs of STCs. Therefore, a better error correction capability could be available because of the damage diffusion in STCs. The analytical and comparison results show that the proposed scheme can well balance robustness and embedding efficiency. Furthermore, due to the compatibility with STCs, the proposed codes can replace STCs to enhance the robustness of those STC framework-based steganographic schemes. Improving the steganographic framework to further enhance the robustness against various attacks, especially the geometrical attacks, is our future work.

Acknowledgements. This work was supported by the Key R&D Program of Guangdong Province (Grant No. 2019B010136003), National Key R&D Program of China (Grant No. 2017YFB0802200), National Natural Science Foundation of China (Grant No. 61802145), Natural Science Foundation of Guangdong Province, China (Grant No. 2017A-030313390, 2018A030313387), Science and Technology Program of Guangzhou, China (Grant No. 201804010428), the Fundamental Research Funds for the Central Universities, the Opening Project of State Key Laboratory of Information Security, the Opening Project of Guangdong Provincial Key Laboratory of Information Security

Technology (Grant No. 2017B030314131), the Opening Project of Guangdong Key Laboratory of Intelligent Information Processing and Shenzhen Key Laboratory of Media Security.

References

1. Filler, T., Judas, J., Fridrich, J.J.: Minimizing additive distortion in steganography using syndrome-trellis codes. IEEE Trans. Inf. Forensics Secur. **6**(3), 920–935 (2011)
2. Li, B., Wang, M., Huang, J., Li, X.: A new cost function for spatial image steganography. In: 2014 IEEE International Conference on Image Processing (ICIP), pp. 4206–4210 (2014)
3. Sedighi, V., Cogranne, R., Fridrich, J.: Content-adaptive steganography by minimizing statistical detectability. IEEE Trans. Inf. Forensics Secur. **11**(2), 221–234 (2016)
4. Holub, V., Fridrich, J.: Digital image steganography using universal distortion. In: Proceedings of the First ACM Workshop on Information Hiding and Multimedia Security, pp. 59–68 (2013)
5. Su, W., Ni, J., Li, X., Shi, Y.Q.: A new distortion function design for JPEG steganography using the generalized uniform embedding strategy. IEEE Trans. Circuits Syst. Video Technol. **28**(12), 3545–3549 (2018)
6. Zhang, Y., Luo, X., Yang, C., Ye, D., Liu, F.: A JPEG-compression resistant adaptive steganography based on relative relationship between DCT coefficients. In: 2015 10th International Conference on Availability, Reliability and Security (ARES), pp. 461–466 (2015)
7. Zhang, Y., Luo, X., Yang, C., Liu, F.: Joint JPEG compression and detection resistant performance enhancement for adaptive steganography using feature regions selection. Multimedia Tools Appl. **76**(3), 3649–3668 (2017)
8. Pereira, S., Voloshynovskiy, S., Madueno, M., Marchand-Maillet, S., Pun, T.: Second generation benchmarking and application oriented evaluation. In: Information Hiding, pp. 340–353 (2001)
9. Liu, W., Liu, G., Dai, Y.: Damage-resistance matrix embedding framework: the contradiction between robustness and embedding efficiency. Secur. Commun. Netw. **8**(9), 1636–1647 (2015)
10. Zhang, X., Wang, S.: Stego-encoding with error correction capability. IEICE Trans. Fundam. Electron. Commun. Comput. Sci. **88**(12), 3663–3667 (2005)
11. Kin-Cleaves, C., Ker, A.D.: Adaptive steganography in the noisy channel with dual-syndrome trellis codes. In: 2018 IEEE International Workshop on Information Forensics and Security (WIFS), pp. 1–7 (2018)
12. Larsen, K.: Short convolutional codes with maximal free distance for rates 1/2, 1/3, and 1/4 (corresp.). IEEE Trans. Inf. Theory **19**(3), 371–372 (1973)
13. Bocharova, I.E., Hug, F., Johannesson, R., Kudryashov, B.D.: On the exact bit error probability for Viterbi decoding of convolutional codes. In: Information Theory and Applications Workshop, pp. 1–6 (2011)
14. Bas, P., Filler, T., Pevný, T.: "Break our steganographic system": the ins and outs of organizing BOSS. In: International Workshop on Information Hiding, pp. 59–70 (2011)
15. Kodovský, J., Fridrich, J.: Calibration revisited. In: Proceedings of the 11th ACM Workshop on Multimedia and Security, pp. 63–74 (2009)
16. Chang, C.C., Lin, C.J.: LIBSVM: a library for support vector machines. ACM Trans. Intell. Syst. Technol. **2**(3), 27 (2011)

Audio Information Hiding Based on Cochlear Delay Characteristics with Optimized Segment Selection

Candy Olivia Mawalim$^{(\boxtimes)}$ and Masashi Unoki

Graduate School of Advanced Science and Technology,
Japan Advanced Institute of Science and Technology,
1-1 Asahidai, Nomi, Ishikawa 923–1292, Japan
{candyolivia,unoki}@jaist.ac.jp

Abstract. Audio information hiding (AIH) based on cochlear delay (CD) characteristics is a promising technique to deal with the trade-off between inaudibility and robustness requirements effectively. However, the use of phase-shift keying (PSK) for blindly detectable AIH based on CD characteristics caused abrupt phase changing (phase spread spectrum), which leads to bad inaudibility. This paper proposed the technique to reduce the spread spectrum from PSK by segment selection process with spline interpolation optimization. Objective evaluation to measure the detection accuracy (BDR) and inaudibility (PEAQ and LSD) was carried out with 102 various genre music clips dataset. Based on the evaluation result, our proposed method could successfully reduce the spread spectrum caused by PSK by having improvement on inaudibility test with adequate detection accuracy up to 1024 bps.

Keywords: Cochlear delay characteristics · Phase shift keying · Spread spectrum · Spline interpolation

1 Introduction

The fast growth of digital technology makes digital data easy to access. This growth, however, also causes a drawback that caused a significant loss for digital production organization [1]. To deal with this drawback, the methods of multimedia information hiding (MIH) (hiding data in digital content) techniques are widely studied as a promising solution. This study focuses on one direction of MIH, which is audio information hiding (AIH). AIH is more challenging than in the image or video information hiding [1,2] since the sensitivity and dynamic level of the human auditory system (HAS) is more than the human visual system (HVS). Consequently, achieving the inaudibility requirement in AIH technique

This work was supported by a Grant-in-Aid for Scientific Research (B) (No. 17H01761) and I-O DATA foundation.

is more difficult than achieving the invisibility in image information hiding technique. Moreover, audio has less signal representation than image signals in a one-time sample which leads to the trade-off between the robustness and the inaudibility requirements in AIH [2].

Several AIH techniques have been proposed by prior researchers [1,3]. For instance, the least significant bit (LSB) modification works by modifying the most insignificant bits of the audio signals [3]. Spread spectrum method works by spreading the pseudorandom signal, which contains the hidden information into the audio data [3]. However, the trade-off between the robustness and the inaudibility issue has remained as the main problem.

Along with the development in AIH research, the psychoacoustics study has been taken into account for improving the inaudibility performance [1]. One of them is cochlear delay (CD) characteristics. Utilizing the CD characteristics for AIH could give promising result [4]. However, the use of phase shift keying (PSK) for blind detection caused the abrupt phase spectrum changes (spread spectrum) [5]. The phase spread spectrum implies to significant audio quality distortion.

This paper studies about the technique of reducing the spread spectrum caused by PSK. Based on the characteristics of the spread spectrum in PSK, the segment selection process with spline interpolation optimization is proposed. We evaluate our proposed method to measure the detection accuracy and inaudibility with comparison with the prior research.

2 Audio Information Hiding Based on Cochlear Delay Characteristics

Cochlear delay (CD) is the characteristic found in HAS. This delay occurs due to the structure of the basilar membrane in the cochlea. When the higher components signal enters, the base of the basilar membrane will be excited; whereas the apex side in the case of the lower frequency components [6]. Figure 1 shows the illustration of the CD characteristics. The study on the affection of cochlear delay to the sounds synchronization perceptual judgment by Alba et al. [7] reported that the HAS is not sensitive with CD characteristics. Consequently, the AIH based on CD characteristics can satisfy the inaudibility requirement.

In signal processing, the CD characteristics were first modelled by Unoki and Hamada [4] using the first order infinite impulse response (IIR) all-pass filter, as follows:

$$H(z) = \frac{-b + z^{-1}}{1 - bz^{-1}} \tag{1}$$

where $0 < b < 1$ is the filter parameter. The filter $H(z)$ caused group delay, as follows:

$$\tau(\omega) = -\frac{d\arg(H(e^{j\omega}))}{d\omega} \tag{2}$$

where $H(e^{j\omega}) = H(z)|_{z = e^{j\omega}}$.

An all-pass filter can pass all the amplitude components; therefore, the phase information can be controlled. For AIH, two-CD IIR filters ($H_0(z)$ and $H_1(z)$)

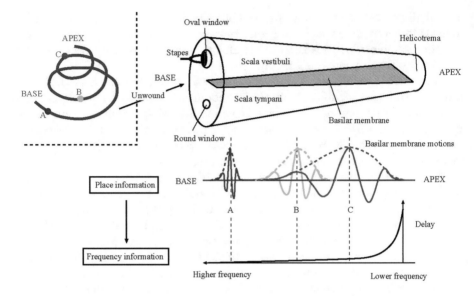

Fig. 1. Cochlear delay characteristics [8]

with filter parameters $b_0 = 0.795$ and $b_1 = 0.865$, respectively were utilized to represent the hidden bit information "0" and "1" (refers as phase-shift keying (PSK) concept). The PSK concept can satisfy robustness criteria, but it caused phase spread spectrum, which results in perceptible distortion [1,3]. To deal with this issue, the weighting function was utilized in the non-blind approach AIH based on CD characteristics [4]. However, for the blind approach [5], the discontinuity in the beginning segments of each frame obtained from PSK is important to obtain adequate detection. This implies to the trade-off between inaudibility and the robustness. Figure 2 shows the phase spread spectrum among frames in the blind detectable AIH based on CD characteristics.

3 Proposed Method

Our proposed method aims to reduce the spread spectrum caused by PSK in blindly detectable AIH based on CD characteristics. In the data embedding process, we conducted the optimization for selecting the embedding segments and spline interpolation to avoid embedding into the perceptible segments. Subsequently, we utilized the CZT in data detection process to achieve the blindness requirement.

3.1 Data Embedding

Figure 3 shows the block diagram for the data embedding scheme. Before passing through the CD filters, the selection process is conducted to reduce the abrupt

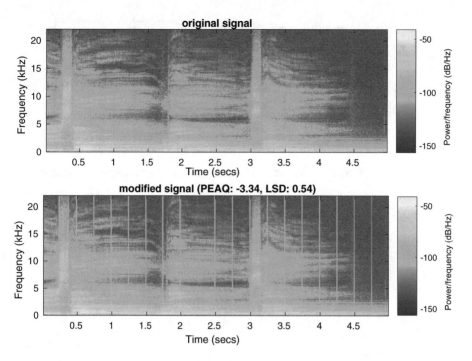

Fig. 2. Spectrogram comparison between original signal (top) and modified signal by blind detectable AIH based on CD characteristics (bottom)

phase changing (spread spectrum) among frames. This process was inspired by considering the advantage of high detection rate in the high payload of the previous works. From the empirical investigation, at least 40 samples are necessary as the size of the embedding frame to obtain adequate detection accuracy in normal condition. As shown in Fig. 4, the spread spectrum also occurred due to the discontinuity between frames. For this reason, the spline interpolation was also applied after filtering to connect each frame smoothly. In detail, the proposed embedding steps are as follows:

1. The original signal was segmented into K frames with specific frame size L.
2. The selection of embedding segments are conducted based on two options of criteria:
   - First, by maximizing the absolute different of higher frequency components (6–8 kHz) power spectrum of the beginning of original signal $x(n)$ (P_x) and artificial modified signal $x'(n)$ (P'_x). The artificial modified signal $x'(n)$ was obtained by filtering the original signal $x(n)$ segment by segment with $H_0(z)$. The higher power in higher frequency was chosen since it is one of the characteristics of spread spectrum by PSK. Mathematically, this process can be expressed as follows:

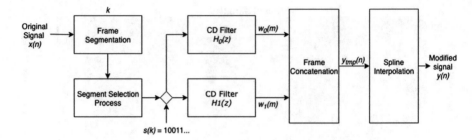

Fig. 3. Block diagram for data embedding procedure in proposed method

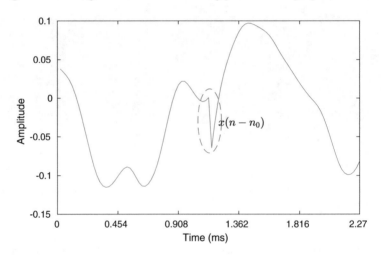

Fig. 4. Discontinuity occurred by PSK

$$n_0 = \arg\max(\Delta(Px(n - n_0), Px'(n - n_0)))$$
$$= \arg\max(|P_x(n - n_0) - P'_x(n - n_0)|) \ (3)$$

where $P_x(n - n_0) = \frac{1}{N^2} \sum_{n=1}^{N} |\text{FFT}(x(n - n_0))|^2$, $0 \leq n_0 \leq N_0$ and N_0 is the upper limitation of the variation of n_0.

- Second, by minimizing the root mean square error (RMSE) of the artificial modified signal $x'(n)$ before and after spline interpolation. From the PSK, the RMSE varies directly with the sound distortion. Mathematically, this minimization process can be expressed as follows:

$$n_0 = \arg\min(\text{RMSE}(x(n - n_o), x'(n - n_0)))$$
$$= \arg\min \sqrt{\frac{1}{L} \sum_{n=1}^{L} (x(n - n_o) - x'(n - n_0))^2} \ (4)$$

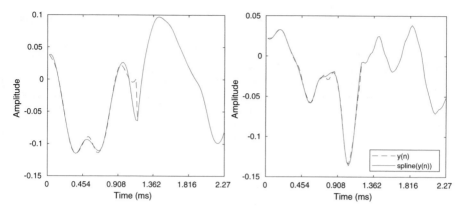

Fig. 5. Illustration of spline interpolation: (a) without RMSE minimization, and (b) with RMSE minimization

3. The original signal $x(m)$ in each segment was filtered simultaneously using two CD filters ($H_0(z)$ and $H_1(z)$) based on the hidden bit stream, $s(k)$, with transfer functions $H_0(z)$ and $H_1(z)$. The outputs of these filters were the intermediate signals, $w_0(m)$ and $w_1(m)$ (as shown in Eqs. 5 and 6).

$$w_0(m) = -b_0(m)x(m) + x(m-1) + b_0(m-1)w_0(m-1) \qquad (5)$$

$$w_1(m) = -b_1(m)x(m) + x(m-1) + b_1(m-1)w_1(m-1) \qquad (6)$$

m is the sample number in each segment range $[(k-1) \times L + 1, k \times L]$ and L is frame size).

4. The intermediate signals from each segment were then concatenated to form the temporary modified signal $y_{tmp}(n)$.
5. After concatenation process, the discontinuity found in each beginning of frame is connected by using spline interpolation. Figure 5 shows example of spline interpolation result (red solid line) from the signal with discontinuity (blue dashed line). The outputs of the spline interpolation are the modified signal $y(n)$.

3.2 Data Detection

As aforementioned, chirp z-transformation (CZT) [9] was utilized to estimate the zero position of the CD filters for the blindly detectable AIH based on CD characteristics [5]. CZT can evaluate the z-transform along contours [10]; therefore the estimation of zero position of CD filters is possible. Mathematically, the following equation represents CZT [11]:

$$\text{CZT}(y(n)) = W^{\frac{m^2}{2}} \times \text{IFFT}[\text{FFT}(y(n) \times r^{-n} \times W^{\frac{n^2}{2}}) \times \text{FFT}(W^{\frac{n^2}{2}})] \qquad (7)$$

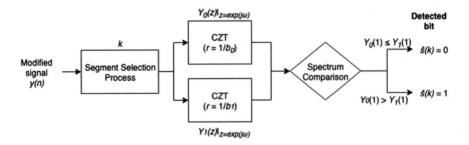

Fig. 6. Block diagram for data detection procedure in proposed method

where W is a complex scalar describing the complex ratio between points on the contour, r is the complex starting point, and m is the length of the transform frame.

Figure 6 shows the block diagram of the detection process. In essence, the payload, the filter parameter $b(n)$, the frame length are required as secret key in the receiver side for detection process. The detail detection steps were implemented as follows:

1. The modified signal $(y(n))$ was segmented into k frames with the corresponding frame length in the embedding method.
2. The segment selection process was conducted with regard to the selection procedure in embedding process (using highest power difference in high frequency components ranged from 6–8 kHz $(\arg\max(\Delta(Py, Py')))$ or lowest RMSE of modified signal before and after spline interpolation segment $(\arg\min(\text{RMSE}(y(n), y'(n)))))$. Order the selected segments based on the criteria and select the first N frames (N = the length of watermark) which refer to the embedded segments.
3. The CZT with $r = 1/b_0$ and $r = 1/b_1$ was conducted for each selected segment.
4. The CZT spectrums were compared to distinguish the hidden bit information (bit "0" or bit "1").
5. Finally, these detected bits were concatenated to form the detected signal.

4   Evaluation and Result

In this evaluation, the detection accuracy and inaudibility performance between the previous blindly detectable AIH based on CD characteristics in LTI system [5] and LTV system [12], the proposed method with segment selection without spline interpolation, the proposed method with optimized spline interpolation (least RMSE) in both LTI system and LTV system are compared. We employed the RWC (Real World Computing) database [13] with 102 audio clips from various music genre for the evaluation. The original clip has 44.1 kHz sampling frequency and 16-bit quantization. Each audio clip was cut into 10 s duration length with no silent part. The hidden information is the random binary bit-stream that embedded into one channel (monaural). The payload ranges from 4 bps to 1024 bps.

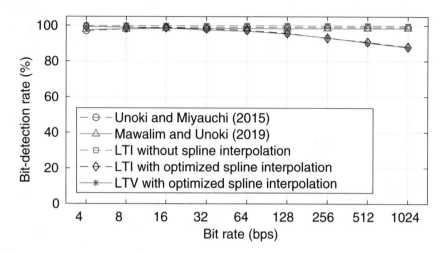

Fig. 7. Evaluation of detection accuracy in term of BDR

4.1   Detection Accuracy

The bit detection rate (BDR) is the measurement for evaluating the detection accuracy. Figure 7 shows the average BDR of each comparative AIH method. Although the detection rate in higher payload is a bit reducing for the proposed methods, it still passes the adequate detection rate threshold (80%) up to 1024 bps in both LTI system and LTV system.

4.2   Inaudibility

The inaudibility test was performed by using the perceptual evaluation of audio quality (PEAQ) [14] and log spectral distance (LSD) [15] measurements. The PEAQ outputs the objective difference grades (ODGs). The ODGs ranges in five scales: 0 (imperceptible), −1 (perceptible but not annoying), −2 (slightly annoying), −3 (annoying), and −4 (very annoying). Generally, the score less than −2 is defined as the required threshold for inaudibility aspect. The LSD measurement aims to evaluate the distance or distortion between two spectral (in AIH case, between the original spectra $X(\omega, k)$ and the modified spectra $Y(\omega, k)$) with k as the frame index. The threshold for LSD is 1 dB. The LSD can be calculated as follows:

$$\text{LSD}(X, Y) = \sqrt{\frac{1}{K} \sum_{k=1}^{K} \left(10\log_{10} \frac{|X(\omega, k)|^2}{|Y(\omega, k)|^2} \right)^2}, \tag{8}$$

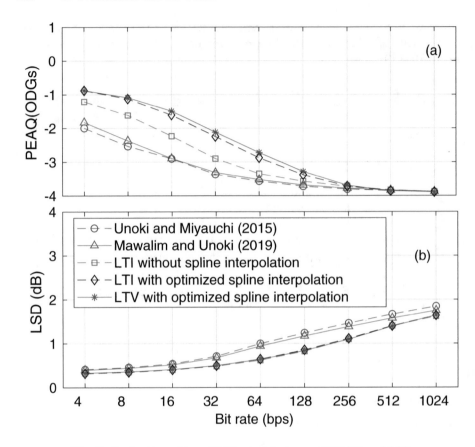

Fig. 8. Evaluation of inaudibility test in term of PEAQ and LSD

Figure 8 shows the inaudibility test of the previous methods and the proposed methods. Based on this result, we can see that the distortion caused by PSK could be reduced significantly by using the proposed methods, especially for the lower payload. The segment selection process played an important role in improving the inaudibility performance (as shown from the green dashed-dot line) in both PEAQ and LSD. In addition, the spline interpolation and implementation in LTV system could reduce the discontinuity and resulted in better inaudibility. Segment selection process by minimizing the RMSE of spline interpolation segments in LTV system could achieve the best inaudibility performance among the proposed methods. Figure 9 shows the example of spectrogram of proposed methods in one specific audio clip as reference to compare with the previous method in Fig. 2. In Fig. 9, the spectrograms show that the spread spectrum has been successfully reduced. Furthermore, the objective evaluation also could be in term of PEAQ and LSD.

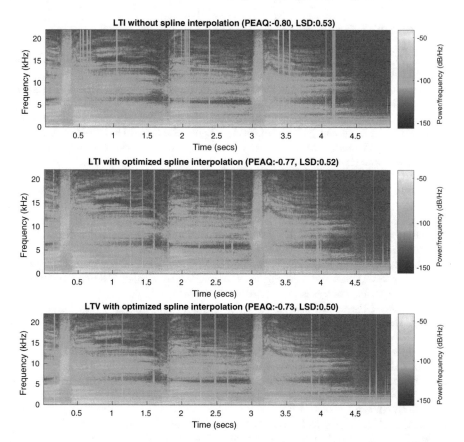

Fig. 9. Spectrogram comparison of modified signals with different criteria: delta power in high frequency components $(\arg\max(\Delta(Py, Py')))$ without spline interpolation in LTI system (top); minimum RMSE of before and after spline interpolation segment $(\arg\min(\text{RMSE}(y(n), y'(n))))$ in LTI system (middle); minimum RMSE of before and after spline interpolation segment $(\arg\min(\text{RMSE}(y(n), y'(n))))$ in LTV system (bottom).

5 Conclusion and Future Works

We investigated the techniques for reducing the spread spectrum caused by PSK in previous blindly detectable AIH based on CD characteristics. The selection of embedding segments and spline interpolation optimization techniques were proposed to deal with the spread spectrum issue. These techniques could solve the spread spectrum issue. As the result, it could significantly improve the inaudibility performance in term of PEAQ and LSD with adequate detection accuracy in term of BDR (up to 1024 bps). As our future direction, the robustness test against several signal processing operations will be carried out and the way to deal with frame synchronization as the common attack will be considered.

References

1. Lin, Y., Abdulla, W.H.: Audio Watermark: A Comprehensive Foundation Using MATLAB. Springer, Heidelberg (2015)
2. Hartung, F., Kutter, M.: Multimedia watermarking techniques. Proc. IEEE **87**(7), 1079–1107 (1999)
3. Hua, G., Huang, J., Shi, Y.Q., Goh, J., Thing, L.L.V.: Twenty years of digital audio watermarking – a comprehensive review. Signal Process. **218**, 222–242 (2016)
4. Unoki, M., Hamada, D.: Method of digital-audio watermarking based on cochlear delay characteristics. Int. J. Innov. Comput. Inf. Control **6**(3B), 1325–1346 (2010)
5. Unoki, M., Miyauchi, R.: Robust, blindly-detectable, and semi-reversible technique of audio watermarking based on cochlear delay. IEICE Trans. Inf. Syst. **E98-D**(1), 38–48 (2015)
6. Dau, T., Wegner, O., Mallert, V., Kollmeier, B.: Auditory brainstem response (ABR) with optimized chirp signals compensating basilar membrane dispersion. J. Acoust. Soc. Am. **107**(3), 1530–1540 (2000)
7. Aiba, E., Tsuzaki, M., Tanaka, S., Unoki, M.: Judgment of perceptual synchrony between two pulses and verification of its relation to cochlear delay by an auditory model. Jpn. Psychol. Res. **50**(4), 204–213 (2008)
8. Unoki, M., Miyauchi, R.: Method of digital-audio watermarking based on cochlear delay characteristics. In: Multimedia Information Hiding Technologies and Methodologies for Controlling Data, pp. 42–70. IGI Global (2013)
9. Rabiner, L.R., Schafer, R.W., Rader, C.M.: The chirp-z transform algorithm. IEEE Trans. Audio Electroacoustics **AU-17**(2), 86–92 (1969)
10. Rabiner, L.R., Schafer, R.W., Rader, C.M.: The chirp-z transform algorithm and its application. Bell Syst. Tech. J. **48** (1969). https://doi.org/10.1002/j.1538-7305.1969.tb04268.x
11. Martin, G.D.: Chirp Z-Transform Spectral Zoom Optimization with MATLAB, Sandia National Laboratories SAND 2005-7084 (2005)
12. Mawalim, C.O., Unoki, M.: Feasibility of audio information hiding using linear time variant IIR filter based on cochlear delay. In: NCSP 2019 Proceedings, pp. 323–326 (2019)
13. Goto, M., Hashiguchi, H., Nishimura, T., Oka, R.: RWC music database: music genre database and musical instrument sound database. In: Proceedings of ISMIR, pp. 229–230 (2003)
14. Kabal, P.: An examination and interpretation of ITU-R BS. 1387: perceptual evaluation of audio quality. Technical report, TSP Lab, McGill University (2003). http://mmsp.ece.mcgill.ca/Documents
15. Gray, A.J., Markel, J.: Distance measures for speech processing. In Proceedings of International Conference on Acoustics Speech and Signal Processing, vol. 24, no. 5, pp. 380–391 (1976)

A Watermarking Method for 3D Game Model Based on FCM Clustering and Density Tag Estimation of Vertex Set

Shengxian Wang[1(✉)], Li Li[1(✉)], Jianfeng Lu[1(✉)],
and Ching-Chun Chang[2(✉)]

[1] School of Computer Science, Hangzhou Dianzi University,
Hangzhou 310018, China
1308427740@qq.com, {lili2008,jflu}@hdu.edu.cn
[2] Department of Computer Science, University of Warwick,
Coventry CV4 7AL, UK
ching-chun.chang@warwick.ac.uk

Abstract. The current watermarking method is robust to several common attacks, but it is difficult to resist attacks of game models, such as cropping, game packaging, game assembly. In order to resist the attacks, this paper proposes a blind multi-region watermarking method based on FCM clustering and density tag estimation of vertex set for 3D game model. In the method, in order to resist the attack of game packaging, we preprocess the game model. In order to resist attacks of cropping and game assembly, we divide the model into multiple regions by FCM clustering and embed watermark into all regions. In order to improve the invisibility of the watermark, we select a vertex set of moderate density by density tag estimation method in each region. In addition, we embed the watermark by changing the positional relationship of the adjacent three vertices in the vertex set. The experimental results show that the proposed method achieved robustness against common attacks such as additive noise, cropping, and can resist attacks of game models, such as game packaging and game assembly.

Keywords: Game packaging · Game assembly · FCM · 3D game model · Density tag

1 Introduce

Digital watermarking is a method for copyright protection and content authentication, which can be classified into non-blind and blind watermarking methods [1]. The non-blind watermarking method requires original carrier during watermark extracting, while the blind watermarking methods can extract watermark correctly without original carrier [2, 3].

With the widespread use of three-dimensional models (3D) on the Internet, the security of 3D models has received great attention, which motivate us to design a watermarking for 3D models. Most of existing 3D watermarking methods can be classified into embedding watermark in spatial domain and embedding watermark in

© Springer Nature Switzerland AG 2020
L. C. Jain et al. (Eds.): SICBS 2019, AISC 1145, pp. 139–156, 2020.
https://doi.org/10.1007/978-3-030-46828-6_13

frequency domain [4, 5]. The first method embeds the watermark by modifying the vertex coordinates, while the second method embeds the watermark by modifying the coefficients in the frequency domain [6].

Zein [7] proposed a method based on FCM clustering to embed watermarks by changing the relationship between vertex coordinates and the average coordinate of adjacent vertices. This method makes it hard to find the location of the watermark embedding, but the capacity of the method is limited to the model topology. Jiang [8] utilizes a reversible watermarking method in the encryption domain (RDH-ED), and after extracting the embedded data, the original content can be reconstructed without loss. Feng [9] proposed a method for embedding two different watermarks into a three-dimensional mesh model. The first watermarking method was based on 3D model feature segmentation, and the second watermarking method was based on 3D model redundant information. In 2012 and 2017, Liu proposed two 3D model watermarking methods [10, 11]. In [10], the model is divided into multiple longitude regions, and the watermark is embedded by changing the longitude region of the vertex. In [11], the vertex coordinates are changed so that the vertex coordinates are divided by a given integer to embed the watermark. These two methods can effectively resist similarity transformation attacks well, but can be corrupted easily by geometric attacks such as cropping.

Nowadays, many existing watermarking methods are designed to obtain robustness against common attacks. However, with the types of attacks increasing, designing a robust 3D watermarking method which can resist special attacks is still a difficult task.

Game packaging refers to a series of operations on 3D models such as format conversion and platform migration. After game packaging, the model will have some isolated points. Game assembly refers to the practice of assembling parts of multiple 3D models together to form a new 3D model. In real-life applications, the game model is usually attacked by cropping, game packaging, game assembly, etc. To address this problem, a new watermarking method is proposed for 3D models. Firstly, we pre-process the 3D model, transform the polygon mesh model into a triangular mesh model and remove the isolated points existing in the model. Secondly, we divide the original model into multiple regions by FCM clustering, and select a vertex set of moderate density in each region to embed the watermark. Then, we construct the centroid of the vertex set as feature vertex with the largest degree in 1-ring neighbors. Finally, according to the distance to the centroid of vertex set, the vertex set is divided into several groups with three vertices. The watermark is embedded into the 3D model by changing the positional relationship of the three vertices in each group. The method has the following advantages: (1) When a region is attacked, the watermark can be extracted since we divide the model multiple region and embed the watermark in multiple region. Hence, the robustness against geometric attacks is improved. (2) If watermark is embedded in the vertex set of sparse density, it causes visible deformation. If watermark is embedded in the vertex set of dense density, the relative position between the vertices may be changed by additive noise attacks. Therefore, select a vertex set of moderate density to embed the watermark, so that the method realizes good invisibility and robustness. (3) This method has a larger watermark capacity than conventional method; (4) After extracting the watermark, the original 3D mesh model can be well-recovered. The flowchart of the proposed methods are shown in Fig. 1.

The rest of paper is organized as follows. The related works are presented in Sect. 2. The method of watermark embedding and extraction are proposed in Sect. 3. The experimental results are discussed in Sect. 4. This paper is concluded in Sect. 5.

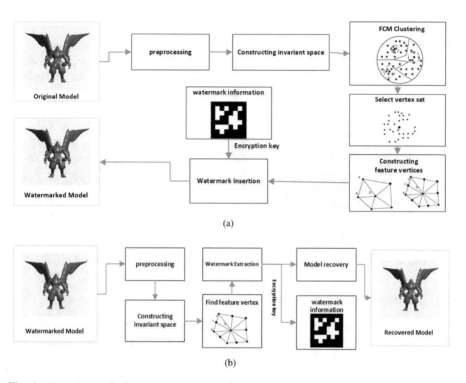

Fig. 1. Flowchart of the proposed method: (a) flowchart for watermark embedding, (b) flowchart for watermark extraction.

2 Overview

Firstly, we preprocess the model, transform the polygon mesh model into a triangular mesh model and remove the isolated points existing in the model. Secondly, we establish a Cartesian coordinate system of the model invariant space to resist similarity transformation attacks. Finally, we divide the model into multiple regions by FCM clustering, select a vertex set of moderate density in each region, and construct the centroid of the vertex set as the feature vertex.

2.1 Preprocess

In order to better construct feature vertex, we need to transform the polygon mesh model into a triangular mesh model. In order to resist the attacks of game packaging, we need to remove isolated vertices.

Triangulation
Most tools for 3D models are based on triangular mesh, but some of the game models are polygon mesh model. In order to better construct feature vertex, we need to transform the polygon mesh model into a triangular mesh model. Figure 2(a) shows the results after Delaunay triangulation [12] for model *cloak*.

Remove Isolated Vertices
After game packaging, some isolated vertices may be added to the model. Therefore, we need to remove isolated vertices to resist game packaging attacks. An example of removing isolated vertices is shown in Fig. 2(b).

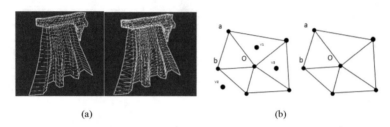

(a)   (b)

Fig. 2. (a) Delaunay triangulation for model *cloak*, (b) Removing isolated vertices

2.2   Establishing a New Coordinate System

In order to obtain robustness against the similarity transformation, such as rotation, translation and scaling, the original model M is transformed into an invariant coordinate system by using the PCA method [13].

(1) Convert M into M_1 to resist transformation attacks:

$$c = (c_x, c_y, c_z) = \frac{1}{n}\sum_{i=1}^{n} v_i = \left(\frac{1}{n}\sum_{i=1}^{n} x_i, \frac{1}{n}\sum_{i=1}^{n} y_i, \frac{1}{n}\sum_{i=1}^{n} z_i\right), \quad v_i \in M \quad (1)$$

$$v_i' = (x_i', y_i', z_i') = (x_i - c_x, y_i - c_y, z_i - c_z), \quad v_i' \in M_1 \quad (2)$$

Where the center of the original Model M is calculated by using Formula (1). v_i' denotes the coordinates of the vertex after moving the origin of the original Model to the center.

(2) Convert M_1 into M_2 to resist rotation attacks:

$$\widehat{v_i} = \left[\widehat{x}_i, \widehat{y}_i, \widehat{z}_i\right] = U \bullet v_i', \quad \widehat{v}_i \in M_2 \quad (3)$$

where the matrix U is the eigenvalue matrix of the matrix Cov.

$$Cov = \begin{bmatrix} \sum_{i=1}^{n} x_i'^2 & \sum_{i=1}^{n} x_i'y_i' & \sum_{i=1}^{n} x_i'z_i' \\ \sum_{i=1}^{n} y_i'x_i' & \sum_{i=1}^{n} y_i'^2 & \sum_{i=1}^{n} y_i'z_i' \\ \sum_{i=1}^{n} z_i'x_i' & \sum_{i=1}^{n} z_i'y_i' & \sum_{i=1}^{n} z_i'^2 \end{bmatrix} \tag{4}$$

Then, we perform eigenvalue decomposition on the matrix Cov.

$$Cov = UHU^T \tag{5}$$

Where U is a orthogonal matrix and the H is a diagonal matrix.

(3) Convert M_2 into M_3 to resist scaling attacks:

$$\hat{v}_i = \frac{s_i}{S_{\max}} \widehat{v}_i, \quad \hat{v}_i \in M_3 \tag{6}$$

$$S_{\max} = \max(s_1, s_2, s_3, \ldots s_i) \tag{7}$$

$$s_i = \sqrt{(\hat{x}_i^2 + \hat{y}_i^2 + \hat{z}_i^2)/3} \tag{8}$$

(4) In our method, we need to move the coordinate origin to the centroid $\hat{v}_m$ in each region, and convert M_3 into M_4:

$$v_i'' = (x_i'', y_i'', z_i'') = (\hat{x}_i - \hat{x}_m, \hat{y}_i - \hat{y}_m, \hat{z}_i - \hat{y}_m), \quad v'' \in M_4 \tag{9}$$

2.3 KNN-Based Density Estimation of the Vertex Set

In our method, there are five density tags for vertex set: sparse, slightly sparse, moderate, slightly dense, dense. Figure 3 shows the vertex set labelled with different tags. we measure the density tag of the model by three features: energy, volume, evenness.

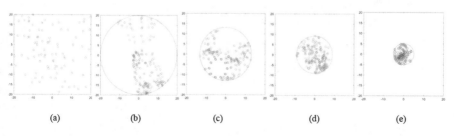

    (a)   (b)   (c)   (d)   (e)

Fig. 3. Five density tags: (a) sparse, (b) slightly sparse, (c) moderate, (d) slightly dense, (e) dense.

(1) Energy

$$En = \sum_{i=1}^{N} \left((x_i - x)^2 + (y_i - y)^2 + (z_i - z)^2\right) \tag{10}$$

Energy is the sum of the squares of all vertices to the centroid, which reflects the distribution of the vertices near the centroid. If the energy is small, it means the vertex set is dense. (x, y, z) are the coordinates of the centroid of the vertex set.

(2) Volume

$$V = Volume(CH(M)) \tag{11}$$

$CH(M)$ represents the convex hull of the model M. If the convex hull is small, it means that the vertex set is dense. Volume represents the volume of the $CH(M)$.

(3) Evenness

$$Ev = \sqrt{\frac{1}{N^2} \sum_{i=1}^{N} \sum_{j=1}^{N} \left((x_i - x_j)^2 + (y_i - y_j)^2 + (z_i - z_j)^2\right)} \tag{12}$$

Evenness reflects the average distance between the vertices. If the evenness is small, it means the vertex set is dense. N represents the sum of the vertices.

Method steps:

(1) Select 100 vertex sets for each tag from the training set. Use the 500 vertex sets as a training set of the KNN method.
(2) Combine the energy, volume, and evenness of the vertex set as a three-dimensional feature vector.
(3) Classify the input vertex set by KNN method ($k = 20$): Calculate the distance between the eigenvector of the input vertex set and the eigenvector of the training set, obtain their tag values of the nearest 20 samples, and select the tag with the most occurrences as the tag for the input vertex set.
(4) Output the density tag of the input vertex set.

To evaluate the feasibility of the density tag estimation method, we test on several vertex sets. Figure 4 shows the results of the classification.

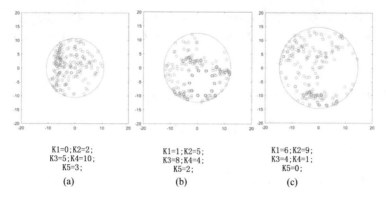

K1=0;K2=2;
K3=5;K4=10;
K5=3;

(a)

K1=1;K2=5;
K3=8;K4=4;
K5=2;

(b)

K1=6;K2=9;
K3=4;K4=1;
K5=0;

(c)

Fig. 4. The results of the classification: (a) slightly dense, (b) moderate, (c) slightly sparse.

2.4 Divide the Model by FCM Clustering

In the method, we divide the model M into λ regions $M_k(k \in (1, \lambda))$ by FCM clustering. After clustering, each vertex has a membership degree for each region. The membership matrix R_i of the ith vertex is shown in Formula (13):

$$R_i = [r_{i1}, r_{i2}, r_{i3}, \ldots r_{i\lambda}], \quad \sum_{k=1}^{\lambda} r_{ik} = 1 \tag{13}$$

Where $r_{ik}(k \in (1, \lambda))$ denotes the membership degree of the ith vertex to the kth region. Then, we divide the vertices into λ regions according to the rules: if $r_{ik}(k \in (1, \lambda))$ is the maximum of R_i, then the vertex v_i belongs to the region M_k.

2.5 Constructing Feature Vertex

In order to find the watermarked vertex set after 3D models are attacked, we need to construct the centroid of vertex set as the feature vertex with the largest degree in 1-ring neighbors.

Suppose that the maximum degree of vertices in a 3D model is D. In our method, we need to increase the degree of the centroid to $D + 1$. We increase the degree of the centroid by adding the triangle connected to the centroid. Figure 5 shows how to construct the centroid as feature points.

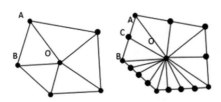

Fig. 5. Constructing feature vertex.

In the triangle (O, A, B), we select the midpoint C of A, B, add a triangle (O, A, C) into 3D model. Increase the degree of the centroid successively until the degree is D + 1.

3 Watermark Embedding and Extraction

This section organized into three parts: watermark embedding, watermark extraction and model recovery.

3.1 Watermark Embedding

Suppose that the watermark sequence is $b(b_1, b_2, b_i, ..b_m)$, where $b_i \in (-1, 1)$, $1 \leq i \leq m$). we generate a pseudo-random sequence $b'(b'_1, b'_2, b'_i, ..b'_m)$, where $b'_i \in (-1, 1), 1 \leq i \leq m$) by an encryption key Kw, the watermark embedding steps are as follows:

Step 1: Divide the 3D model into λ regions $M^{(k)} (k \in (1, \lambda))$ by FCM clustering.
Step 2: Select a vertex V_i in the kth region. Select V_i and 3 * m vertices closest to V_i which does not contain the 1-ring neighbor of V_i as a vertex set, where V_i is the centroid of the vertex set. Evaluate the density of the vertex set. If the density is moderate, embed watermark in the vertex set; otherwise, select V_{i+1} and 3 * m vertices closest to V_{i+1} as the next vertex set and evaluate its density tag.
Step 3: Construct the centroid of the vertex set as the feature vertex.
Step 4: Convert the 3D model coordinate system O_{xyz} into the new coordinate system $\bar{O}_{xyz}$, and move the coordinate system origin to the feature vertex $v^{(k)}$ of the region. Convert the Cartesian coordinate system of the kth region into a spherical coordinate system $(\rho_i^{(k)}, \phi_i^{(k)}, \theta_i^{(k)})$.
Step 5: Divide the 3 * m vertices $v_{3\kappa}^{(k)} (\kappa \in (1, m), k \in (1, \lambda))$ into m groups by distance. Each group has 3 vertices $(v_{\kappa 1}^{(k)}, v_{\kappa 2}^{(k)}, v_{\kappa 3}^{(k)}) (\kappa \in (1, m), k \in (1, \lambda))$, and one bit is embedded into each group.

Embed one bit by changing the position of the middle vertex in each group. The embedding formula is as follows.

$$\hat{\rho}_{\kappa 2}^{(k)} = \begin{cases} \rho_{\kappa 1}^{(k)} \times (1 - a) + \rho_{\kappa 2}^{(k)} \times a, & a \in (0, 0.5) \quad if \quad b'_\kappa = -1 \\ \rho_{\kappa 2}^{(k)} \times a + \rho_{\kappa 3}^{(k)} \times (1 - a), & a \in (0, 0.5) \quad if \quad b'_\kappa = 1 \end{cases} \quad (14)$$

Assuming ρ_{middle} is the middle point of $\hat{\rho}_{\kappa 1}^{(k)}$, $\hat{\rho}_{\kappa 3}^{(k)}$, the statement that $\hat{\rho}_{\kappa 2}^{(k)} \in (\rho_{\kappa 1}^{(k)}, \rho_{middle})$ after embedding bit -1 is proved as follows.

$$\hat{\rho}_{\kappa2}^{(k)} = \rho_{\kappa1}^{(k)} \times (1-a) + \rho_{\kappa2}^{(k)} \times a$$
$$< \rho_{\kappa1}^{(k)} \times (1-a) + \rho_{\kappa3}^{(k)} \times a \qquad (15)$$
$$< \rho_{\kappa1}^{(k)} \times 0.5 + \rho_{\kappa3}^{(k)} \times 0.5 = \rho_{middle}$$

$$\hat{\rho}_{\kappa2}^{(k)} = \rho_{\kappa1}^{(k)} \times (1-a) + \rho_{\kappa2}^{(k)} \times a$$
$$> \rho_{\kappa1}^{(k)} \times (1-a) + \rho_{\kappa1}^{(k)} \times a = \rho_{\kappa1}^{(k)} \qquad (16)$$

Therefore, after embedding bit -1, the relationship that $\hat{\rho}_{\kappa2}^{(k)} \in (\rho_{\kappa1}^{(k)}, \rho_{middle})$ is preserved. Similarly, the relationship that $\hat{\rho}_{\kappa2}^{(k)} \in (\rho_{middle}, \rho_{\kappa3}^{(k)})$ is kept after embedding bit 1.

Figure 6(a) shows the change in the position of the middle vertex in each group. Figure 6(b) shows the position range of original middle vertex and the position range of middle vertex after embedding a bit. If a approach 0.5, the relative values of $\hat{\rho}_{\kappa2}^{(k)}$ and ρ_{middle} are easy to change, and thus robustness is poor. If a approach 0, $\rho_{\kappa2}^{(k)}$ changes greatly, which causes visual distortion. It is proved by experiments that when $a = 0.3$, the robustness and transparency of the method are near optimal.

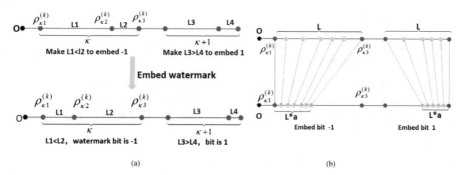

Fig. 6. Change of the middle vertex: (a) Change of the position, (b) Change of the position range

Repeat Step 5 until all regions are embedded with watermarks. Figure 7 shows the watermarked points cloud game model. Three copies of watermarks are embedded into three different regions in the model, respectively. Different colors represent different regions, and the vertices within the circle belong to a vertex set of moderate density.

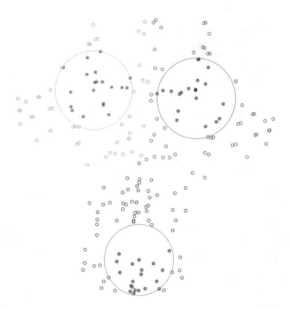

Fig. 7. The watermarked points cloud game model

3.2 Watermark Extraction

Watermark extraction requires encryption key Kw, and watermark can be extracted correctly by the following steps.

Step 1: Find the feature vertices in the model. For each feature vertex, select $3 * m$ vertices closest to feature vertex that do not contain the 1-ring neighbors.

Step 2: Extract the watermark at the kth feature point, and the extraction rule is as follows:

$$\bar{b}_\kappa^{(k)} = \begin{cases} = -1 & if \quad \hat{\rho}_{\kappa 2}^{(k)} - \hat{\rho}_{\kappa 1}^{(k)} < \hat{\rho}_{\kappa 3}^{(k)} - \hat{\rho}_{\kappa 2}^{(k)} \\ = 1 & if \quad \hat{\rho}_{\kappa 2}^{(k)} - \hat{\rho}_{\kappa 1}^{(k)} > \hat{\rho}_{\kappa 3}^{(k)} - \hat{\rho}_{\kappa 2}^{(k)} \end{cases} \tag{17}$$

Step 3: Repeat Step 1 and Step 2 until all but the last watermarks are extracted from the feature vertices, and then extract the last watermark by comparing all other watermarks. The watermark extraction formula is as follows.

$$\bar{b}_\kappa = \begin{cases} = -1 & if(N_{i0} \geq N_{i1}) \\ = 1 & if(N_{i0} < N_{i1}) \end{cases} \tag{18}$$

Where N_{i0} represents the number of '-1' in all watermark bits, N_{i1} represents the number of '1' in all watermark bits.

3.3   Model Recovery

When embedding a watermark, the centroid of the vertex set is constructed as the feature vertex. After all the watermarks are extracted, the feature points can be further restored. If V_a, V_b and V_c is in the 1-ring neighbors of the feature vertex, and V_a, V_b and V_c satisfy the following equation:

$$V_b = (V_a + V_c)/2 \qquad (19)$$

then, it implies that V_a, V_b, and V_c are collinear, and V_b is the midpoint of V_c and V_a. Hence, we need to delete the vertex V_b and the triangle containing the vertex V_b to restore the feature vertex.

After extracting watermark, we should restore the coordinate of middle vertex for each group. The process is as follows:

$$\breve{\rho}_{\kappa 2}^{(k)} = \begin{cases} (\hat{\rho}_{\kappa 2}^{(k)} - \hat{\rho}_{\kappa 1}^{(k)} \times (1 - a)) \div a & if \quad b_\kappa'^{(k)} = -1 \\ (\hat{\rho}_{\kappa 2}^{(k)} - \hat{\rho}_{\kappa 3}^{(k)} \times (1 - a)) \div a & if \quad b_\kappa'^{(k)} = 1 \end{cases} \qquad (20)$$

4   Experimental Comparison and Results

To test the invisibility and robustness of the proposed method, we compare the experiment results with the methods in [10] and [11]. Figure 10(a) shows five 3D game models. In the method, 60 bits are embedded into each 3D model, and the robustness is tested by performing the attacks of additive noise, cropping, game assembly, and game packaging. To evaluate the invisibility of the proposed method, the root mean square (*RMSE*) is given by

$$RMSE = \sqrt{\sum_{i=1}^{N} (x_i' - x_i)^2 + (y_i' - y_i)^2 + (z_i' - z_i)^2} \qquad (21)$$

Moreover, the signal-to-additive noise ratio (*SNR*) is also used for evaluating the invisibility, which is defined as

$$SNR = 10 \log_{10} \left(\frac{\sum_{i=1}^{N} (x_i^2 + y_i^2 + z_i^2)}{\sum_{i=1}^{N} ((x_i' - x_i)^2 + (y_i' - y_i)^2 + (z_i' - z_i)^2)} \right) \qquad (22)$$

Where N denotes the number of the vertices, (x_i, y_i, z_i) are the original coordinates and (x_i', y_i', z_i') are the modified coordinates respectively. A smaller *RMSE*, or a higher *SNR*, means better invisibility of the watermark. Moreover, in order to test the robustness of the method, the correlation coefficient (*Corr*) between the extracted watermark **w** and the embedded w is given by

$$Corr(w, w') = \frac{\sum\limits_{i=1}^{M} w_i \times w_i'}{\sqrt{\sum\limits_{i=1}^{M} (w_i')^2}\sqrt{\sum\limits_{i=1}^{M} (w_i)^2}} \qquad (23)$$

The values of *Corr* range between 0 and 1. In addition, the bit error rate (*BER*) also is used to measure the robustness, which represents the error rate of the extracted watermark bits.

4.1   Optimal Number of Clusters

The FCM divides the model into multiple regions. When a region is attacked, the watermark can still be extracted from other regions, which enhance the robustness of our scheme. If we embed watermarks into more regions, we can achieve higher robustness. On the other hand, however, the visual quality of the carrier model will degrade. We test on five game models shown in Fig. 10(a) and calculate the average. Figure 8 shows the trade between invisibility and robustness as the number of regions increase.

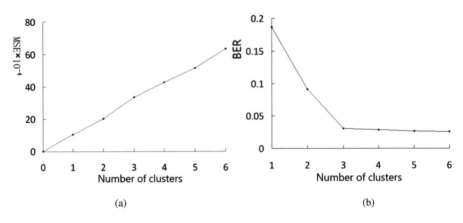

(a)   (b)

Fig. 8. The invisibility and robustness: (a) the invisibility of watermark, (b) the robustness attacked by 5% cropping

When the number of regions is 3, the method achieved good invisibility and robustness under the 5% cropping attack. Therefore, in the following experiment, we let the model to be divided into 3 regions.

4.2   Capacity Comparisons of the Methods

In our method, a group of three vertices can be embedded with a bit. In the case where the number of vertices is the same, our method has a larger capacity than the method [10] and [11]. Figure 9 shows the watermark capacity of three method as the number of the vertices increasing.

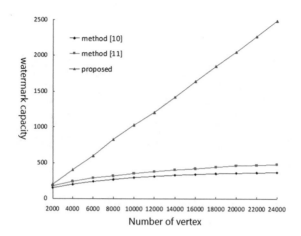

Fig. 9. The watermark capacity of three method as the number of the vertices increasing

4.3   Imperceptibility Comparisons of the Methods

The five original models and five watermarked models by using our method are shown in Fig. 10. Embedding a bit only needs to change the coordinate of middle vertex for each group, and selecting a vertex set of moderate density improves the transparency of the watermark. As shown in Fig. 10(b), the watermarked models have no obvious distortion compared the original models.

From Table 1, we can find that the proposed method attained lower *RMSE* and higher *SNR* than the methods in [10] and [11].

Table 1. Invisibility of the proposed method compared to method in [10] and [11] in terms of *RMSE* $\times$ 10^{-4} and *SNR*

Models	Methods					
	Method in [10]		Method in [11]		The proposed method	
	RMSE	SNR(dB)	RMSE	SNR(dB)	RMSE	SNR(dB)
Model 1	72	39.2	70	39.2	31	81.7
Model 2	69	41.9	69	40.5	28	83.2
Model 3	65	43.1	64	42.6	27	84.1
Model 4	59	45.2	55	44.2	25	85.2
Model 5	58	45.5	53	44.7	24	85.8

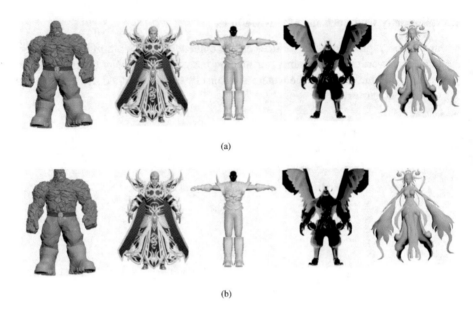

Fig. 10. (a) Original model, (b) Watermarked model.

4.4   Robustness Comparisons of the Methods

In order to test the robustness against the attacks of additive noise, cropping, game packaging and game assembly, we perform different kinds of attacks on five models. Figure 11 shows the five models under different attacks, Table 2 illustrates the values of *Corr* and Fig. 12 shows the *BER* average values respectively.

Additive Noise Attack
The robustness against additive noise attack is tested by performing different degree (0.05%, 0.1%, 0.2%) on 3D models. The *Corr* values in Table 2 and the BER shown in Fig. 12(a) indicate that the proposed method can resist additive noise attack. From Fig. 12(a) we can see that the proposed method has obtained better robustness than the method in [10] and [11].

In our method, we select a vertex set with the degree tag of moderate to embed watermark in order to avoid the change of relative position of the vertices under the attacks of additive noise. We can further decrease the value a to improve the robustness against additive noise attacks.

Cropping Attack
The robustness against cropping attack is tested by performing different degree (5%, 10%, 20%) on 3D models. From Table 2 and Fig. 12(b), it indicates that the proposed method can resist cropping attack. It can be seen that the *Corr* and the *BER* of the proposed method are higher than the methods in [10] and [11], which shows the

proposed method has stronger robustness. When the strength of the attack is less than 20%, the values of *Corr* are greater than 0.83, which shows the proposed method can obtain good robustness.

Our method constructs the centroid of the vertex set as the feature vertex. The degree of feature vertex is unchanged under the cropping attack. We divide the model into multiple regions, and embed watermark into each region. When a region is attacked, the watermark can be extracted from other regions. Therefore, our method is robust against cropping attack.

Game Assembly Attack
Game assembly refers to assembling parts of multiple 3D models together to form a new 3D model. The third three columns of Table 2 shows *Corr* values when the assembled component accounts for 5%, 10%, and 20% of the 3D model, and Fig. 12(c) shows the performance against the game assembly attacks. We observe that when assembled component exceeds 20% of the original model, the values (*Corr*) is less than 0.65 and the values (*BER*) is greater than 25%. And compared to the robustness against cropping, the robustness against game assembly attack is poor.

The proposed method can obtain the robustness against game assembly attack by embedding watermark into multiple regions. Compared to cropping attack, however, the robustness against game assembly attack is poor. The underlying reason is that in the process of assembling, the maximum degree of 3D model may be changed. And it is also possible that the vertex set closest to the feature vertex is changed.

Game Packaging Attack
Game packaging refers to performing a series of operations on 3D models such as format conversion and platform migration. The third last line of Table 2 shows the robustness against game packaging attacks. After game packaging, the model would have some isolated vertices. Before embedding watermark, we remove the isolated vertices via the preprocessing in order to resist the game packaging attacks.

Similarity Transformation
In our method, we establish a Cartesian coordinate system of the model invariant space to resist similarity transformation attacks. If the signal operations do not change the relative position of vertex coordinates, the constructed invariant space is unchanged under the attacks, such as translation, scaling and rotation. The last two lines of Table 2 shows that the proposed method can resist similarity transformation attacks.

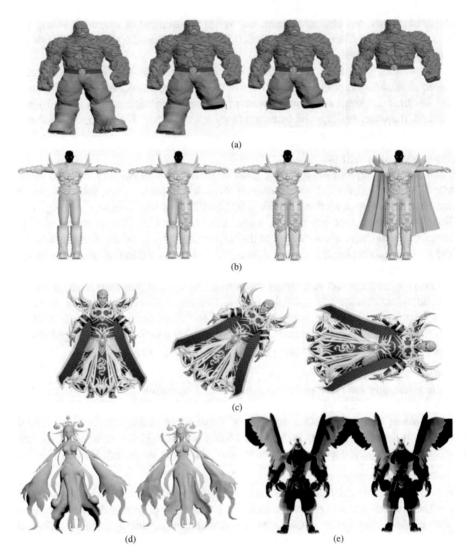

Fig. 11. Attacks: (a) the cropping strength is 0, 10%, 20%, 40% from left to right, (b) the game assembly strength is 0, 5%, 10%, 20% from left to right, (c) the rotating angle is 0, 45°, 90° from left to right, (d) the original model and the model under additive noise attack (0.2%), (e) the original model and the model under game packaging attack.

Table 2. The *Corr* of five 3D models under different kinds of attacks

Models	Additive noise			Cropping			Game assembly			Pack.	Scal.	Rota.
	0.05%	0.1%	0.2%	5%	10%	20%	5%	10%	20%	/	*0.5	90°
Model 1	0.922	0.853	0.713	0.955	0.886	0.832	0.902	0.754	0.625	1	1	1
Model 2	0.927	0.864	0.734	0.962	0.891	0.846	0.924	0.768	0.627	1	1	1
Model 3	0.931	0.876	0.754	0.974	0.895	0.851	0.920	0.775	0.630	1	1	1
Model 4	0.937	0.881	0.767	0.985	0.896	0.852	0.927	0.785	0.658	1	1	1
Model 5	0.931	0.878	0.741	0.990	0.913	0.864	0.932	0.808	0.661	1	1	1

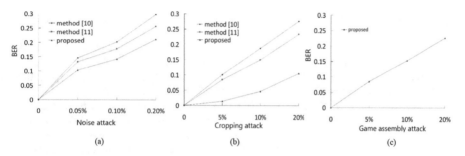

Fig. 12. Comparison of the *BER*: (a) robustness against additive noise, (b) robustness against cropping, (c) robustness against game assembly

5 Conclusion

In this paper, a novel robust 3D models watermarking method is proposed for game models. Firstly, we preprocess the model to resist the attack of game packaging; Secondly, we divide the original model into multiple regions by FCM clustering, and select a vertex set of moderate density in each region to embed watermark. According to the above experiments, our method achieved good watermark transparency and improved the robustness of the 3D game watermarking method against attacks of cropping and additive noise. Furthermore, our method can resist the attacks of game packaging and game assembly.

In the future, we will investigate the following two possible research directions: (1) Design a more efficient way to construct feature vertex (2) Combine powerful machine learning techniques to resist more kinds of attacks.

Acknowledgement. This work was partially supported by National Natural Science Foundation of China (No. 61370218, No. 61971247), and Public Welfare Technology and Industry Project of Zhejiang Provincial Science Technology Department (No. LGG19F020016).

References

1. Deng, C., Gao, X., Li, X., Tao, D.: A local Tchebichef moments-based robust image watermarking. Signal Process. **89**(8), 1531–1539 (2009)

2. Feng, X., Liu, Y., Fang, L.: Digital watermark of 3D CAD product model. Int. J. Secur. Appl. **9**(9), 305–320 (2015)

3. Soliman, M., Hassanien, A., Onsi, M.: An adaptive watermarking approach based on weighted quantum particle swarm optimization. Neural Comput. Appl. **27**(2), 469–481 (2016)

4. Hachani, M., Ouled, Z., Puech, W.: Feature-based image watermarking algorithm using SVD and APBT for copyright protection. Future Internet **9**(13), 1–15 (2017)

5. Soliman, M., Hassanien, A., Onsi H.: A robust 3D mesh watermarking approach using genetic algorithms. In: Advances in Intelligent Systems and Computing, vol. 323, pp. 731–741. Springer, Warsaw (2014)

6. Zhang, Y., Wang, C., Wang, X., Wang, M.: Robust mesh data hiding based on irregular wavelet transform. In: Proceedings of European Signal Processing Conference, vol. 12, no. 1, pp. 1–5 (2013)

7. El Zein, O., El Bakrawy, L., Ghali, N.: A robust 3D mesh watermarking algorithm utilizing fuzzy C-Means clustering. Future Comput. Inform. J. **2**(2), 148–156 (2017)

8. Jiang, R., Zhou, H., Zhang, W., Yu, N.: Reversible data hiding in encrypted three-dimensional mesh models. IEEE Trans. Multimed. **20**(1), 55–67 (2018)

9. Feng, X., Zhang, W., Liu, Y.: Double watermarks of 3D mesh model based on feature segmentation and redundancy information. Multimed. Tools Appl. **68**(3), 497–515 (2014)

10. Liu, J., Wang, Y., Li, Ye., Liu, R., Chen, J.: A robust and blind 3D watermarking algorithm using multiresolution adaptive parameterization of surface. Neurocomputing **237**, 304–315 (2017)

11. Liu, J., Yang, Y., Ma, D., Wang, Y., Pan, Z.: A watermarking method for 3D models based on feature vertex localization. IEEE Access **6**, 56122–56134 (2018)

12. George, P., Borouchaki, H.: Delaunay triangulation and meshing: application to finite elements. Paris (1998)

13. Kalivas, A., Tefas, A., Pitas, I.: Watermarking of 3D models using principal component analysis. In: Proceedings of IEEE International Conference on Acoustic, Speech, and Signal Processing 2003, pp. 676–679. IEEE, New York (2003)

A Reversible Information Hiding Scheme Based on Index Encryption Strategy in VQ-Image

Cheng-Ta Huang[1], Chi-Yao Weng[2], Cheng-Hsing Yang[2(✉)], and Cheng-You Guo[3]

[1] Department of Information Management, Oriental Institute of Technology, New Taipei City 220, Taiwan
[2] Department of Computer Science, National Pingtung University, Pingtung 90003, Taiwan
chyang@mail.nptu.edu.tw
[3] Institute of Information Security, Nation Tsing Hua University, Hsinchu 30013, Taiwan

Abstract. In 2017, Yan et al. proposed a method of image cryptosystem based on VQ and cross chaotic map. In their method, they applied index table to swap the index with randomization as the method of image cryptosystem. Swapping the index can achieve the effect of image encryption, but it could be suffered from lower security in image while the statistical analysis is used. This paper improves the Yan et al. scheme and proposes a novel image cryptosystem based on encryption of index-table and codebook. Our scheme uses two methods, confusion on index table and diffusion on codebook, to encrypt the images and make the encrypted image more security. Experimental result shows that the proposed scheme can efficient prevent the attack of histogram analysis and pay fewer effort on preprocessing. The proposed scheme not only has more security against statistical analysis but makes our scheme more flexibility on information hiding.

Keywords: VQ · Index encryption · Reversible data hiding · Diffusion

1 Introduction

With rapid revolution of communications and multimedia technology, multimedia data plays an important role for our daily lives. However, the demand of new applications for people via the public network is quickly promoting. For solving the people requirement, apply image compression and encryption scheme, which can reduce the storage and image size and easy to transmission and to protect the confidentiality of image, have been proposed in recent years. It is worthy to achieve this goal that is different from the traditional image-encryption and image-compression [1–10]. The existing method of traditional image-encryption and image-compression can be classified into three categories: first category, the compression-oriented is to insert the encryption method into the compression algorithm [1–3]. In the second category, encryption-oriented [4, 5] intends to merge compression into cryptosystem. The last category regards to consider that bother of compression and encryption are very importance to each other [6], this work belongs to.

© Springer Nature Switzerland AG 2020
L. C. Jain et al. (Eds.): SICBS 2019, AISC 1145, pp. 157–169, 2020.
https://doi.org/10.1007/978-3-030-46828-6_14

Reversible data hiding (RDH) [13–18] is an import issue of information hiding technique. The data hiding technology embeds secret data into the cover media, such as image, text, and audios. However, RDH scheme not only extract the secret data but also recover the cover media with completely from the hidden media. The RDH technology is useful to some special application where the cover media is not allowed to be destroyed, for example, medical image, military map, and low text. Many RDH methods have been proposed to embed the secret message into image. According to the development schemes, the cover image could be transferred into three domains, such as spatial domain [17], frequency domain [19], and compressed domain [12, 20], and then, select the hiding strategies, for instance, histogram shifting, differencing expansion, or prediction error, to conceal secret information. Basically, these RDH schemes make and seek more accommodation to embed secret data. If an image has been encrypted, the encrypted image will lose a lot of redundancy information, resulting in missing the task of lossless recovery. To avoid the missing task, the RDH method in encrypted image is thus proposed [21, 22].

The method of encrypted image can achieve the recovery cover image with lossless, but the hiding payload are limited. Some researches focus on compressing the image to preserve the space for image encryption and data hiding. Yan et al. in 2017 proposed a new image cryptosystem based on vector quantization (VQ). In their method, the cover image is compressed to a codebook and index by using traditional VQ [11], and the indexes will be encrypted by cross chaotic map. Although the index can achieve the effect of image encryption, but their method could be suffered the vulnerable of statistical analysis. In addition, Yan et al.'s method could not hide secret message into. In this article, we improve the security of index encryption and make the proposed method to achieve information hiding.

The rest of this article is organized as follows. In Sect. 2, we will briefly show the preliminary works based on Vector Quantization approach and Yan et al.'s method. The proposed scheme in detail is introduced in Sect. 3. Experimental results and comparison are illustrated in Sect. 4. Finally, Sect. 5 concludes the whole paper.

2 Related Works

This section gives some basic of VQ for image and describe the related work of Yan et al.'s method. In addition, the technology of reversible data hiding will be introduced briefly.

2.1 VQ Encoding

The vector quantization (VQ) image compression method is proposed by Lind et al. in 1980 [11]. VQ is an efficient lossy image compression technique where the compressed ratio is depended by the size of codebook. Generally speaking, the smaller codebook size will have good compression effect, but the bad image quality will be made. In VQ method, the cover image is firstly divided into nonoverlapping blocks, and use a codebook to find the neighboring codewords, then, the block index is thus outputted.

The conventical VQ method can be divided into three phases, including codebook generation, image encoding, and image decoding. The codebook generating is related with the image quality while image encoding is executed. Therefore, a better method for codebook generating will make high image quality. Many methods have been proposed for codebook generating, such as LBG algorithm, genetic algorithm. After codebook generating, we can obtain many codewords and collect all the codewords to produce a codebook which owns index to assign the codewords.

After receiving the codebooks, for each block in an image is thus compressed. The compress method for each block u is to estimate the distance that the pixels within a block are closed to codewords v. The estimated distance function is shown in Eq. (1).

$$d(u, v) = \sum_{i=1}^{k} (u_i - v_i) \tag{1}$$

where d is Euclidean distance.

According to Euclidean distance d, the closer index between a block and codewords could be found. For each block, we can apply the same method to seek all the indexes, called as index table, the VQ encoding is finished.

The third phase apply the image decoding to reconstruct the image. Image coding is a faster than image encoding because the time complexity of searching index is short than codewords computing and comparison. The image reconstruction fetches the index from the index table and find the codewords from codebook. Finally, reconstruct the pixel in each block by replacing the codewords of codevector.

2.2   Yan et al.'s Image Encryption Method VQ

In 2017, Yan et al. proposed an encryption algorithm based on VQ [12]. They apply cross chaotic map to achieve index encryption. The index encryption is based on two methods of codebook diffusion and index table confusion. For the index table confusion, the sequence of index produced by VQ encoding and chaotic sequence generated by cross chaotic map are obtained. Then, rearrange the index by the chaotic sequence and to get the swapping index sequence by following equation:

$$q(i) \leftrightarrow q(X_i), i = 1, 2, \ldots, n \tag{2}$$

where $\leftrightarrow$ is the swapping function between $q(i)$ and $q(X_i)$, X_i is the element within the chaotic sequence.

Codebook diffusion indicates that the codewords in a codebook are shuffled with chaotic sequence. However, the property of codewords distribution keeps unchanged before and after shuffling operation. The Eq. (3) shows that each codeword are operated with diffusion sequence using XOR operation.

$$C_i' = \left(\left(c_i + \left(x_i' \oplus y_i' \right) \right) \oplus c_{i-1}' \right) mod\, 256 \tag{3}$$

where C_i' and C_i are the codebook with diffusion and permuted, respectively.

2.3 Reversible Data Hiding

The concept of reversible data hiding (RDH) is first introduced in 2003. The requirement of RDH is that the hidden message and stego image can be completely extracted and recovered respectively. So far, two methods of RDH, including differencing expansion (DE) [14] and histogram shifting [17], are applied into the research of RDH. The DE method is first mentioned by Tian in 2003 [14]. The Tian's method used the differencing between two neighboring pixels and expand the differencing for information hiding. In this method, the total embedding payload is near to 0.5 bpp (bits per pixel). DE method can hide more message, but the image distortion in the stego image is very huge. In 2006 [17], Ni et al. proposed a novel RDH based on histogram shifting (HS) for overcoming the huge image distortion.

The HS approach uses the statistical character to generate the pixel distribution, and then, shift a vast room for concealing the secret message. In HS method, firstly, input a cover image and generate the pixel distribution. From this distribution, seek a highest peak value and zero (or lowest) peak value. Next, shift all the pixels belonging to the range [peak point, zero point] by 1 unit. Finally, hide the message into when the pixel value is equal to peak point. As the result, the hiding capacity is limited on the altitude of peak point. If more and more high peak points are selected, the hiding capacity will be enlarger.

3 The Proposed Method

This research aims to improve the index encryption based on Yan et al.'s method and make the proposed method to achieve information hiding. The Fig. 2 shows the flowchart of our proposed method. In our proposed method, the cover image is firstly encoded using VQ-encoding, then, the index table is obtained. The index table is used to sort by ascendant order for information hiding. The operations of index table confusion and codebook diffusion can be paralleled processing. We use the random generator to generate a secret bit string and apply the XOR operation with codebook and secret bit string, called as codebook diffusion. Although Yan et al.'s method has a good method for index encryption, the index encryption strategy is vulnerable to statistical analysis. For enhancing the security of index encryption, we thus proposed another index encryption method to diffuse the index table. The index table is divided into non-overlapping blocks. For each block, the random string key_1 is generated, and then, do the operation of XOR operation with key_1 and each block. The proposed approaches in terms of index encryption, information hiding, image decryption and message extracting, are mentioned in detail as below.

3.1 Image Encryption

In the index encryption phase, the index table is firstly divided by non-overlapping blocks with the size of 2×2. Assume that the size of the index table is $m \times n$. The following steps are introduced the index diffusion processing.

Step 1: Apply the *key_T* to generate the secret sequence $r_{i,j}$, where $i = 1, 2, 3,...,$ $m-1, j = 1, 2, 3,..., n-1$, and $r_{i,j} \in [0, k-1]$. Additionally, k is the size of codebook. The size of the secret sequence is $((m \times n))/4$.

Step 2: For each index in a block, is encrypted by the following Equation.

$$\begin{cases} TE_{i,j} = \left(T_{i,j} + r_{i,j}\right) mod\, k \\ TE_{i,j+1} = \left(T_{i,j+1} + r_{i,j}\right) mod\, k \\ TE_{i+1,j} = \left(T_{i+1,j} + r_{i,j}\right) mod\, k \\ TE_{i+1,j+1} = \left(T_{i+1,j+1} + r_{i,j}\right) mod\, k \end{cases} \tag{4}$$

Figure 2 shows the example of the index in a block (Fig. 1).

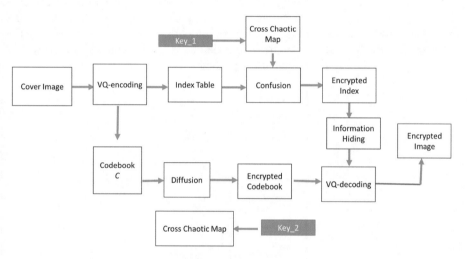

Fig. 1. The flowchart of our proposed approach

$T_{i,j}$	$T_{i,j+1}$
$T_{i+1,j}$	$T_{i+1,j+1}$

Fig. 2. The example of index in a block

3.2   Information Hiding

After finishing the procedure of index diffusion, the index differencing between two neighboring index keeps unchanged, called as differencing preserving. Inspect the property of differencing preserving, the secret bits can be hidden into. Here, we apply the histogram shifting (HS) of prediction error to achieve the reversible data hiding. Using HS method will encounter the exception of overflow and underflow. We should

162   C.-T. Huang et al.

preprocess each index while the index is suffered the exception of underflow/overflow. The following Eq. (5) shows the preprocess procedure.

$$\begin{cases} TE'_{i,j} = 1, d_n = 1 & if\ TE_{i,j} = 0 \\ TE'_{i,j} = 1, d_n = 0 & if\ TE_{i,j} = 1 \\ TE'_{i,j} = k - 2, d_n = 1 & if\ TE_{i,j} = k - 1 \\ TE'_{i,j} = k - 2, d_n = 0 & if\ TE_{i,j} = k - 2 \end{cases} \tag{5}$$

where $TE'_{i,j}$ is a new index after preprocessing, k is the size of the codebook. The total length of preprocess is $d_n,\ n = 1, 2, \dots$. The d_n is concatenated to the secret message, and then, run the message hiding procedure.

For pursuing high embedding payload, the embedding procedure is divided into two parts. The first part of embedding procedure applies the fixed index to predict the neighboring index, shown in Fig. 3. Then, collect all the prediction errors and hide the message into the highest errors. The first hidden message is mentioned as below.

Step 1: Use the Eq. (6) to obtain the prediction error $e_{i,j}$.

$$\begin{cases} e_{i,j+1} = TE'_{i,j} - TE'_{i,j+1} \\ e_{i+1,j+1} = TE'_{i,j} - TE'_{i+1,j+1} \\ e_{i+1,j} = TE'_{i,j} - TE'_{i+1,j} \end{cases} \tag{6}$$

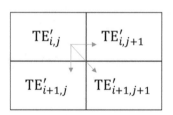

Fig. 3. The example of first part prediction

Step 2: Collect all of prediction error $e_{i,j}$ and generate the error distribution. Find two pairs information of (pr, zr) and (pl, zl) from the error distribution and hiding the secret message into two pairs, then, obtain new error distribution $e'_{i,j}$ according to the following Eq. (7) and Eq. (8). Notably, the pair information of (pr, zr) and (pl, zl) indicate that the peak points and zero points in the right-side of the error distribution and in the left-side of the error distribution, respectively.

$$\begin{cases} e' = e, if\ b = 0\ and\ e = pr \\ e' = e + 1, if\ b = 1\ and\ e = pr \\ e' = e + 1, if\ zr > e > pr \\ e' = e, else \end{cases} \tag{7}$$

$$\begin{cases} e' = e, \text{ if } b = 0 \text{ and } e = pl \\ e' = e - 1, \text{ if } b = 1 \text{ and } e = pl \\ e' = e - 1, \text{ if } zl < e < pl \\ e' = e, \text{ else} \end{cases} \tag{8}$$

Step 3: Inverse the new error distribution $e'_{i,j}$ and obtain new hidden index $TEH_{i,j}$ according to the following equation.

$$\begin{cases} TEH_{i,j+1} = TE'_{i,j} - e'_{i,j+1} \\ TEH_{i+1,j+1} = TE'_{i,j} - e'_{i+1,j+1} \\ TEH_{i+1,j} = TE'_{i,j} - e'_{i+1,j} \end{cases} \tag{9}$$

The second part of embedding procedure is similar with the first part of embedding procedure except for prediction computing. The prediction value is computed by average of the neighboring pixels, depicted in Fig. 4. The hidden message in detail is shown as below.

Step 1: Use the Eq. (10) and Eq. (11) to obtain the prediction error $e_{i,j}$.

$$avg = \frac{TEH_{i,j+1} + TEH_{i+1,j+1} + TEH_{i+1,j}}{3} \tag{10}$$

$$e = avg - TE'_{i,j} \tag{11}$$

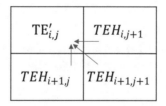

Fig. 4. The example of second part prediction

Step 2: The method is the same as the first part embedding of Step 2. Use the Eq. (7) and Eq. (8) to obtain new error distribution $e'_{i,j}$.

Step 3: Inverse the new error distribution $e'_{i,j}$ and obtain new hidden index $TEH_{i,j}$ according to the following equation.

$$TEH_{i,j} = avg - e' \tag{12}$$

After finishing all the operation of first part and second part, the hidden index $TEH_{i,j}$ is received. Where $TEH_{i,j}$ is encrypted index. The encrypted stego image is reconstructed when $TEH_{i,j}$ is decoded by VQ.

3.3   Image Decryption and Information Extracting

This process can extract the message and recover the host index of VQ encoding. Image decryption and information extract can be regarded as inverse process of image encryption and information hiding. In addition, the image decryption and information extract can be paralleled manipulation. The method of our codebook recovery is the same as Yan et al.'s method except for the Eq. (3). Our codebook recovery phase applies the Eq. (13) to recover the codebook.

$$C_i = \left(\left(c_i' \oplus c_{i-1}' \right) - \left(x_i' \oplus y_i' \right) \right) mod\, 256 \tag{13}$$

In the embedding phase, the index is first encrypted using confusion manner, then, hide the message into the confused index. It indicates that the data extracting phase should be executed after index decryption. Assume that the message is extracted before index decryption, nothing will be extracted.

After receiving the decrypted index, the hidden message and index can be completely extracted and recovered respectively. The first phase of data extracting applies the Eq. (10) to compute the predicted index, then, use Eq. (11) to obtain the prediction error e'. Collect all of the prediction error e' and generate the error distribution. From the error distribution, find two pairs of (pr, zr) and (pl, zl), then, extract the secret bit b from two pairs and recover the prediction error e according to the following Eq. (14) and Eq. (15).

$$\begin{cases} b = 0\,and\,e = e',\,if\,e' = pr \\ b = 1\,and\,e = e' - 1,\,if\,e' = pr + 1 \\ e = e' - 1,\,if\,zr \geq e' > pr + 1 \\ e = e',\,else \end{cases} \tag{14}$$

$$\begin{cases} b = 0\,and\,e = e',\,if\,e' = pl \\ b = 1\,and\,e = e' + 1,\,if\,e' = pl - 1 \\ e = e' + 1,\,if\,zl \leq e < pl - 1 \\ e = e',\,else \end{cases} \tag{15}$$

Next, use the Eq. (10) to compute the average index avg and apply the Eq. (16) to get the index $TE'_{i,j}$. That is the first phase of data extracting.

$$TE_{i,j}' = avg - e \tag{16}$$

After finishing the first phase of data extracting, the second phase is prepared to operate. Take the index $TE'_{i,j}$ and fetch the its neighboring index to calculate the prediction error e' using Eq. (17).

$$\begin{cases} e_{i,j+1}' = TE_{i,j}' - TEH_{i,j+1} \\ e_{i+1,j+1}' = TE_{i,j}' - TEH_{i+1,j+1} \\ e_{i+1,j}' = TE_{i,j}' - TEH_{i+1,j} \end{cases} \tag{17}$$

Apply the two pairs information of (pr, zr) and (pl, zl) as the same as the first phase of data extracting. Extract the secret bit b from two pairs and recover the prediction error $e_{i,j}$ according to the following Eq. (14) and Eq. (15). In the end, take the prediction error $e_{i,j}$ to recover the index $TE'_{i,j}$ using Eq. (18). Run all the processing of above-mention, secret messages have been extracted. But the index remains not to recover because of the underflow/overflow exception. The d_n information of exception is received from separated the secret message. Recover the index $TE'_{i,j}$ by applying the Eq. (19).

$$\begin{cases} TE'_{i,j+1} = TE'_{i,j} - e_{i,j+1} \\ TE'_{i+1,j+1} = TE'_{i,j} - e_{i+1,j+1} \\ TE'_{i+1,j} = TE'_{i,j} - e_{i+1,j} \end{cases} \tag{18}$$

$$\begin{cases} TE_{i,j} = 0 & \text{if } TE'_{i,j} = 1, d = 1 \\ TE_{i,j} = 1 & \text{if } TE'_{i,j} = 1, d = 0 \\ TE_{i,j} = k & \text{if } TE'_{i,j} = k - 1, d = 1 \\ TE_{i,j} = k - 1 & \text{if } TE'_{i,j} = k - 1, d = 0 \end{cases} \tag{19}$$

For the index decryption phase, we should recover the index $TE_{i,j}$ to original index $T_{i,j}$. The secret sequence $r_{i,j}$ is generated by key_T. For each index in a block, is decrypted by the following Equation.

$$\begin{cases} T_{i,j} = \left(TE_{i,j} - r_{i,j}\right) \bmod k \\ T_{i,j+1} = \left(TE_{i,j+1} - r_{i,j}\right) \bmod k \\ T_{i+1,j} = \left(TE_{i+1,j} - r_{i,j}\right) \bmod k \\ T_{i+1,j+1} = \left(TE_{i+1,j+1} - r_{i,j}\right) \bmod k \end{cases} \tag{20}$$

where k is the size of codebook.

4 Experimental Result

Several experimental results are presented in this section to illustrate the performance of our proposed. Our proposed method improves the Yan et al.'s method and makes the proposed method more adequately to be in place of information hiding. In our experiments, the five commonly used standard with 512×512 gray-level images, including, Lena, Jet, Baboon, Peppers, and Boat, are used to demonstrate the result of our proposed method. Here, we used the standard Lena image to discuss the comparison the performance in terms of index encryption and hiding capacity between Yan et al. and ours. Figure 5 shows the VQ encoding simulation using Lena image, including the index distribution, pixel distribution, and codebook distribution.

From the Fig. 5 and Fig. 6, it is obviously to know that the index encryption and index distribution using Yan et al.'s method can be detected by HVS (Human Vision

166   C.-T. Huang et al.

System). The index distributions without encryption and with encryption are displayed in Fig. 5(a) and Fig. 6(a), respectively. It indicates that index distribution with encryption using Yan et al.'s method has vulnerable to statistical analysis. Our proposed method of index encryption has more secure than Yan et al., shown in Fig. 7. Compare to the index distributions displayed in Fig. 5(a), Fig. 6(a), and Fig. 7(a), index distributions using our encryption approach is different to others, means that our scheme has better powerful in the statistical analysis.

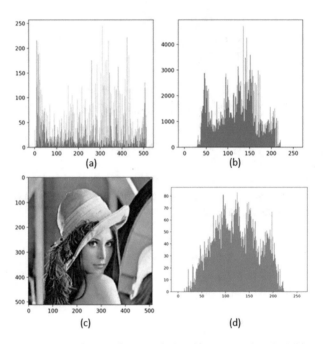

Fig. 5. Lena image using VQ encoding; (a) index histogram; (b) pixel histogram; (c) VQ encoding; (d) histogram of pixels in codebook.

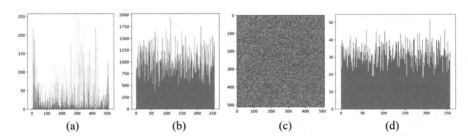

Fig. 6. Lena image using Yan et al.'s method; (a) index histogram; (b) pixel histogram; (c) encrypted image; (d) histogram of codebook.

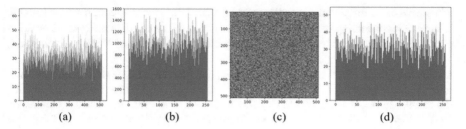

Fig. 7. Lena image using our encryption method; (a) index histogram; (b) pixel histogram; (c) encrypted image; (d) histogram of codebook diffusion

For embedding capacity, Table 1 shows the performance of our proposed approach without using index encryption. From this table, 2 × 2 block size has more hiding capacity than that of block size with 2 × 1. The reason is that a block has more accurate prediction and more hiding payload is gained. Table 2 illustrates the performance of our proposed approach using index encryption. Compare to Table 1 and Table 2, the Table 2 has the hiding capacity as similar as Table 1, but the average of processing bits is less than Table 1. Additionally, the preprocessing bits used the Peppers image has drastically improved since the encryption method is applied for index encryption. It indicates that the index encryption will make all of indexes with uniform distribution, leading to fewer index should be suffered from overflow/underflow.

Table 1. The hiding capacity and the payload of preprocessing using VQ encoding

Blok size	Cover image	Lena	Jet	Baboon	Peppers	Boat	Average
2 × 2	Hiding capacity (Bits)	2255	3136	612	2835	4208	2609.2
	Preprocessing (Bits)	106	0	19	1032	195	270.4
2 × 1	Hiding capacity (Bits)	1951	4028	630	3165	5317	3018.2
	Preprocessing (Bits)	106	0	19	1032	195	270.4

Table 2. The hiding capacity and the payload of preprocessing using index encryption

Blok size	Cover image	Lena	Jet	Baboon	Peppers	Boat	Average
2 × 2	Hiding capacity (Bits)	2227	3134	598	2784	4189	2586.4
	Preprocessing (Bits)	112	107	134	104	134	118.2
2 × 1	Hiding capacity (Bits)	1934	4032	631	3127	5301	3005
	Preprocessing (Bits)	151	128	144	113	135	134.2

5   Conclusion

In this paper, we have proposed a method of differencing preserving which is index encryption based on VQ. In our scheme, we improve the Yan et al.'s method and apply the index differencing to conceal the information. Our method only not have the robustness of security analysis as the same as Yan et al.'s method but provide a suitable information concealing method which is not offered by Yan et al.'s method. Simulation results show that our scheme has a high performance than that of Yan et al.'s method.

Acknowledgement. This work was partially supported by the Ministry of Science and Technology of the Republic of China under the Grant No. MOST 107-2221-E-153-001, 107-2813-C-153-050-E, 108-2221-E-153-004-MY2, 108-2221-E-153-006 and the Taiwan Information Security Center at National Sun Yat-sen University (TWISC@NSYSU) and at Tsing Hua University (TWISC@NTHU).

References

1. Xiang, T., Qu, J., Yu, C., Fu, X.: Degradative encryption: an efficient way to protect SPIHT compressed images. Optics Commun. **285**(24), 4891–4900 (2012)
2. Ji, X.Y., Bai, S., Guo, Y., Guo, H.: A new security solution to JPEG using hyper-chaotic system and modified zigzag scan coding. Commun. Nonlinear Sci. Numer. Simul. **22**(1–3), 321–333 (2015)
3. Zhang, Y., Xiao, D., Liu, H., Nan, H.: GLS coding based security solution to JPEG with the structure of aggregated compression and encryption. Commun. Nonlinear Sci. Numer. Simul. **19**(5), 1366–1374 (2014)
4. Chang, H.K., Liu, J.L.: A linear quadtree com-pression scheme for image encryption. Sig. Process. Image Commun. **10**(4), 279–290 (1997)
5. Wong, K.W., Yuen, C.H.: Embedding compression in chaos-based cryptography. IEEE Trans. Circuits Syst. II Express Briefs **55**(11), 1193–1197 (2008)
6. Zhu, H., Zhao, C., Zhang, X.: A novel image encryption compression scheme using hyper-chaos and Chinese remainder theorem. Sig. Process. Image Commun. **28**(6), 670–680 (2013)
7. Wang, L., Ye, Q., Xiao, Y., Zou, Y., Zhang, B.: An image encryption scheme based on cross chaotic map. In: Proceeding of 2008 Congress on Image and Signal Processing, pp. 22–26. IEEE, China (2008)
8. Tong, X.J.: The novel bilateral-diffusion image encryption algorithm with dynamical compound chaos. J. Syst. Softw. **85**(4), 850–858 (2012)
9. Tong, X.J., Wang, Z., Zhang, M., Liu, Y.: A new algorithm of the combination of image compression and encryption technology based on cross chaotic map. Nonlinear Dyn. **72**, 229–241 (2013)
10. Zhang, M., Tong, X.: A new algorithm of image compression and encryption based on spatiotemporal cross chaotic system. Multimed. Tools. Appl. **71**(1–2), 1–25 (2014)
11. Linde, Y., Buzo, A., Gray, R.M.: An algorithm for vector quantizer design. IEEE Trans. Commun. **28**(1), 84–95 (1980)
12. Yan, B., Bai, S.: Design of image confusion-diffusion cryptosystem based on vector quantization and cross chaotic map. In: Proceeding of 2017 2nd International Conference on image Vision and Computing (ICIVC), pp. 639–644. IEEE, China (2017)
13. Xuan, G., Yao, Q., Yang, C., Gao, J., Chai, P., Shi, Y.Q., Ni, Z.: Lossless data hiding using histogram shifting method base on integer wavelets. In: Proceeding of International Workshop on Digital Watermarking, Lecture Notes in Computer Science, vol 4283, pp. 323–332. Springer, Berlin (2006)
14. Tian, J.: Reversible data embedding using a difference expansion. IEEE Trans. Circuits Syst. Video Technol. **13**(8), 890–896 (2003)

15. Huang, Y., Kuo, J., Hsieh, W.: Capacity controllable location map free reversible watermarking. Chin. J. Electron. **24**(3), 519–523 (2015)
16. Thodi, D.M., Rodriguez, J.J.: Expansion embedding techniques for reversible watermarking. IEEE Trans. Image Process. **16**(3), 721–730 (2007)
17. Ni, Z., Shi, Y.Q., Ansari, N., Su, W.: Reversible data hiding method in encrypted images based on block shifting. IEEE Trans. Circuits Syst. Video Technol. **16**(3), 354–362 (2006)
18. Liu, Z.L., Pun, C.M.: Reversible image reconstruction for reversible data hiding in encrypted images. Sig. Process. **161**, 50–62 (2019)
19. Liu, Y.J., Chang, C.C.: Reversible data hiding for JPEG images employing all quantized non-zero AC coefficients. Displays **51**, 51–56 (2018)
20. Rahmani, P., Dastghaibyfard, G.: Novel reversible data hiding scheme for two-stage VQ compressed image based on search-order coding. J. Vis. Commun. Image Represent. **50**, 186–198 (2018)
21. Fu, Y.J., Kong, P., Yao, H., Tang, Z.J., Qin, C.: Effective reversible data hiding in encrypted image with adaptive encoding strategy. Inf. Sci. **494**, 21–36 (2019)
22. Wu, H.T., Cheung, Y.M., Yang, Z.Y., Tang, S.H.: A high-capacity reversible data hiding method for homomorphic encrypted images. J. Vis. Commun. Image Represent. **62**, 87–96 (2019)

Introduction of Reversible Data Hiding Schemes

Tzu-Chuen Lu$^{(\boxtimes)}$ and Thanh Nhan Vo

Department of Information Management, Chaoyang University of Technology,
168, Jifeng East Road, Wufeng District, Taichung 41349, Taiwan R.O.C.
tclu@cyut.edu.tw

Abstract. The hiding techniques proposed and published recently that have particularly focused on using Reversible Data Hiding (RDH) methods because they satisfied many requirements simultaneously that they are safe and reversible. RDH method gained more attention in the past few years. Because of its increasing applications in the areas such as military, law enforcement, and health care. This study lists some recently RDH methods such as Difference Expansion scheme (DE), the Histogram Shifting scheme (HS), the Pixel-Value Ordering scheme (PVO), the Dual-image based scheme, and the Interpolation-based scheme. The readers who were interested in this field can comprehend data hiding methods especially RDH technique.

Keywords: Reversible steganography · Difference expansion · Histogram shifting · Pixel-value ordering scheme · Dual-image · Interpolation

1 Introduction

In a general hiding scheme, a sender tries to embed some secret data into a host image to generate the stego-image. The stego-image then sent to the receiver for different kind of purposes, such as secret sharing, image authentication, temper detection. This kind of schemes try to hide huge data into the image. However, more hiding payload may cause more image distortion. Hence, how to achieve an acceptable balance between image quality and information load is an important research topic in the field of information hiding.

Reversible data hiding technique (RDH) is an application of the information hiding. The method uses the image feature to hide the secret data into the host image. In the hiding procedure of RDH, the recovery information, such as image features, is extracted and concatenated with the secret message to hide in the host image for generating the stego-image. In the extraction procedure, the concealed information, the secret message and image features, are extracted from the stego-image to restore its original status. The image feature is used to recover the host image. The reproducibility of RDH is particularly suitable for processing military or medical images. The basic concept of the embedding and extraction procedure of RDH is shown in Fig. 1.

© Springer Nature Switzerland AG 2020
L. C. Jain et al. (Eds.): SICBS 2019, AISC 1145, pp. 170–183, 2020.
https://doi.org/10.1007/978-3-030-46828-6_15

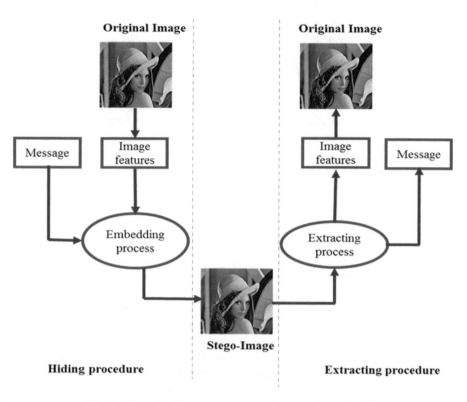

Fig. 1. The embedding and the extraction procedure of RDH

Teklap et al. [1] divided the RDH into two categories. The first type of RDH directly adds the secret data into the host pixel and subtracts the original pixel from the stego-pixel to retrieve and restore the original pixel. However, such a method often has overflow or underflow problems. Rabbani et al. [2] use the modulo-module (Modulo Arithmetic) technique to solve the overflow problem. However, this can be linked to the serious problem of "salt and pepper" noise, whereby the stego-images exhibit many black spots and white spots, similar to salt and pepper, scattered in the image. Furthermore, the hiding payload of such methods is very low. For example, the methods proposed by Vleeschouwer et al. [3] do not have the problem of "salt and pepper", yet the hiding capacity is very low.

The second type of RDH performs data hiding by changing image features, such as the Least Significant Bit (LSB) or high-frequency coefficient values after wavelet transform. For example, Goljan et al. [4] used the undistorted compression method to compress the LSB portion of each pixel of the host image in order to generate a compressed code. Following this, the compressed code is hidden in the LSB of the hidden image, along with the secret data. The receiver uses the compression code to replace the part of the original LSB of the pixel. The image feature used with this method is the LSB of the pixel, and the amount of information loading is limited by the compression of the LSB. However, the compression results of the LSB are usually of poor quality. Thus the pure hiding payload that used to hide the secret data is very low.

In 1996, Morimoto et al. [5] proposed an RDH based on Patchwork Algorithm. In their method, a host image is first divided into several blocks. Each block can be used to hide one single bit. The pixel values in each block are then randomly divided into two sets, denoted as set X and set Y. If the secret data to be hidden is 0, the pixel value in set X is increased, and that in set Y is reduced. Conversely, if the hidden information is 1, the pixel value in set X is reduced, and the pixel value in set Y is increased. In the process of decoding, the average of set X and set Y in each block is analyzed in order to estimate the hidden information.

According to the techniques and methodology, RDH can be further divided into several categories, such as Difference Expansion (DE), Histogram-Shifting (HS), Pixel-Value-Ordering (PVO), Dual-image based, Interpolation-based and so on. In the next sections, we will describe these schemes more detail.

2 Difference Expansion (DE) Scheme

Tian [6] proposed the first DE-based scheme in 2003. In his scheme, two neighboring pixels can be used to conceal one secret bit. The average hiding payload is $1/2 = 0.5$ bit per pixel (bpp). However, the scheme has overflow and underflow problems. Hence, the scheme requires an additional information, called location map, for further recovering procedure. In order to increase the payload, Tian performed the scheme multiple times, called the Multi-levels Information Hiding method. Thus the visual quality is not good enough.

Let consider two pixels as an example. Suppose that $p_1 = 35$, $p_2 = 30$ are two neighboring pixels, where $0 \leq p_1, p_2 \leq 255$. Assume the secret bit is $s = 1$. First, the scheme calculates the distance between p_1 and p_2 by $d = p_1 - p_2 = 35 - 30 = 5$ and calculates a fixed value $\alpha = \left[\frac{p_1 + p_2}{2}\right] = \left[\frac{35 + 30}{2}\right] = 32$ as a base point. The secret bit is hidden into the distance by $d' = 2 \times d + s = 2 \times 5 + 1 = 11$. The final stego-pixels are $p_1' = \alpha + \left[\frac{d'+1}{2}\right] = 32 + \left[\frac{11+1}{2}\right] = 38$ and $p_2' = \alpha - \left[\frac{d'}{2}\right] = 32 - \left[\frac{11}{2}\right] = 27$.

In the extraction process, the receiver computes the expanded difference by $d' = p_1' - p_2' = 38 - 27 = 11$ and calculates the base point by $\alpha' = \left[\frac{p_1' + p_2'}{2}\right] = \left[\frac{38 + 27}{2}\right] = 32$. The original distance is $d = \left[\frac{d'}{2}\right] = \left[\frac{11}{2}\right] = 5$. The secret bit is extracted by $b = d' - \left[\frac{d'}{2}\right] \times 2 = 11 - 10 = 1$. The original pixels are restored by $p_1 = \alpha' + \left[\frac{d+1}{2}\right] = 32 + \left[\frac{5+1}{2}\right] = 35$ and $p_2 = \alpha' - \left[\frac{d}{2}\right] = 32 - \left[\frac{5}{2}\right] = 30$.

In the embedding process, if the stego-pixel satisfies $0 \leq \alpha - (2 \times d + s) \leq 255$ and $0 \leq \alpha + (2 \times d + s) \leq 255$, then the stego-pixel is expandable. Otherwise, if the pixel does not satisfy the rules, the pixel is non-expandable. The scheme tries to use another strategy to embed the data. The strategy is called changeable strategy.

The scheme changes the hiding equation to $d' = 2 \times \left[\frac{d}{2}\right] + s$. If the stego-pixel fits the rules $0 \leq \alpha - \left(2 \times \left[\frac{d}{2}\right] + s\right) \leq 255$ and $0 \leq \alpha + \left(2 \times \left[\frac{d}{2}\right] + s\right) \leq 255$, then the pixel is changeable which can be used to hide one secret bit. But, the receiver does not know that the stego-pixel is computed by $d' = 2 \times d + s$ or $d' = 2 \times \left\lfloor \frac{d}{2} \right\rfloor + s$. Hence, a location map is used to record whether the pixel is expandable or not. If the pixel is expandable, the location map is recorded with 1. Otherwise, the location map is recorded with 0. Moreover, the original least significant bit (OLSB) of the distance also needed to be recorded. The location map and OLSB are compressed and concatenated along with the secret data to be conceal into the host image.

The weakness of the DE-based scheme is low image quality and low hiding payload. In order to improve the image quality of DE-based schemes, Ansari et al. [8] proposed the Histogram Shifting scheme to perform RDH.

3 Histogram Shifting (HS) Scheme

In 2006, Ansari et al. [8] proposed a HS-based RDH. In their scheme, the occurrence of all pixel values in the host image are counted and plot them on a histogram, then based on the magnitude of the occurrence of pixel values, the pixel with the most occurrence is determined from the histogram, denoted as the peak point (P point), while the pixel value with zero occurrence is called the zero point (Z point).

Next, the pixel value of bin P + 1 is shifted to that of the bin P + 2, and the total number of bin P + 1 reduced to zero in order to generate a space to hide the secret data. The shifting operator shifts the bins between P + 1 to Z − 1 to the right hand side by 1 unit. All the pixel values in the range [P + 1,Z - 1] are increased by 1. Then, the values between the range [P + 1,Z - 1] will shift to the range [P + 2,Z].

The main drawback of HS-based scheme is also overflow and underflow problems. If the pixel is equal to 255 or 0 and the pixel is shifted to the right hand side or the left hand side, then the value is changed to 256 or -1. Ni et al. change the values of 0 to 1 and 255 to 254 for preventing the underflow or overflow problem. An extra information is needed to record whether the pixel is changed or not. The location map is concatenated with the secret data to conceal into the peak points.

The hiding capacity of HS-based scheme is very few, since the number of the extra information needed to be subtracted from the total number of hiding bits. Besides, the hiding payload is limited by the size of the peak points. The average peak signal to noise ratio (PSNR) value of the HS scheme is approximately 48 dB. In 2013, Li et al. [9] enhanced the HS-based method by proposing a Pixel Value Ordering (PVO) RDH.

4 Pixel Value Ordering (PVO) Scheme

In 2013, Li et al. [9] proposed a pixel-value-ordering (PVO) based-predictor method. The purpose of PVO method is to reduce the number of shifted pixels and obtain high image quality. Firstly, Li et al. divided the original image into several non-overlapped blocks. The pixel values in the block are $(p_1, p_2, \ldots, p_n)$. Then, the scheme sorted the pixel values in ascending order to archive $(p_{\sigma(1)}, p_{\sigma(2)}, \ldots, p_{\sigma(n)})$. Here $\sigma_{(n)}$ is the hierarchy of pixel value p_n in the value range after sorted. If there are many pixels with the same value that follow the hierarch and the position of $p_{n-1} < p_n$ the order follows $p_{\sigma(n-1)} < p_{\sigma(n)}$.

If $i < j$ and $p_i = p_j$ to derive $p_{\sigma(i)} < p_{\sigma(j)}$. Then they used the second largest value $p_{\sigma(n-1)}$ to predict its maximum $p_{\sigma(n)}$. (or used the second smallest value $p_{\sigma(2)}$ to predict its minimum $p_{\sigma(1)}$). The prediction-error PE_{max} is calculated by $PE_{max} = p_{\sigma(n)} - p_{\sigma(n-1)}$.

There are three cases occur in each block, and the stego-pixel values are calculated follow these rules

If $PE_{max} = 0$, then the stego-pixel values equal the original pixel values respectively, secret bit cannot be embedded in the block, and $\widetilde{PE}_{max} = PE_{max}$.

If $PE_{max} = 1$, then bit b is embedded into the maximum by $y_{\sigma(n)} = p_{\sigma(n)} + b$, the remnants of the block are unchanged, and (PE) $\widetilde{PE}_{max} = PE_{max} + b$.

If $PE_{max} > 1$, then $y_{\sigma(n)} = p_{\sigma(n)} + 1$, the remnants of the block are unchanged, and $\widetilde{PE}_{max} = PE_{max} + 1$. This case secret bit b is not embedded into the block.

In the equation $b \in \{0, 1\}$ is a secret bit which is embedded into the block, the prediction-error $\widetilde{PE}_{max}$ is used for extracting process and $\widetilde{PE}_{max} = y_{\sigma(n)} - y_{\sigma(n-1)}$.

For example, the original image is divided into the blocks sized 2×2. Each block has four pixels (p_1, p_2, p_3, p_4) that derive from these pixels by ordering from left to right, top to bottom in each block. Finally, the stego-image is collected from all stego-blocks, then it is sent to the receiver. The embedding process is illustrated in Fig. 2.

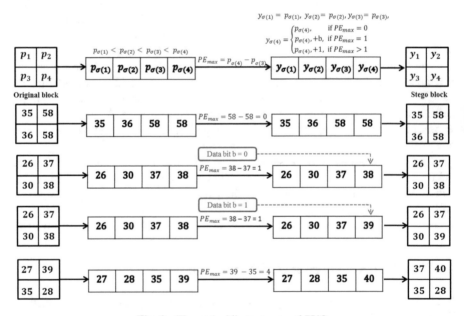

Fig. 2. The embedding process of PVO

Similarly, the receiver uses the extracting process to extract the secret data and reconstructs the original image as follows. The receiver divides the stego-image into non-overlapped blocks. Then the extracting process is implemented in each stego-block. The receiver sorts the pixel values in ascending order and derives

$(y_{\sigma(1)}, y_{\sigma(2)}, \ldots, y_{\sigma(n)})$. Next, the scheme calculates the prediction-error $\widetilde{PE}_{max} = y_{\sigma(n)} - y_{\sigma(n-1)}$. Hence, there are three cases can be occurred.

If $\widetilde{PE}_{max} = 1$, the stego-pixel maximum is unchanged and the values of the original block equal the values of the stego-block, respectively. Secret bit $b = 0$ is extracted.

If $\widetilde{PE}_{max} = 2$, the maximum of the block is calculated by $p_{\sigma(n)} = y_{\sigma(n)} - 1$, the remnants in the block are attributed equal the remnants of the stego-block, respectively. Secret bit $b = 1$ is extracted.

If $\widetilde{PE}_{max} > 2$, the maximum of the original block is calculated $p_{\sigma(n)} = y_{\sigma(n)} - 1$, the remnants in the block are attributed equal the remnants of the stego-block, respectively.

The original image is restored and the scheme obtains the secret data simultaneously. Figure 3 shown the extracting example.

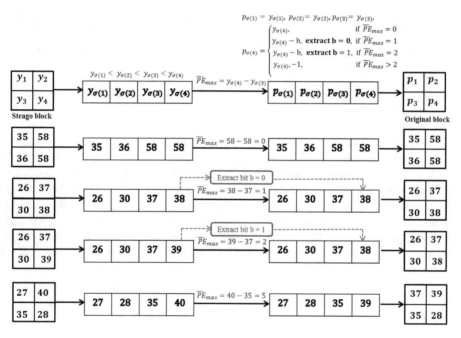

Fig. 3. The extracting process of PVO

In the PVO-based scheme, the middle pixels do not require shifting operator. Hence, the image quality of the PVO-based schemes is better than that of the HS-based schemes. In Li et al.'s PVO scheme, if the difference is equal to 0 or 1, the pixel is embeddable, thus the embeddable range is very small.

The PVO-based scheme also needs a location map to record the boundary pixel which may has underflow or overflow problem. Hence, the pure hiding payload is still limited. Therefore, the scholars proposed the dual-images based scheme to solve the problem.

5 Dual-Image Based Scheme

In 2006, Zhang et al. [21] proposed a method namely Exploiting Modification Direction (EDM). The scheme transferred the secret data is the binary numeral system into $(2n+1)$ numeral system. Then they embedded n secret digits in $(2n+1)$ numeral system into each group including n cover pixels. This method could embed more secret bits than the other methods at that time. However, this method obtained higher distortion image. Because the pixels were modified (i.e. increases or decreases by 1) by the value of the $(2n+1)$ number system, it is too large change in the binary numeral system.

In 2007, Chang et al. [22] applied EDM method to propose their method. In the scheme, two secret digits in the quinary-based numeral system are embedded into two cover pixels by using two steganography images. Chang et al. creates a new branch in data hiding techniques that is dual-images data hiding technique. Their scheme not only increase the hiding capacity but also image quality. The security issue is guaranteed because the receiver must have both of two stego-images that could extract the secret data. After Chang et al., many scholars improved it and achieved the significant results in their experiments.

Lee et al. (2009) proposed a dual-mage scheme that embedded data into the set of pixels including a pixel as a center position and four neighbor pixels of four directions, there are the upper, lower, left and right pixels correspond to center pixel. Chang et al. (2013) converted the secret data into decimal secret symbol and embedded into the pixels in the right diagonal. Qin et al. (2014) embedded secret message into two cover images in two different ways, they embedded data into the first cover image using EMD method and the second one using the three rules belong to the first cover image. Lu et al. (2015) proposed center folding dual-image strategy (CSF) to reduce the value of the secret data for reducing the distortion between the stego-image and the original image. The method not only could enhance embedding capacity but also maintained high image quality. In Lu et al.'s scheme, an original image is presented as $p_1, p_2, \ldots, p_{w \times h}$, where w is the width and h is the height. In the embedding process of CSF, the secret data is concealed in the cover images as the following procedure.

The secret data with k bits is converted from binary numeral system to decimal numeral system. For example, the secret data is $\{101111011110\}$ and k is set to be 3. The decimal numbers of the secret data are $\{5, 7, 3, 6\}$. Each decimal number of the secret message is embedded into a cover image alternately. Before the embedding processing, the secret data was folded by using CFS.

With $k = 3$, there are $2^k = 8$ different kinds of values in the value range, where $d = \{0, 1, 2, 3, 4, 5, 6, 7\}$. The center value of the value range is $2^{k-1} = 4$. The center value was subtracted from the secret data to generate the folded value. An example folded values with $k = 3$ are shown in Table 1. We can see that the folded value $\tilde{d}$ can be separated into two parts, negative part and positive part.

Table 1. CFS folded values with $k = 3$

d	0	1	2	3	4	5	6	7
$\tilde{d}$	-4	-3	-2	-1	0	1	2	3

Then, the scheme divides $\tilde{d}$ into two values, $\tilde{d}_1$ and $\tilde{d}_2$ by

$$\begin{cases} \tilde{d}_1 = \left\lfloor \frac{\tilde{d}}{2} \right\rfloor, \\ \tilde{d}_2 = \left\lceil \frac{\tilde{d}}{2} \right\rceil. \end{cases} \tag{1}$$

The values are added to the cover pixel $p_{i,j}$ to generate the stego-pixels by

$$\begin{cases} p'_{i,j} = p_{i,j} + \tilde{d}_1, \\ p''_{i,j} = p_{i,j} - \tilde{d}_2, \end{cases} \tag{2}$$

In the embedding process, the scheme might has overflow or underflow problem. A simple way to prevent the problem is to limit the embedding range within $[2^{k-1}, 256 - 2^{k-1}]$. If the pixel is within the range then the pixel is embeddable. Otherwise, the pixel is non-embeddable and the stego-pixel is the same with the original pixel.

Consider an example, a cover image I = {35, 41, 75, 60} and the secret symbols {101 010 011 110} and $k = 3$. Firstly, the first three secret data is transformed to a decimal number $d = (101)_2 = (5)_{10}$. The value is folded by $\tilde{d} = 5 - 2^{k-1} = 5 - 4 = 1$. The value is divided into two parts $\tilde{d}_1 = \frac{1}{2} = 0$ and $\tilde{d}_2 = \frac{1}{2} = 1$. The first two stego-pixels are computed by $p'_{1,1} = 35 + 0 = 35$ and $p''_{1,1} = 35 - 1 = 34$. Next, the second secret data $d = (010)_2 = (2)_{10}$ is folded and divided by $\tilde{d}_1 = \frac{-2}{2} = -1$ and $\tilde{d}_2 = \frac{-2}{2} = -1$. The second stego-pixels are $p'_{1,2} = 41 + (-1) = 40$ and $p''_{1,2} = 41 - (-1) = 42$.

In the extracting process, the receiver must have both two stego-images to extract the secret messages. Firstly, the receiver determines whether the pair of pixels of two stego-pixels $p'_{i,j}$ and $p''_{i,j}$ are embeddable. If $p'_{i,j} = p''_{i,j}$) and the pixel value is not within $[2^{k-1}, 256 - 2^{k-1}]$, then the pixel is non-embeddable. Conversely, if $p'_{i,j} \neq p''_{i,j}$, then the pixel is embeddable. For the embeddable pixel, the scheme subtract $p'_{i,j}$ from $p''_{i,j}$ to restore the data symbol $\tilde{d}$

$$\tilde{d} = p'_{i,j} - p''_{i,j} \tag{3}$$

Then, the extracted $\tilde{d}$ is added with 2^{k-1} to recover the original d by

$$d = \tilde{d} + 2^{k-1}. \tag{4}$$

Lastly, the original pixel is recovered by

$$p_{i,j} = \left\lceil \frac{p'_{i,j} + p''_{i,j}}{2} \right\rceil \tag{5}$$

For example, the stego-pixels are $p'_{1,1} = 35$ and $p''_{1,1} = 34$. Because $p'_{1,1} \neq p''_{1,1}$ the pixel is embeddable. The data symbol $\tilde{d}$ is $\tilde{d} = 36 - 35 = 1$ and the original d is $d = 1 + 4 = (5)_{10} = (101)_2$. Here, secret message is $(101)_2$. The original pixel is $p_{1,1} = \frac{34+35}{2} = 35$.

Consider an overflow example, suppose that the pixel value is $p_{i,j} = 255$ and $d = (111)_2 = (7)_{10}$. The value is larger than $256 - 2^{3-1} = 252$. The folded value is $\tilde{d} = 7 - 4 = 3$, then $\tilde{d}$ is divided into two parts $\tilde{d}_1 = 1$, and $\tilde{d}_2 = 2$ to generate the stego-pixels $p'_{i,j} = 255 + (1) = 256$ and $p''_{i,j} = 255 - 2 = 253$. The pixel $p'_{i,j}$ has overflow problem. Hence, the stego-pixels are set to be $p'_{i,j} = 255$ and $p''_{i,j} = 255$. There are no secret embedded in the pixel.

In the extracting process, the scheme determines that $p'_{i,j} = p''_{i,j}$ and both of the pixels are not within $\left[2^{k-1}, 256 - 2^{k-1}\right]$. Hence, the pixel is non-embeddable.

6 Interpolation-Based Scheme

Interpolation technique is used to compute the reasonable value of the intermediate pixels when the cover image is stretched and the secret message is embedded into the intermediate pixels with less distortion quality [23]. There are many interpolation methods have been proposed. The nearest neighbor interpolation uses the closet pixel $p_{i,j}$ from the cover image to generate the prediction image pixel $p'_{i,j}$. This interpolation type can be used to enlarge and reduce images. The bilinear interpolation, a type of linear interpolation, determines the intensity of pixel by calculating the weighted average of four neighboring pixels.

In 2009, Jung and Yoo [23] used interpolation method in data hiding scheme, they proposed Neighbor Mean Interpolation (NMI) method. In NMI, the cover image can be generated by using an input image sized $w \times h$. This method not only increases the capacity but also guarantees the security. They scaled the size of the input image down to become the original image sized $w/2 \times h/2$. The original image just equals a quarter of the input image size. Subsequently, the cover image is generated by using the interpolation method to return the original image size back to $w \times h$. The cover image has high-resolution pixels which these interpolated pixels are just the intermediate pixels. The intermediate pixel values are calculated by referencing the value of the neighboring pixels, then they are allocated the intermediate pixel positions of the cover image. The diagram of NMI is shown in Fig. 4.

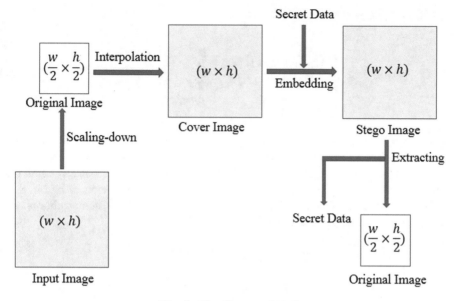

Fig. 4. The diagram of NMI

In order to increase the computing speed, this method used the interpolation method called neighbor mean interpolation to generate the intermediate pixels and fill in the cover image when the original image was scaled up to become the cover image. The result of the interpolation that the cover image has $w \times h$ size and more secret message can be embedded in the intermediate pixels. The interpolation rules were presented for all pixels as follows.

The embedding NMI rules are

$$
p'_{i,j} = \begin{cases}
p_{i,j}, & \text{if } i = K \times m, j = K \times n, \\
\left(p_{i,j-1} + p_{i,j+1}\right)/K, & \text{if } i = K \times m, j = K \times n + 1, \\
\left(p_{i-1,j} + p_{i+1,j}\right)/K, & \text{if } i = K \times m + 1, j = K \times n, \\
\left(p_{i-1,j-1} + p'_{i-1,j} + p'_{i,j-1}\right)/(K+1), & \text{otherwise.}
\end{cases} \tag{6}
$$

In the equation, $0 \leq j \leq i$ and $m, n = 0, 1, \ldots, 127$. K is presented a coefficient value that was scaled up.

The extracting NMI rules are

$$
b = \begin{cases}
p'_{i,j} - \left[\frac{p'_{i,j} + p'_{i,j}}{K}\right], & \text{if } i = K \times x, j = K \times y, \\
p'_{i,j} - \left[\frac{p'_{i,j} + p'_{i,j+1}}{K}\right], & \text{if } i = K \times x, j = K \times y + 1 \\
p'_{i,j} - \left[\frac{p'_{i,j} + p'_{i+1,j}}{K}\right], & \text{if } i = K \times x + 1, j = K \times y \\
p'_{i,j} - \dfrac{\left(\frac{K \times p'_{i,j} + p'_{i,j+2} + p'_{i+2,j}}{K}\right)}{(K+1)}, & \text{otherwise.}
\end{cases} \tag{7}
$$

In the equation, b is secret data, $0 \leq j \leq i$ and , $y = 0, 1, \ldots, 127$.

The convenient way of NMI with $K = 2$ is

$$p'_{i,j} = \begin{cases} p_{i,j}, & \text{if } i = 2m, j = 2n, \\ \left(p_{i,j-1} + p_{i,j+1}\right)/2, & \text{if } i = 2m, j = 2n+1, \\ \left(p_{i-1,j} + p_{i+1,j}\right)/2, & \text{if } i = 2m+1, j = 2n, \\ \dfrac{p_{i-1,j-1} + \frac{p'_{i-1,j}}{2} + p'_{i,j-1}}{6}, & \text{otherwise.} \end{cases} \tag{8}$$

Figure 5 shows an example of the original image sized 2×2 expanded to the cover image sized 3×3. The pixel values of the original image are $p_{0,0} = 52$, $p_{0,1} = 61$, $p_{1,0} = 85$, $p_{1,1} = 88$. Four corners pixels of the cover image are respectively equal the initial pixels of the original image. The other virtual pixels are computed by Eq. 8.

Next, the scheme calculates the differences d_1, d_2, and d_3 from the neighbor pixels and the corner pixel $p'_{0,0}$ in the cover image as follows.

$$d_1 = p'_{0,1} - p'_{0,0} = 56 - 52 = 4,$$

$$d_2 = p'_{1,0} - p'_{0,0} = 68 - 52 = 16,$$

$$d_3 = p'_{1,1} - p'_{0,0} = 58 - 52 = 4.$$

The scheme then calculates how many secret bits s_1, s_2, and s_3 can be embedded into $p'_{0,1}, p'_{1,0}$, and $p'_{1,1}$. The equations are

$$s_1 = \log_2 |d_1| = \log_2 |4| = 2,$$

$$s_2 = \log_2 |d_2| = \log_2 |16| = 4,$$

$$s_3 = \log_2 |d_3| = \log_2 |6| = 2.$$

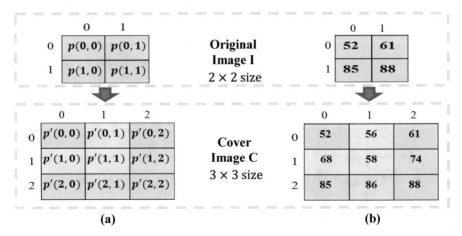

(a)   **(b)**

Fig. 5. An example for scaling up the original image using NMI method

Assume that the secret message is $b = (10110101)_2 = (b_1b_2b_3)_2$, where $b_1 = (10)_2, b_2 = (1101)_2, b_3 = (01)_2$ corresponding to the number of bits s_1, s_2, and s_3.

Thirdly, the scheme converts the sub-messages from binary form to decimal number, such that $b_1 = (10)_2 = (2)_{10}, b_2 = (1101)_{10} = (13)_{10}, b_3 = (01)_2 = (1)_{10}$ and embed them into $p'_{0,1}, p'_{1,0}$, and $p'_{1,1}$ respectively by

$$p''_{i,j} = p'_{i,j} + b_k. \tag{9}$$

The stego-pixels are $p''_{0,1} = 56 + 2 = 58, p''_{1,0} = 68 + 13 = 81$, and $p''_{1,1} = 58 + 1 = 59$. The rest pixel values of the cover image are unchanged. The stego-image is shown in Fig. 6.

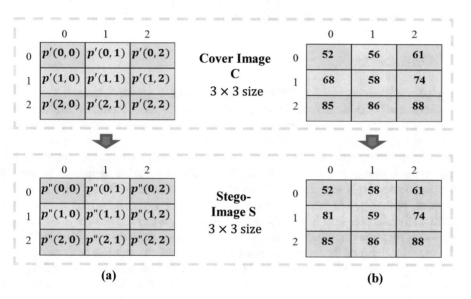

(a)   (b)

Fig. 6. The stego-image of Fig. 4

In the extracting process, the equation is used to extract the secret message

$$b = \begin{cases} p''_{i,j} - \left[\frac{p''_{i,j} + p''_{i,j}}{2}\right], & \text{if } i = 2x, j = 2y, \\ p''_{i,j} - \left[\frac{p''_{i,j} + p''_{i,j+1}}{2}\right], & \text{if } i = 2x, j = 2y + 1, \\ p''_{i,j} - \left[\frac{p''_{i,j} + p''_{i+1,j}}{2}\right], & \text{if } i = 2x + 1, j = 2y, \\ p''_{i,j} - \frac{2 \times p''_{i,j} + \frac{p''_{i,j+2}}{2} + p''_{i+2,j}}{6}, & \text{otherwise.} \end{cases} \tag{10}$$

In the equation, b is the secret data, $0 \le j \le i$ and $x, y = 0, 1, \dots, 127$.

In Fig. 5, $p''_{0,0} = 52, p''_{0,1} = 58, p''_{1,0} = 81$, and $p''_{1,1} = 59$. The differences are

$$d'_1 = p''_{0,1} - p''_{0,0} = 58 - 52 = 6,$$

$$d'_2 = p''_{1,0} - p''_{0,0} = 81 - 52 = 29,$$

$$d'_3 = p''_{1,1} - p''_{0,0} = 59 - 52 = 7.$$

Next, the scheme determines the amount of secret bits which embedded in each pixel. The lengths of the secret bits are $s_1 = 2, s_2 = 4$, and $s_3 = 1$. Then the secret bits can be computed by $b_1 = (2)_{10} = (10)_2$, $b_2 = (13)_{10} = (1101)_2$, $b_3 = (1)_{10} = (01)_2$. The final secret message are $b = (10110101)_2$.

The NMI method has the high computing speed and high capacity. This method utilizes to embed large secret data, and maintains a high visual quality.

7 Conclusions

This paper introduces several RDH methods that applied to grayscale images. The paper presented the basic ideas from the typical articles that focus on a detailed explanation. RDH methods have shown many advantages such as receiver can extract both secret message and original data. This feature is very important in the application and many scholars are interested in RDH methods. The paper presented and arranged according to the research order, gradually improving the quality of hidden images in practice. Scholars increasingly exploit the features of images to create many new methods with superior advantages in image hiding techniques. For example, in the dual-image method, the receiver must have both of two stego-images that can be able to extracting the secret message and reconstructing the original image. Besides, the PSNR (dB) and the capacity (bpp) of this technical image is higher than the techniques to use single images. The interpolation-based method (NMI) can hide a large amount of confidential data that still ensures high security. Because NMI method does not use the input image directly for the embedding process. The cover image is generated from the input image goes through a transformation process, called interpolation technique. Also ensuring stego-image quality is relatively good that makes this method is difficult for third parties can be hard to penetrate and stealing secret data which hidden in the cover image.

References

1. Tekalp, A.M., Celik, M.U., Sharma, G., Saber, E.: Lossless generalized-lsb data embedding. IEEE Trans. Image Process. **14**, 253–266 (2005)
2. Rabbani, M., Honsinger, C. W., Jones, P.W., Stoffel, J.C.: Lossless recovery of an original image containing embedded data. US Patent Appl. **6**, 791 (2001)
3. Vleeschouwer, C.D., Delaigle, J.F., Macq, B.: Circular interpretation of bijective transformations in lossless watermarking for media asset management. IEEE Trans. Multimedia **5**, 97–105 (2003)

4. Goljan, J., Fridrich, J., Du, R.: Invertible authentication watermark for jpeg images. In: Proceedings of the SPIE Conference on Security and Watermarking of Multimedia Content, pp. 223–227. IEEE, Las Vegas (2001)
5. Morimoto, N., Bender, W., Gruhl, D., Lu, A.: Techniques for data hiding. IBM Syst. J. **35**, 313–336 (1996)
6. Tian, J.: Reversible data embedding and content authentication using difference expansion. IEEE Trans. Circ. Syst. Video Technol. **13**, 831–841 (2003)
7. Alatar, A.M.: Reversible watermark using the difference expansion of a generalized integertransform. IEEE Trans. Image Process. **13**, 1147–1156 (2004)
8. Ansari, N., Ni, Z., Shi, Y.Q., Su, W.: Reversible data hiding. IEEE Trans. Circ. Syst. Video Technol. **16**, 354–362 (2006)
9. Li, X., Li, J., Li, B., Yang, B.: High-fidelity reversible data hiding scheme based onpixel-value-ordering and prediction-error expansion. Sig. Process. **93**, 198–205 (2013)
10. Li, X., Peng, F., Yang, B.: Improved pvo-based reversible data hiding. Dig. Signal Process **25**, 255–265 (2014)
11. Hsiao, J.Y., Chang, C.C., Chan, C.S.: Finding optimal least-significant-bit substitution in image hiding by dynamic programming strategy. Pattern Recogn. **36**, 1583–1595 (2003)
12. Lee, C.F., Huang, Y.L.: Reversible data hiding scheme based on dual stegano-images using orientation combinations. Telecommun. Syst. **52**, 2237–2247 (2013)
13. Chang, C.C., Lee, C.F., Wang, K.H., Huang, Y.L.: A reversible data hiding scheme based on dual steganographic images. In: Proceedings of the 3rd International Conference on Ubiquitous Information Management and Communication, ICUIMC, vol. 9, pp. 228–237 (2009)
14. Tseng, C.Y., Lu, T.C., Wu, J.H.: Dual imaging-based reversible hiding technique using lsb matching. Sig. Process. **108**, 77–89 (2015)
15. Wu, J.H., Lu, T.C., Huang, C.C.: Dual-image-based reversible data hiding method using center folding strategy. Sig. Process. **115**, 195–213 (2015)
16. Chen, K.C., Lee, C.F., Weng, C.Y.: An efficient reversible data hiding with reduplicated exploiting modification direction using image interpolation and edge detection. Multimedia Tools Appl. **76**, 9993–10016 (2017)
17. Chang, C.C., Qin, C., Hsu, T.J.: Reversible data hiding scheme based on exploiting modification direction with two steganographic images. Multimedia Tools Appl. **74**, 5861–5872 (2014)
18. Verma, H.K., Malik, A., Sikka, G.: An image interpolation based reversible data hiding scheme using pixel value adjusting feature. Multimedia Tools Appl. **76**, 13025–13046 (2017)
19. Lu, T.C.: Adaptive (k, f1) interpolation-based hiding scheme using center folding strategy. Multimedia Tools Appl. **76**, 1827–1855 (2017)
20. Jung, K.H., Yoo, K.Y.: Data hiding method using image interpolation. Comput. Stan. Interfaces **31**, 465–470 (2009)
21. Zhang, X.P., Wang, S.Z.: Efficient steganographic embedding by exploiting modification direction. IEEE Commun. Lett. **10**(11), 1–3 (2006)
22. Chang, C.C., Kieu, T.D., Chou, Y.C.: Reversible data hiding scheme using two steganographic images. In: Proceedings of IEEE Region 10 International Conference TENCON 2007, pp. 1–4. IEEE, Taipei (2016)
23. Jung, K.H., Yoo, K.Y.: Data Hiding method using image interpolation. Comput. Stan. Interfaces **31**(2), 465–470 (2009)
24. Lee, C.F., Huang, Y.L.: An efficient image interpolation increasing payload in reversible data hiding. Expert Syst. Appl. **39**, 6712–6719 (2012)

Study on the Factors Affecting Sharing Behavior of Social Media Platform Users

Hsiu-Ching Chang[1], Ming-Yueh Wang[2], Ci-Jhan He[1], Jia-Pei Lin[1], and Chi-Yueh Hsu[1(✉)]

[1] Chaoyang University of Technology, Taichung City, Taiwan
jj0418s@gmail.com, polo443311@yahoo.com.tw,
{s10826607, cyhsu}@gm.cyut.edu.tw
[2] National University of Kaohsiung, Kaohsiung City, Taiwan
mywang@nuk.edu.tw

Abstract. User's behavior on social media platform becomes so complicate that only linear approach is hard to examine the structural issue. Therefore, the back propagation neural network is employed in this study since this method applies to the issue computed with inner corresponding rule. SIPS is used to investigate the complex sharing willingness of users on social media platform through back propagation neural network. 280 questionnaires were collected and finally 254 efficient questionnaires were received in 90.7% efficiency. The factors affecting sharing behavior on social media included user attributes (privacy, experience, check-in, information source, and reward activity), habits, and sharing behavior on social media platform. The results showed that the whole social media platform has significant impact on privacy.

Keywords: Artificial neural network · Back propagation neural network · Consumer behavior · Reward activity

1 Introduction

The rapid development of the internet gives that people have not only access to obtain information but also delivers message and furthermore becomes a social platform. People with common topics, interests or ideas gather together in social media community. The internet development is out of imagination. Over 82% of world population (about 1.2 billion people) have visited social media websites (comScore 2011). Institute for Information Industry (2014) point outed that up to 96.2% of netizens in Taiwan tracked contents, fan pages, brands or celebrities on social media websites. Social media websites play an important role in our daily life. Divol et al. (2012) recognized that new complexity exists on these social media and believed that research on marketing of social media needs to be re-conceptualized. There are many advantages on social media marketing. It could connect businesses with consumers, develop and maintain customer relationships in a timely manner (Kaplan and Haenlein 2010). In 2016, Facebook owned the most users which was expected to reach 160 million then, and one minute of seven minutes on the internet surfing was spent on Facebook (Smart 2015; comScore 2011). Such a considerable users count as marketing advantage

© Springer Nature Switzerland AG 2020
L. C. Jain et al. (Eds.): SICBS 2019, AISC 1145, pp. 184–193, 2020.
https://doi.org/10.1007/978-3-030-46828-6_16

appeals to enterprise. Smart (2015) pointed out 63% of users use Facebook to acquire information. In China, despite censorship of social networks like Facebook and Twitter, social media like Sina Weibo and Wechat have become popular because they could immediately provide user experience and interactivity (Heggestuen 2013). The user's willingness of sharing plays a crucial role in communication on the social media websites scholars have devoted themselves for better understanding. Because of the rapid internet development, interaction on the internet becomes more complicate and marketing on social media also becomes more difficult for business on social media website. Burnett (2000) distinguished the participation behavior of social media users into interactive and non-interactive. Brodie et al. (2013) classified the participation behavior of social media users into five types such as learning, sharing, advocacy, socializing and co-creation. They believed that the higher the level of social user participation, the easier it is to increase satisfaction, trust, loyalty, emotional loyalty, and cohesion in the community. Dentsu in Japan built up a consumer behavior mode (SIPS) in 2011 in which there is zero-time-difference sharing for a social media website. Casaló et al. (2008) proposed that social media is a powerful communication tool for brand companies. The social media is not only cheap to set up and relatively easy to update content, but also uses the information that netizens get in social media to share. The reposting in your own social circle is easier to attract the attention of the target customers than the horizontal one-way network link between friends, compared to the previous one-way vertical communication between consumers and consumers.

Shang et al. (2006) and Jensen Schau and Gilly (2003) proposed consumers are willing to upload and publish the comment in the community. In addition to being a channel for users to express themselves, build their personal image and market their ideas, it may also be the reason why they like to share their experiences with people who have the same beliefs about the same brand. Sundara et al. (1998) and McAlexander et al. (2002) pointed out that the psychology of publishing comes from helping others make better purchasing decisions, and by publishing and sharing their own experience in using products or services, they can give others faster and more accurate selection suggestions. According to De Vries et al. (2012) believe that the relevant content published by the brand in social media marketing activities, its vividness, entertainment, information, update and positive and negative evaluations of netizens would have impact on the popularity of the brand. Factors affecting user's willingness of sharing consist of relationship quality, information sharing, website quality, member of social media community, information value, the internet interaction and word of mouth (Foster et al. 2010; Fan et al. 2012; Fan et al. 2013a; Fan et al. 2013b; Wu and Lo 2014). In the present study, we attempt to explore the factors of affecting users' willingness of sharing on social media website.

2 Methodology

2.1 Research Framework

In previous studies, researchers usually used regression analysis to study the determinants of affecting user's willingness of sharing. Artificial neural network could replicate the information processing system of biological neural network. Artificial neural network could resolve non-linear and complicate structure and has excellently predictive ability with few assumptions. The rapid development on internet and the rise of social media websites make interaction of the internet users complicate. The attribute of artificial neural network was expected to understand how social media platform affects user's willingness of sharing.

The questionnaires in our study consist of two portions. The first portion including 18 written questions, which describe user habits and factors affecting behavior of sharing on social media platform. The second portion is the possibility of user's information sharing.

2.2   Sampling Design

This study adopted judgmental sampling to collect data of Facebook users who share, pay attention to, follow or click like on information on the social media platform. Online survey was conducted on Facebook fan pages, various groups, and walls, and was shared by the users.

We investigated the determinants of sharing behavior on social media platform. To strengthen the stability and reliability of the questionnaire, we designed a multi-index method. The questionnaire was structured with closed-ended questions that users answered by 5-point Likert scale, except background variables. Gorsuch (1983) suggested that the samples be at least 5 times the questions and more than 100. The survey was conducted during the periods from April 29th, 2017 to May 15th, 2017 on the social media platform, Facebook, to ask users for answering the questionnaire through messages. 300 questionnaires were sent out and 280 were returned. The 254 efficient questionnaires were received. Three experts and scholars reviewed the questionnaire on January 3, 2017. After the review, two experts and scholars repeated the review on June 28 to confirm the validity of the study.

3   Empirical Results

3.1   Parameter of Back Propagation Neural Network

In this study, we investigate the impact of user attributes on willingness of sharing. The input variables include user privacy, information source, experience, check-in, reward activity and commenting habit. The output variable is user's willingness of sharing. The key parameters of back propagation neural network consist of number of neurons in the hidden layers, number of hidden layers, learning rate, and momentum parameter, is stated as follows:

Number of Hidden Layers
The convergence will reach if there is only one or two hidden layers. Without hidden layer, it fails to reflect the interaction among input variables and could lead to bigger error. More layers, however, make the whole network too complicated, the convergence decreases and more local minimum generates that causes it easier to fall into local minimum of error function when the network weights amends.

Number of Neurons in the Hidden Layers
It fails to describe efficiently the question with few neurons in the hidden layers, while it takes more time to converge but reach a smaller error with more neurons. Generally speaking, the number of neurons in the hidden layers could be chosen as follows:

(1) The number of neurons in the hidden layers should be half of the sum or the product of the neurons in the input and the output layer.
(2) If the question has more noise, the number of neurons in the hidden layers be less. On the contrary, it should increase.
(3) If testing sample error is higher than training sample error, the number of neurons in the hidden layers be less. On the contrary, it should increase.

Learning Rate
Too high or too low learning rate has a negative impact on network convergence: too low learning rate takes more time and might fail to avoid relative minimum, while too high one makes oscillation and hard to converge. According to past experience, a learning rate which is 0.5 or between 0.1 and 1.0 usually brings about great convergence.

Momentum Factor
Momentum factor includes partial update weight expression to improve the oscillation and fasten converging. Although the momentum does not cause the influence as critical as learning rate does, it usually leads to the network convergence as long as the parameter is under 0.9.

This study sets 1 as hidden layer, 2, 4, or 6 as neurons, 0.1, 0.5, or 1 as learning rate. The initial values of momentum factor are 0.001, 0.1, 0.3, 0.5, 0.7, or 0.9 respectively. The default of a learning cycle is set 1,000 and 5,000 to implement an experiment with the training data and testing data in the ratio of 8 to 2 (Table 1).

Table 1. Parameters of back propagation neural network

Parameter	Value
Number of input neurons	8
Number of output neurons	1
Number of hidden layer	1
Number of neurons in the hidden layer	2, 4, 6
Initial learning rate	0.1, 0.5, 1
Momentum factor	0.001, 0.1, 0.3, 0.5, 0.7, 0.9
Learning cycle	5,000 (Pre-clustering), 1,000 (Post-clustering)

3.2   Back Propagation Neural Network Model

The values are set according to the answers of questionnaire. The positive or negative sign is stated as follows:

Privacy

The questionnaire on privacy is "I leave personal information on Facebook" and "I use my real name on Facebook". Answering "I leave personal information and use real name on Facebook" is set as 1, and "I neither leave personal information nor use real name on Facebook" as −1. If the value is positive, it represents that the user leaves their personal information and uses their real name on Facebook, who value privacy less. If it is negative, it represents that subjects never leave their personal information nor use their real name on Facebook, who emphasize their privacy.

Information Source

The questionnaire on information source states "the access to acknowledge official Facebook fan pages". Answering official post on news feed is set to 0, fan page link on official website to 1, post showing on news feed after friends click like to 2, media such as magazine, television ad, or television program to 3, self-searching result to 4, and advertisement on platform or App to 5. Since the subjects have no bias toward any source, the coefficient value helps to learn the proportion of user information source on the social media website.

Experience

The questionnaire on experience is "I have recommended or shared the fan page posts after clicking the fan page like and becoming one of the fans in current two years" and "I have received the service or products after clicking the fan page like and becoming one of the fans in current two years". Answering never is set to 0, 1–4 times to 1, 5–9 times to 2, and above 10 times to 3. The coefficient value represents user experience on the social media platform.

Tagging Habit

The questionnaire on tagging habit is "I have tagged other users as sharing". Answering yes is set to 1 and otherwise to 0. The value represents user tagging habit.

Check-in Habit

The questionnaire on check-in habit states "I have checked in". Answering almost never/once a month is set to 0, seldom/once or twice a week as 1, sometimes/3–4 times a week to 2, and usually/5–7 times a week to 3. The coefficient stands for check-in habit of social media platform users.

Reward Activity

The questionnaire on reward activity is "what kind of reward activity promotes my sharing willingness". Answering raffle is set to −1, coupon to 0, and cash to 1. If the coefficient is negative, it stands for the bias toward raffle; if the value is positive, it is decided by whether it is close to 0 or 1 to learn the reward activity bias of social media users.

Commenting Habit

The questionnaire on commenting habit states "I write a comment on the information shared by Facebook member". Answering almost never/once a month is set to 0, seldom/once or twice a week to 1, sometimes/3–4 times a week to 2, usually/5–7 times a week to 3. The value shows the commenting habit of social media platform users.

3.3   Back Propagation Neural Network Model

As shown on Table 2, it reached the highest predictive accuracy in one hidden layer with six neurons. The learning rate is 1. The momentum parameter is 0.7. The predictive accuracy with training data was 68.2% while the one with testing data was 67.3%, and overall predictive accuracy was 67.4% with the smallest MSE (0.539).

Table 2. Results based on the back propagation neural network model

Input/output layer	Hidden layer	Learning rate	Momentum parameter	Learning cycle	Accuracy of training data	Accuracy of testing data	Total accuracy	MSE
8/1	1(2)	0.1	0.001	5000	54.4%	69.0%	57.3%	0.659
8/1	1(2)	0.1	0.1	5000	61.9%	54.8%	60.0%	0.655
8/1	1(2)	0.1	0.3	5000	62.4%	64.2%	60.1%	0.622
8/1	1(2)	0.1	0.5	5000	59.0%	68.8%	60.1%	0.622
8/1	1(2)	0.1	0.7	5000	60.7%	59.5%	60.2%	0.622
8/1	1(2)	0.1	0.9	5000	61.3%	66.7%	60.2%	0.622
8/1	1(2)	0.5	0.001	5000	60.2%	70.7%	60.2%	0.637
8/1	1(2)	0.5	0.1	5000	58.6%	65.1%	60.2%	0.632
8/1	1(2)	0.5	0.3	5000	59.4%	59.4%	60.3%	0.627
8/1	1(2)	0.5	0.5	5000	57.6%	63.5%	60.4%	0.640
8/1	1(2)	0.5	0.7	5000	59.5%	64.3%	60.4%	0.647
8/1	1(2)	0.5	0.9	5000	63.3%	64.9%	60.4%	0.650
8/1	1(2)	1	0.001	5000	59.9%	60.8%	60.5%	0.655
8/1	1(2)	1	0.1	5000	60.9%	59.4%	60.4%	0.625
8/1	1(2)	1	0.3	5000	59.5%	68.7%	60.5%	0.621
8/1	1(2)	1	0.5	5000	62.3%	60.6%	60.5%	0.633
8/1	1(2)	1	0.7	5000	61.7%	61.6%	60.5%	0.621
8/1	1(2)	1	0.9	5000	59.2%	67.6%	60.4%	0.641
8/1	1(4)	0.1	0.001	5000	62.5%	58.0%	60.5%	0.655
8/1	1(4)	0.1	0.1	5000	62.6%	63.2%	60.3%	0.622
8/1	1(4)	0.1	0.3	5000	62.8%	56.7%	60.3%	0.621
8/1	1(4)	0.1	0.5	5000	59.5%	63.8%	60.3%	0.621
8/1	1(4)	0.1	0.7	5000	59.1%	64.6%	60.3%	0.621
8/1	1(4)	0.1	0.9	5000	60.1%	57.5%	60.1%	0.627
8/1	1(4)	0.5	0.001	5000	61.1%	66.1%	60.2%	0.628
8/1	1(4)	0.5	0.1	5000	65.3%	55.7%	60.3%	0.628
8/1	1(4)	0.5	0.3	5000	61.8%	73.3%	60.1%	0.649
8/1	1(4)	0.5	0.5	5000	57.7%	59.5%	59.8%	0.641
8/1	1(4)	0.5	0.7	5000	60.6%	68.7%	60.7%	0.619
8/1	1(4)	0.5	0.9	5000	58.2%	65.4%	60.5%	0.619
8/1	1(4)	1	0.001	5000	59.6%	64.6%	60.3%	0.643
8/1	1(4)	1	0.1	5000	60.9%	63.8%	60.9%	0.632
8/1	1(4)	1	0.3	5000	60.0%	63.4%	60.5%	0.629
8/1	1(4)	1	0.5	5000	59.7%	63.6%	60.1%	0.617
8/1	1(4)	1	0.7	5000	59.0%	66.6%	60.7%	0.616

(*continued*)

Table 2. (*continued*)

Input/output layer	Hidden layer	Learning rate	Momentum parameter	Learning cycle	Accuracy of training data	Accuracy of testing data	Total accuracy	MSE
8/1	1(4)	1	0.9	5000	56.3%	67.3%	60.8%	0.627
8/1	1(6)	0.1	0.001	5000	55.4%	69.9%	57.7%	0.650
8/1	1(6)	0.1	0.1	5000	59.0%	57.6%	57.8%	0.650
8/1	1(6)	0.1	0.3	5000	58.1%	54.9%	57.7%	0.651
8/1	1(6)	0.1	0.5	5000	57.3%	56.0%	58.2%	0.645
8/1	1(6)	0.1	0.7	5000	60.5%	57.9%	58.2%	0.645
8/1	1(6)	0.1	0.9	5000	59.3%	57.0%	58.2%	0.646
8/1	1(6)	0.5	0.001	5000	59.3%	60.7%	58.6%	0.646
8/1	1(6)	0.5	0.1	5000	58.6%	58.7%	58.6%	0.646
8/1	1(6)	0.5	0.3	5000	60.7%	59.6%	59.2%	0.633
8/1	1(6)	0.5	0.5	5000	61.4%	65.9%	59.6%	0.630
8/1	1(6)	0.5	0.7	5000	59.8%	51.6%	59.7%	0.630
8/1	1(6)	0.5	0.9	5000	62.4%	58.6%	60.1%	0.655
8/1	1(6)	1	0.001	5000	61.4%	61.2%	60.1%	0.634
8/1	1(6)	1	0.1	5000	61.3%	62.0%	60.2%	0.638
8/1	1(6)	1	0.3	5000	65.3%	57.5%	66.3%	0.557
8/1	1(6)	1	0.5	5000	67.2%	83.1%	67.3%	0.545
8/1	1(6)	1	0.7	5000	68.2%	67.3%	67.4%	0.539
8/1	1(6)	1	0.9	5000	70.2%	75.9%	67.3%	0.538

3.4   Sensitivity Analysis

Sensitivity analysis was employed to assess the impact of each attribute on the user's sharing behavior. The user's attributes consist of privacy, information source, experience, tagging habit, check-in habit, reward activity, and commenting habit. The crucial factor affecting the user's behavior of sharing is check-in habit (0.5817), followed by privacy using real name on Facebook (−0.51965). The rest of factors are arranged as commenting habit (−0.4856), tagging habit (−0.2264), and information source (0.0562) in order (Table 3, Fig. 1).

Table 3. Estimates based on sensitivity analysis

	Value
Privacy (personal information)	0.2726
Privacy (real name)	−0.51965
Information source	0.0562
Experience	0.2576
Check-in habit	0.5817
Commenting habit	−0.4856
Reward activity	−0.25115
Tagging habit	−0.2264

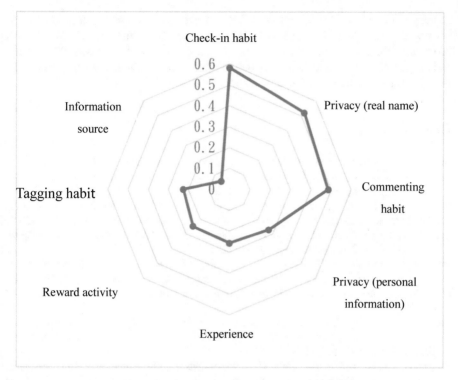

Fig. 1. Radar chart based on sensitivity analysis

4   Concluding Remarks

Artificial neural network replicating the biological neural network is an information processing system. It resolves not only non-linear and complicate structure but it has excellent prediction with few limitation. This study aims to explore the behavior of users on social media platform by employing artificial neural network. The predictive accuracy of the model reaches 67.4%.

As for users with high usage frequency, the behavior of common post or check-in would promote user's sharing willingness Open comment and discussion could also raise their sense of engagement and encourage their sharing. On the contrary, privacy and information source have less significant effects on sharing behavior. For users with low usage frequency, their usage sharing experience or used experience have a great impact on their sharing that private sharing behavior fits this kind of users because they do not like to post. The users with high usage frequency prefer to reward activities, while users with low usage frequency has no specific preference for reward activities. As for businesses, the ability to get information from a wide range of users and disseminate it to a wider range of users, and the ability to integrate different channels as part of their marketing project are critical in developing a successful social media brand strategy (Kim and Ko 2012; Rapp et al. 2013). Once the social media campaign begins. ers un- Instead, they listen carefully because they know that consumers not only "consume" the campaign, but can comment on it ("create"), share it with their friends and anyone else ("connect") and provide their uncensored thoughts about it ("control") for any and all to view (Hoffman and Fodor 2010).

References

Burnett, G.: Information exchange in virtual communities: a typology. Inf. Res. **5**(4) (2000)

Brodie, R.J., Ilic, A., Juric, B., Hollebeek, L.: Consumer engagement in a virtual brand community: an exploratory analysis. J. Bus. Res. **66**(1), 105–114 (2013)

Casaló, L.V., Flavián, C., Guinalíu, M.: Promoting consumer's participation in virtual brand communities: a new paradigm in branding strategy. J. Mark. Commun. **14**(1), 19–36 (2008)

Chiu, S.-T., Hsiao, C.-H.: Viral marketing campaign on the Facebook fan page of the sports events. Q. Chin. Phys. Educ. **28**(1), 11–20 (2014)

De Vries, L., Gensler, S., Leeflang, P.S.: Popularity of brand posts on brand fan pages: an investigation of the effects of social media marketing. J. Interact. Mark. **26**(2), 83–91 (2012)

Divol, R., Edelman, D., Sarrazin, H.: Demystifying social media. McKinsey Q. **2**(12), 66–77 (2012)

Fan, T.-Y., Liu, F.-M., Ding, J.-D.: The impact of website interactivity and word-of-mouth effect on social networking website loyalty-a case study of group buying communities. J. Yu Da Univ. Sci. Technol. (32), 157–174 (2012)

Fan, T.-Y., Liu, F.-M., Lin, Y.-Y., Wu, C.-S.: The impact of website quality, relationship quality and information sharing intension on social networking website loyalty-a case study of Mobiles01. J. Sci. Technol. Humanit. Transworld Univ. (17), 73–89 (2013a)

Fan, T.-Y., Liu, F.-M., Wang, H.-C.: The impact of relationship quality and information sharing intension on user' social website loyalty-a case study of social networking game. J. Cardinal Tien Coll. Nurs. (11), 24–32 (2013b)

Foster, M.K., Francesucci, A., West, B.C.: Why user participate in online social network. Int. J. e-Bus. Manag. **4**(1), 3–19 (2010)

Gorsuch, R.L.: Factor Analysis, 2nd edn. LEA, Hillsdale (1983)

Heggestuen, J.: Confused by China's social networks? Here is a simple infographic showing their US-based equivalents. Business Insider (2013)

Hoffman, D.L., Fodor, M.: Can you measure the ROI of your social media marketing? MIT Sloan Manag. Rev. **52**(1), 41 (2010)

Jensen Schau, H., Gilly, M.C.: We are what we post? Self-presentation in personal web space. J. Consum. Res. **30**(3), 385–404 (2003)

Kaplan, A.M., Haenlein, M.: Users of the world, unite! The challenges and opportunities of social media. Bus. Horiz. **53**(1), 59–68 (2010)

Kim, A.J., Ko, E.: Do social media marketing activities enhance customer equity? An empirical study of luxury fashion brand. J. Bus. Res. **65**(10), 1480–1486 (2012)

McAlexander, J.H., Schouten, J.W., Koenig, H.F.: Building brand community. J. Mark. **66**(1), 38–54 (2002)

Rapp, A., Beitelspacher, L.S., Grewal, D., Hughes, D.E.: Understanding social media effects across seller, retailer, and consumer interactions. J. Acad. Mark. Sci. **41**(5), 547–566 (2013)

Shang, R.A., Chen, Y.C., Liao, H.J.: The value of participation in virtual consumer communities on brand loyalty. Internet Res. **16**(4), 398–418 (2006)

Wu, H.-H., Lo, W.-F.: Why do you want to do "like", "Comment" or "Share" on Facebook: the study of antecedent on Facebook user's behavioral intentions. Mark. Rev. **11**(2), 107–131 (2014)

Institute for Information Industry: Taiwan Consumer Mobile Device and APP Use Behavior Research Survey Report (2014). http://www.find.org.tw/market_info.aspx

Smart M.: Report on data on Taiwan's Internet and Action Survey for the third quarter of 2015 (2015). https://www.smartm.com.tw/Article/31373431cea3

Intelligent Computing

A Survey of Bus Arrival Time Prediction Methods

Mao-Lun Chiang[1], Chin-Feng Lee[2]($\boxtimes$) (ID), Ting-You Lin[1],
and Somya Agrawal[2]

[1] Department of Information and Communication Engineering,
Chaoyang University of Technology, Taichung City 41349, Taiwan ROC
mlchiang@cyut.edu.tw, mjnhb7510@gmail.com
[2] Department of Information Management, Chaoyang University of Technology,
Taichung City 41349, Taiwan ROC
lcf@cyut.edu.tw, asomya@gm.cyut.edu.tw

Abstract. Traffic flow is often affected by several stochastic factors and the timely availability of precise travel time data of moving automobiles is valuable for both drivers and the commuters. Public transportation system gives people an option of choosing busses to the passengers to commute which reduces their own vehicle use. However, it often encounters inaccurate arrival time, which is affected by many factors such as vehicle congestion, road conditions, weather conditions, vehicle breakdowns. To plan their travel itinerary in a better way, the passengers often need real-time information. They don't want to spend their time waiting for the bus. In this paper, we have done a literature survey to study the existing methods used to predict the arrival time of buses. Different models that have been proposed to reduce the travel time of passengers have been presented and discussed.

Keywords: Bus arrival time · Historical model · Dynamic model · Machine learning · Artificial neural network · Kalman filter · Prediction model

1 Introduction

Traffic flow is often affected by various stochastic variables such as road conditions, the total number of moving automobiles at a particular time, the time of travel, climate and the driving patterns of vehicles, etc. [1]. The availability of precise travel time data of moving automobiles in real-time is useful for both drivers and the commuters. This information is more crucial when dispatching is based on the prediction of presumed commuters waiting along the travel route rather than the predetermined schedule. Public transportation system gives people another option of choosing busses to commute which reduces their own vehicle use. To know the arrival time, the passenger company provides an estimated arrival schedule and an instant arrival system provided by the Transportation Bureau. However, it often encounters inaccurate arrival time affected by the driving speed, climatic conditions, driver habits, passenger flow, delays or unpredictable conditions, which indirectly affects the local people and the tourists'

© Springer Nature Switzerland AG 2020
L. C. Jain et al. (Eds.): SICBS 2019, AISC 1145, pp. 197–206, 2020.
https://doi.org/10.1007/978-3-030-46828-6_17

impressions of the city. Therefore, prediction and calculation of the arrival time for buses have been explored by many scholars using various models [1].

In the past few years, many studies attempted to use historical average travel time and several other inputs to estimate the travel time of the route accurately [3, 5, 10, 23, 24]. In this paper, we have used them for the estimation of bus arrival time. Researchers often use archived data or data available online using the Global Positioning System (GPS) as the primary source of information. This available data is then analyzed using different analysis methods such as Kalman filter, regression models, artificial neural network (ANN), support vector machine (SVM), radial basis functions, hybrid models, and dynamic models, etc. [1]. In this paper, we have done a literature survey to study the existing methods used to estimate the arrival time of busses, and different models that have been proposed to reduce the travel time of passengers.

2   A Model Discussion on Predicting the Arrival of Traffic Between Destinations

2.1   Historical Model

Future and the current travel times of the busses are estimated using the historical model. This model is based on previous historical travel time within the same time period [3]. The parameters of average travel time and average speed were used to make this prediction. When the congestion is minimum in areas with stable traffic, the model gives more accurate results.

2.2   Dynamic Model

The researchers designed a dynamic model for estimating bus arrival time by using the next station's GPS data [5, 8, 13]. Taking GPS data as input, prediction can be the parameters such as departure time based on arrival time, station's distance, and historical departure time. This method was found to be more accurate when the prediction time was compared to the actual arrival time. The dynamic model proved to be superior compared to the historical model [23].

2.3   Statistical Model

Future and the current travel times of the busses are estimated using the historical model. This model is based on previous historical travel time within the same time period [3]. The parameters of average travel time and average speed were used to make this prediction. When the congestion is minimum in areas with stable traffic, the model gives more accurate results.

2.3.1   Regression Model

Regression model predicts the interrelationships between different explanatory variables (independent and dependent) using linear functions, where linear functions were formulated by using a set of independent variables.

Lin et al. proposed that the bus arrival time is impacted by the driver behavior, traffic signs, and jams, etc. Maiti et al. proposed the use of stay time, the total distance, number of stops, climatic descriptors, etc. as independent variables [15, 20]. These factors were analyzed by many researchers to study their impact on the dependent variable of bus travel time using regression models [2].

2.3.2   Kalman Filter

The researchers proposed a Kalman filter model that used stay time, run time, speed, and position as the parameters to estimate the bus arrival time [6, 22]. It divided the complete travel time into two main parts, running and stay time. Zaki et al. also proposed the prediction of travel time using Kalman filter algorithm. The residence time is estimated by calculating the stay time and the gain factor. When the estimated arrival time was compared to the actual value, the model showed an error rate of 16.7% [14].

Tantawy and Zorkany proposed two techniques, neural network and Kalman filter, to estimate the real-time arrival time based on timings of the day, such as morning, peak, evening and weekdays, weekends, etc. [17].

A Kalman filter model was proposed by Shalaby and Farhan that used Automatic Passenger Counting (APC) system to predict bus arrival and departure times. Both of them were used for route travel time to estimate the arrival time of the passenger. The run time, stay time and stop model were used to predict Automatic Vehicle Location (AVL) and APC data [4].

2.3.3   Polynomial Model

Yu et al. proposed a model using GPS data, distance, speed and traffic related data. This model was based on the techniques of polynomial fitting and cluster analysis for estimating the bus run time. The interval prediction model was based on historical data, while GPS measured the location and time of the bus. Using the real-world data, the model was efficiently implemented in terms of prediction [13].

2.3.4   Other Model

Chandurkar et al. proposed Real-Time Passenger Information Systems (RTPIS) by installing GPS on the bus. Passengers used the mobile app or web app to find bus schedules and RTPIS. The study met needs of the passengers, drivers and transport department heads. Location of all the buses were tracked to estimate the arrival time of different routes. The estimated time was updated each time and a message was displayed to the passenger [9].

2.4   Machine Learning Model

Machine Learning (ML) deals with artificial intelligence which is related to the construction and research of systems that can learn from data that is fed to the computer machine [19]. ML has been widely used in solving problems in various fields for several years. It consists of two stages, the selection and candidate stages, which predicts the parameters of the model through learning the process of existing data.

These models can be used to predict travel time without having to deal with traffic flow implicitly. Models including Artificial Neural Network (ANN) and Support Vector Machine (SVM) and Deep Neural Network (DNN) models are presented next under these categories.

2.4.1 Artificial Neural Network

Researchers have used models such as ANN models, regression models, and statistical models based on historical data, to predict arrival times using automatic vehicle positioning (AVL) data [3]. The forecast was based on stay time and traffic jam. The parameters used were arrival time, stay time, progress compliance, travel time and distance.

The route arrival time prediction model used Android platform which was proposed by Zhou et al. It included three components such as query users, shared users, and servers on Android phones. The system depended on the resources of the shared users, and the proposed predictive model performed well compared to the bus arrival time prediction using a GPS-enabled solution. The proposed system provided a solution that did not cost a lot of money [7].

Historical data and seasonal recurrent neural networks were used together to predict arrival time. The ANN model was used to estimate the arrival time based on different climatic conditions. Compared to the ML or HD model, the ANN model performed better in terms of Mean Absolute Percentage Error (MAPE) [15].

ANN and Kalman filter were used to estimate bus arrival times for comparison. Time indicators, bus delays, arrival times, and travel times were considered to be the main parameters for predicting the arrival time in these models [23]. The ANN model was used for each collected data cluster. Finally, everything was integrated into the layered model. The conclusion was that ANN performed better than the Kalman filter [11].

Zaki et al. developed a hybrid model that combined the characteristics of neural networks with the reliability of Kalman filter. The model predicted bus stay time and route. The data acquisition combined with the time position received by the GPS receiver, and the average speed was recorded, and the travel speed was used to predict the arrival time of the bus. The results showed satisfactory prediction accuracy [14].

Maiti et al. compared historical data models (HD), ANN and SVM. The route location and time stamp were the main features. The location of the route took into account the road conditions and the length of stay. The timestamp feature captured vehicle speed, which also included traffic. Studies have shown that the performance of the HD model was 2.5 times superior to the ANN model, 2 times superior than the SVM model, and the accuracy was 75.65% [15].

Tantawy and Zorkany proposed the use of techniques such as Kalman filter and NN to estimate time based on daily patterns, such as morning, peak, evening, weekends, working days, weekend traffic conditions [24]. In field tests, low-complexity traffic did a better prediction of arrival time for neural networks, while in daily heavy strokes, Kalman filter were superior to neural networks [17].

2.4.2   Support Vector Machine

Support Vector Machine (SVM) is a supervised algorithm with classification and regression analysis. It is often used in face recognition, handwriting recognition, and text categorization. It showed good learning ability for time series prediction and statistical analysis.

Researchers have proposed several SVM models for estimating bus arrival times [20]. Three indicators were used in the model, called GPS coverage, accuracy and release rate. The multi-indicator evaluation was performed using SVM. These three indicators played an important role in the forecast. The researchers obtained measurements of each indicator, and the SVM learned from the statistical learning evaluation model of the training data, which was later used in the multi-indicator model [12].

Yu et al. proposed a method using SVM to eliminate outliers through the Grubbs test to estimate future bus travel times. A current data generated by the variables. The results showed that comparing the Kalman filter with Grubbs test, the proposed SVM with forgetting factor predicted more accurate travel time [16].

2.4.3   Deep Neural Network

Deep Neural Network (DNN) model is an ANN which has multiple hidden layers that exist between the input and output to handle huge and complex variable relationships. This makes the model more complicated. No exact rule of thumb exists to determine the optimal number of hidden layers and their nodes. The most common rule is that "the optimal size of the hidden layer is usually between the input size and the output layer."

Treethidtaphat et al. proposed a method using only one route data to construct the DNN model [21]. The study explored a method to improve bus travel time prediction performance at any distance along the route. Due to cloud computing, more data can lead to deeper and more complex DNN.

2.4.4   K-nearest

In 2012, Sinn et al. used three algorithms in their study such as linear regression, K-nearest neighbor prediction, and kernel regression [8]. Kernel regression model showed the relationship between bus arrival time and location updates. These algorithms were based on travel time, location, distance between stops, and delay. The predicted arrival time was used for the bus electronic signage. Dong et al. proposed a method using K-nearest neighbor or ANN to predict the arrival time of the bus. The model used the bus's site link, speed, travel time and stay time as the main parameters along with K nearest neighbors. Short-range prediction were based on actual traffic conditions, while very large distance predictions were based on the nearest neighbor K. The final result was that ANN or KNN alone provided higher accuracy (reduced error rate over a 60-min time span 12%) [10].

2.4.5   Radial Basis Function

Wang et al. proposed a Radial Basis Function (RBF) neural network to predict bus arrival time. This method was completed in two stages, one using historical data and the second using online data. RBF was used with historical data and trained by variables such as stay time, delay, distance and travel speed. The Kalman filter was used for online data and travel speed adjustment. The proposed model was more accurate when online RBF was used without online adjustment [18].

2.4.6   Hybrid Models

Some researchers proposed a hybrid framework that integrated two or more models for arrival time estimation. As and Mine proposed an ANN dynamic model based on historical data to distinguish time periods. The results indicated that ANN was superior to historical model [23]. Yamaguchi et al. proposed a bus delay prediction model between the bus stops. The results demonstrated the superiority of the prediction model based on Gradient Lift Decision (GBDT) and the effect of considering travel time over the previous interval. Zaki et al. proposed a hybrid scheme that combined NN with decision rules from historical data and Kalman Filter. The proposed algorithm relied on real-time location and included historical travel time, traffic conditions and spatial changes [14].

All models presented here are mentioned in Fig. 1. The use of various techniques has been discussed in preliminary work such as historical models, dynamic models, statistical models, machine learning and hybrid models.

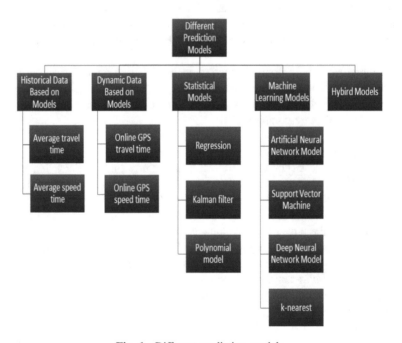

Fig. 1. Different prediction models

3   Affecting the Arrival Factor

The prediction of arrival time depends on a number of factors and variables. The following factors were taken into account to show the impact of them on bus arrival time predictions, as shown in Fig. 2.

Table 1. Affecting factor code

Code	Affect factor
TT	Travel Time
SD	Street Distance
GPS	GPS
S	Speed
TP	Time Period
E	Event
ND	Network Delay
PD	Path Direction

Fig. 2. Arrival time affecting factors

The literature survey highlights different methods, theories, and different models. Researchers have designed many predictive models having significant features to estimate the vehicle arrival time (Table 1).

4   Comparison Method

The accuracy of the model is evaluated by using three measurement parameters: Mean Absolute Error (MAE; sec), Root Mean Square Error (RMSE; sec), and Mean Absolute Percentage Error (MAPE; %). The formulas are as follows:

$$MAE = \frac{\sum \left| t^{actual} - t^{predict} \right|}{N}$$

$$RMSE = \sqrt{\frac{\sum \left(t^{actual} - t^{predict} \right)^2}{N}}$$

$$MAPE = \frac{1}{N} \cdot \sum \frac{\left| t^{autual} - t^{predict} \right|}{t^{autual}} \times 100\%$$

Where t^{actual} is the actual observed bus travel time, $t^{predict}$ is the predicted bus travel time, and N is the test number.

Table 2 compares the methods used to predict the bus arrival time based on different annual prediction models. The annual models used by different researchers are shown in the table below.

Table 2. Compares the methods based on different annual prediction models

YEAR	REFERENCE	MODEL	ANN	DYNAMIC MODEL	HISTORICAL MODEL	REGRESSION	KALMAN FILTER	K-NEAREST	SVM	DNN	POLY FITTING	RBF	DATA	AFFECTING FACTOR	ACCURATE
2004	[3]	D	✓										2000/06~11 (six month)	[TT] [GPS][S] [TP]	MAPE (1.96%, 8.36%)
	[4]	C					✓						2001/05 (one month)	[APC]	
	[2]	C				✓							2002/01~06 (one month)	[SD][TP][ND] [PD]	RMSE 8.24
2009	[5]	B	✓											[TT]	
	[6]	C					✓							[TT]	MAPE 19.7%
2012	[8]	E	✓	✓	✓								2011/06 (one month)	[GPS]	
	[7]	A		✓										[SD][S][ND][PD]	
2013	[10]	D					✓							[SD][GPS]	
	[11]	D	✓										2010/11 (one month)	[SD][GPS]	
	[12]	D					✓							[GPS]	
	[13]	E	✓									✓	2012/08~10 (two month)	[SD] GPS]	MAPE 22.57%
	[14]	E	✓			✓								[TT][GPS]	
	[9]	C							✓					[SD][GPS]	
	[17]	C					✓						Two month	[GPS]	RMSE (1.28, 6.8)
2014	[15]	A		✓									08:00~23:00	[SD][S][E]	MAPE 24.44%
	[16]	D					✓						2006/09~10 (one month)	[TT]	
	[18]	D									✓			[GPS]	MAPE 15.98%
2015	[19]	D	✓												
	[20]	D					✓							[E]	
2017	[21]	D								✓			2017/02~03 (one month)	[SD][TP]	MAE 343
	[22]	E	✓		✓	✓	✓							[SD][GPS]	MAPE (16.35%, 35.78%)
2018	[23]	E	✓	✓									2013/11~12 (one month)	[TT] [GPS][TP]	MAPE <15%
	[24]	E	✓										201311~12 (one month)	[TT][GPS][TP] [ND]	MAE 27
total			7	4	3	3	5	2	3	1	2	1			

The reason we used three metrics was due to their different presentation. In general, if we wanted to know what the average error of the prediction was, MAE could tell the magnitude of this error. If the magnitude of the error was large, the RMSE was sensitive to the error. Finally, MAPE was easy to read. It determined how big the 10 s error was, which was the total percentage of the actual travel time.

5 Observation and Discussion

Different models were showcased in the paper to predict the exact and effective arrival time of bus arrivals. Accuracy was divided into three presentation methods: MAPE, MAE, and RMSE, depending on the evaluation method. We compared the final accuracy with an average MAPE of less than 30% and concluded that the number of machine learning models used by the researchers increased gradually and that the performance was better than the bus arrival time predictions of all other models.

6 Conclusion

In this paper we studied various methods such as ANN, SVM Kalman filter, DNN, dynamic model to predict the bus arrival time. The literature survey included papers on bus arrival time forecasts from 2004 to 2015. As can be seen from the results of this study, no fuzzy logic was deployed to solve GPS data.

Regression models, ANN, Kalman filter and historical data models were used to predict the bus arrival time under a number of factors such as arrival time, dwell time, distance, and run time. Artificial neural network were also used to predict bus arrival times by using unpredictable factors (such as traffic load and seasonal variations). This research supports researchers who used historical data-based models, statistical models, and machine learning models to further investigate bus arrival time prediction.

References

1. Altinkaya, M., Zontul, M.: Urban bus arrival time prediction: a review of computational models. Int. J. Recent Technol. Eng. (IJRTE) 2(4), 164–169 (2013)
2. Patnaik, J., Chien, S., Bladikas, A.: Estimation of bus arrival times using APC data. J. Public Transp. 7(1), 1 (2004)
3. Jeong, R., Rilett, R.: Bus arrival time prediction using artificial neural network model. In: Proceedings of the 7th International IEEE Conference on Intelligent Transportation Systems (IEEE Cat. No. 04TH8749), pp. 988–993, October 2004
4. Shalaby, A., Farhan, A.: Prediction model of bus arrival and departure times using AVL and APC data. J. Public Transp. 7(1), 3 (2004)
5. Zhang, J., Yan, L., Han, Y., Zhang, J.J.: Study on the prediction model of bus arrival time. In: 2009 International Conference on Management and Service Science, pp. 1–3, September 2009
6. Padmanaban, R.P.S., Vanajakshi, L., Subramanian, S.C.: Estimation of bus travel time incorporating dwell time for APTS applications. In: 2009 IEEE Intelligent Vehicles Symposium, pp. 955–959, June 2009

7. Zhou, P., Zheng, Y., Li, M.: How long to wait?: predicting bus arrival time with mobile phone based participatory sensing. In: Proceedings of the 10th International Conference on Mobile Systems, Applications, and Services, pp. 379–392, June 2012

8. Sinn, M., Yoon, J.W., Calabrese, F., Bouillet, E.: Predicting arrival times of buses using real-time GPS measurements. In: 2012 15th International IEEE Conference on Intelligent Transportation Systems, pp. 1227–1232, September 2012

9. Chandurkar, S., Mugade, S., Sinha, S., Misal, M., Borekar, P.: Implementation of real time bus monitoring and passenger information system. Int. J. Sci. Res. Publ. 3(5), 1–5 (2013)

10. Dong, J., Zou, L., Zhang, Y.: Mixed model for prediction of bus arrival times. In: 2013 IEEE Congress on Evolutionary Computation, pp. 2918–2923, June 2013

11. Lin, Y., Yang, X., Zou, N., Jia, L.: Real-time bus arrival time prediction: case study for Jinan, China. J. Transp. Eng. 139(11), 1133–1140 (2013)

12. He, Z., Yu, H., Du, Y., Wang, J.: SVM based multi-index evaluation for bus arrival time prediction. In: 2013 International Conference on ICT Convergence (ICTC), pp. 86–90, October 2013

13. Yu, H., Xiao, R., Du, Y., He, Z.: A bus-arrival time prediction model based on historical traffic patterns. In: 2013 International Conference on Computer Sciences and Applications, pp. 345–349, December 2013

14. Zaki, M., Ashour, I., Zorkany, M., Hesham, B.: Online bus arrival time prediction using hybrid neural network and Kalman filter techniques. Int. J. Mod. Eng. Res. 3(4), 2035–2041 (2013)

15. Maiti, S., Pal, A., Pal, A., Chattopadhyay, T., Mukherjee, A.: Historical data based real time prediction of vehicle arrival time. In: 17th International IEEE Conference on Intelligent Transportation Systems (ITSC), pp. 1837–1842, October 2014

16. Yu, B., Ye, T., Tian, X.M., Ning, G.B., Zhong, S.Q.: Bus travel-time prediction with a forgetting factor. J. Comput. Civil Eng. 28(3), 06014002 (2012)

17. Tantawy, M., Zorkany, M.: A suitable approach for evaluating bus arrival time prediction techniques in Egypt. In: Proceedings of the 2014 International Conference on Communications, Signal Processing and Computers (2014)

18. Wang, L., Zuo, Z., Fu, J.: Bus arrival time prediction using RBF neural networks adjusted by online data. Procedia-Soc. Behav. Sci. 138, 67–75 (2014)

19. Khamparia, A., Pandey, B.: Knowledge and intelligent computing methods in e-learning. Int. J. Technol. Enhanced Learn. 7(3), 221–242 (2015)

20. Yang, M., Chen, C., Wang, L., Yan, X., Zhou, L.: Bus arrival time prediction using support vector machine with genetic algorithm. Neural Netw. World 26(3), 205 (2016)

21. Treethidtaphat, W., Pattara-Atikom, W., Khaimook, S.: Bus arrival time prediction at any distance of bus route using deep neural network model. In: 2017 IEEE 20th International Conference on Intelligent Transportation Systems (ITSC), pp. 988–992, (2017)

22. Kumar, B.A., Vanajakshi, L., Subramanian, S.C.: Bus travel time prediction using a time-space discretization approach. Transp. Res. Part C: Emerg. Technol. 79, 308–332 (2017)

23. As, M., Mine, T.: Dynamic bus travel time prediction using an ANN-based model. In: Proceedings of the 12th International Conference on Ubiquitous Information Management and Communication, p. 20 (2018)

24. Yamaguchi, T., As, M., Mine, T.: Prediction of bus delay over intervals on various kinds of routes using bus probe data. In: 2018 IEEE/ACM 5th International Conference on Big Data Computing Applications and Technologies (BDCAT), pp. 97–106 (2018)

Applying Talent Quality-Management System (TTQS) and Information Visualization to e-Learning Partner Project

Hsing-Yu Hou[1], Somya Agrawal[2]($\boxtimes$), Yu-Lung Lo[2],
and Chin-Feng Lee[2]

[1] National Taichung University of Science and Technology,
Taichung City, Taiwan
hsingyuhou@nutc.edu.tw
[2] Chaoyang University of Technology, Taichung City, Taiwan
{asomya,lcf}@gm.cyut.edu.tw, yllo@cyut.edu.tw

Abstract. In order to reduce the gap of digital skills between urban and rural, the ministry of education in Taiwan carried out an e-learning partner project in rural schools. The quality management of education in such schools can improve the performance of training. In the present study, Talent Quality-management System was used to explore the effect of the training with the help of 18 indexes. Data consisted of 276 records from the teaching side in the behaviour assessment survey in the case university which was collected during the three semesters in the academic years of 2018–2019. The questionnaire included the variables of self-evaluation, administration assessment and information technology support. Using Tableau software and further analysis, the hidden issues could be highlighted to the decision maker. Basic statistical analyses, ANOVA, post-hoc analysis and information visualization were used to carry out analysis in this study. The results demonstrated that regular process management and retention were two main factors affecting the performance in the e-learning partner project. We contend that using insights from this study, the cost of human resources training could be declined further.

Keywords: e-Learning partner project · Information visualization · Quality management · Retention · TTQS

1 Introduction

Recently, the government of Taiwan has started paying more attention to the information technology (IT) related competency level in individuals. There are four stages for the Executive Yuan, which is the executive branch of the government in Taiwan to reduce the gap of digital skills between urban and rural. In the first stage the aim was to improve education, economy, society and culture to reduce the digital divide between 2005 to 2007. In the second stage the aim was to set up digital opportunity centers and create fair e-learning opportunities between 2008 to 2011. In the third stage the aim was to strengthen the information infrastructure, improve the literacy rate of rural students and diverse information application services to deepen the digital care between 2012 to 2015. And in the fourth stage, the aim was to improve digital skills, enrich the digital applications (e-commerce, e-marketing), enjoy mobile services and applications, improve digital marketing competency level of rural enterprises, and strengthen the skills of farmers between 2016 to 2019. According to the project, the ministry of education [1] carried out digital application promotion plan in rural schools [1]. The e-learning partner is one part of the plan. Based on companionship and learning, undergraduates used the medium of Internet to overcome barriers between urban and rural areas. It was the objective of partner universities to train college students to use IT to integrate learning, and help improve the motivation and interest of rural students. The purpose was to promote equal learning opportunities for both rural and urban places.

Till date, many researchers have discussed the effects of digital service and e-learning partner projects. For example, Shuai and Jhou [2] found students' attitude, teachers' personality and internet speed as the main factors affecting online tutoring quality [2]. About social care and social engagement, the e-learning partner can develop the concern of humanity and improve the social interactive experience of undergraduates [3, 4]. For rural students, they can develop characteristics such as learning attitude and learning performance through the support and guidance of e-learning [5–9]. About the evaluation of teaching, the variables of teaching efficacy and teaching satisfaction were discussed [10, 11]. However, the quality management of teaching and counselling was not given too much importance. In the present study, Talent Quality-management System (TTQS) was carried out to explore the effect of undergraduates in the checklist form. A diagnosis was done to investigate the relationship between self-evaluation, administration assessment and IT support of undergraduates during three semesters at the case University. Finally, the visualization results can highlight the problem and offer suggestions for projects hosts in the management process. Therefore, the main objectives of the present study include: (1) To set up the quality management checklist with the TTQS to enhance the undergraduates'

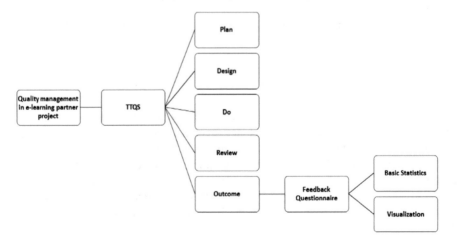

Fig. 1. Research structure

skills and abilities; (2) To find the teaching outcomes of undergraduates through feedback to understand the trend among three semesters; (3) To visualize the items among self-evaluation, administration assessment and IT support to search for the gaps. The research structure is showed in Fig. 1.

2 Literature Review

2.1 Introduction of e-Learning Partner Plan

According to MOE (2016), the core values of e-learning partner plan include "life accompanying life, and living teaching living". The outcome of the plan should be that undergraduates cultivate a spirit of social service and digital care through IT to integrate learning and enhance rural students' interest in learning, and thereby improve the education culture in remote areas. There are two main targets: one is the learning side in primary school, and the other is the teaching side in universities. The teaching side can provide information applications and do consultation. The teaching side (which includes the undergraduate students) and the learning side (which includes the primary students) can be placed on a regular basis, through video equipment and online learning platform in the centralized management computer class. There are totally 20 weeks (first and the second semester each 10 weeks), twice a week, each time 2 lessons (Chinese, English or Mathematics), each class for 45 min with one-on-one online companion and learning.

2.2 TTQS

TTQS is the system which focuses on the quality of service. The Ministry of Labor (formerly the Labor Council) had cited ISO10015, the European vocational training policy, the British IIP talent investment plan, and the Australian active vocational training policy since 2005. Thereafter, the Workforce Development Agency developed a Training Quality System (Taiwan Train Quali System, TTQS) for the training program, design, implementation, check, and results evaluation [12]. In recent years, the Workforce Development Agency has developed a Talent Development Quality Management System (TTQS) for the program to ensure reliability and validity of the training process. The structure of TTQS mainly follows Plan, Design, Do, Review and Outcome (PDDRO) with 18 items for evaluation. TTQS effectively enhances the institutional human capital and this strengthens it to be competitive internationally [13–15]. In e-learning partner project, the teaching quality from the university's side should be guided by a stable system. Therefore, the checklist of TTQS will be applied into the process management.

2.3 Information Visualization

Visual Information systems usually appeal to active methodologies that develop logical reasoning, critical thinking and problem solving skills. An evidence-based decision making system is widely using Tableau visual system such as healthcare, business, accounting, government, science, security or disaster management, taking the right decision based on the available data at the right time matters much in ensuring its success [16–18]. According to some literatures [19–22], visualization tools and complex analyses can be performed in fast decision making. In the e-learning partner project, the host of the project can detect the feedback easily and efficiently through the Business intelligent software of Tableau. After the survey during the three semesters, the trend and problem can be found in visualization analyses.

3 Methodology

3.1 Samples and Procedure

In the first step, the TTQS checklist in e-learning partner project was created with the five PDDRO indexes. In the second step, data was collected from 276 undergraduates from the teaching side in the case university during three semesters for the academic years of 2018–2019 using the on-line feedback survey. The questionnaire which was designed by MOE included self-evaluation, administration assessment and IT support. When the participants agreed to participate in the MOE survey, they were required to fill the questionnaire. The government declared that it will be only for the use of academic institutions and not for business purpose on the front page of the survey. All the participants were blind to the hypotheses of the study and MOE platform questionnaire did not provide us with the personal details of the participants, so their names were kept anonymous. Therefore, it fitted the ethical rules. Using Tableau, evidence-based results were offered with visualization to the host of the e-learning partner. Further, we carried out basic statistical analyses, ANOVA, post-hoc analysis and information visualization in this study.

3.2 Measures and Variables

The undergraduates answered the survey using a five-point scale; the options were "strongly agree," "agree," "neutral," "disagree," and "strongly disagree." The scoring order was 5, 4, 3, 2 and 1 point, respectively. The independent variable was the semester. The dependent variable were the scores of self-evaluation, administration assessment and IT support from the behaviour assessment survey after 10 weeks. Three academic years were coded as 1 (2018/07/31), 2 (2019/01/31) and 3 (2019/07/31). The 28 independent variables were coded as follows: Willing (W1–W7); Ability (A1–A7); Staff (S1–S9); Information technology support (IT1–IT5).

4 Results

According to TTQS operating standard of training quality from Bureau of Employment and Vocational Training [23], TTQS checklist can be shown in Table 1. In the fourth column, there are relative factors between the 18 TTQS items and the management indicators in the e-learning partner project following the MOE rules in Taiwan. The star (*) implies the weakness or shortage of the project with respect to the case university for decision makers. In Table 2, only the mean of IT3 and IT4 was lower than 4.0. But the support level was growing gradually from the first semester to the third semester.

Table 1. TTQS checklist in e-learning partner project

Aspects	Emphatic facets	Evaluation items	Relative factors in this research
Plan	Focus on relativity and the practice of training plan and institution's goal	1.Explicitness 2.Systematic 3.Connections 4.Ability	1. MOE goal: life accompanying life, and living teaching living 2.Standard Operational Process of the project 3.KPI of the project 4.Competency of Host, Assistant, Tutors, Administrative and Academic Support
Design	Focus on the design of training system (including the involvement of beneficial party, standard of course selections, purchase standard)	5.Selection standard of training products and services 6.The involvement of beneficial party 7.Combination of training and targets 8.Systematic design of training	5.Human Resources from Educational Program, IT 6.Tutors experience sharing 7.Post requirement questionnaire (*) 8.SOP of training schedule and instructors
Do	Emphasize the process of training, records and the standardization of management	9.Purchase process and standardization of training products and services 10.Whether the training follows the plan 11.Records and information system	9.1Undergraduates evaluations 9.2 Teaching materials 9.3 Instructors 9.4 Pedagogy 9.5 IT platform 10.Satisfactory Questionnaire Survey (*) 11.Records in e-tutor platform in MOE
Review	Focus on periodical analysis, monitoring and reaction to accidents	12.Evaluation reports and periodical analysis 13.Monitoring and reactions	12.1 Monthly reports 12.2 Middle requirement survey (*) 13.Teaching video sampling and daily record reply
Outcome	Focus on layers, completeness, and sustained improvement of training evaluation	14.Diversity and completeness of training evaluation 15.Results of trainees 16.Extension Effects of training 17.Results of special training 18.Others	14.Behavior Assessment 15.Retention (*) 16.1 Social Responsibility 16.2 Economy 16.3 Humanity (*) 17. Stories in real meeting 18. School Popularity, Marketing

4.1   Basic Statistics

See Table 2.

Table 2. Basic statistics of behavior assessment

Variable	Item	Academic year					
		1		2		3	
		Mean	SD	Mean	SD	Mean	SD
W1	I am willing and actively involved in education and training	4.59	0.60	4.33	0.68	4.52	0.70
W2	I am willing to prepare lessons according to the needs of school children	4.70	0.56	4.59	0.56	4.73	0.49
W3	I am willing to abide by the intellectual property rights to make teaching materials	4.87	0.37	4.79	0.44	4.86	0.43
W4	I will take classes on time and take the initiative to take time off	4.92	0.32	4.82	0.46	4.88	0.38
W5	I am willing to actively participate in the teaching and counseling discussion meeting	4.69	0.49	4.54	0.61	4.69	0.56
W6	I am willing to actively participate in the event	4.65	0.61	4.66	0.59	4.71	0.61
W7	I am willing to continue to participate in several student programs	4.45	0.84	4.24	0.90	4.23	1.05
A1	The project improved my IT application and literacy (information security, intellectual property rights)	4.40	0.74	4.21	0.99	4.32	0.85
A2	The project improved my teaching material production ability	4.65	0.65	4.54	0.62	4.55	0.64
A3	The project improved my communication ability	4.60	0.58	4.67	0.56	4.55	0.64
A4	The project improved my problem solving ability	4.57	0.59	4.54	0.58	4.39	0.71
A5	The project improved my morality (punctuality, responsibility, altruism, good deeds)	4.67	0.52	4.54	0.62	4.69	0.58
A6	The project improved my empathy	4.63	0.53	4.56	0.60	4.61	0.60
A7	The project improved my social participation ability	4.58	0.61	4.56	0.56	4.55	0.67
S1	The education and training planned by our school team helped me	4.63	0.49	4.47	0.67	4.50	0.66
S2	The needs of the schoolchildren provided by the school team helped me	4.52	0.59	4.33	0.73	4.62	0.62
S3	The preparation materials provided by the school team helped me	4.43	0.73	4.31	0.73	4.64	0.61

(*continued*)

Table 2. (*continued*)

Variable	Item	Academic year					
		1		2		3	
		Mean	SD	Mean	SD	Mean	SD
S4	The teaching guidance provided by the school team helped me	4.53	0.65	4.46	0.64	4.61	0.62
S5	The teacher with the class really helps to confirm the status of the elementary school	4.76	0.46	4.71	0.53	4.83	0.40
S6	The class teacher does help with network obstacles	4.73	0.47	4.71	0.55	4.82	0.41
S7	The teacher with the class really assists in the troubleshooting of hardware and software equipment	4.74	0.49	4.70	0.53	4.84	0.40
S8	The class teacher did reply to the teaching diary	4.74	0.47	4.66	0.58	4.77	0.47
S9	The class teacher does counsel the teaching problems and tracking	4.71	0.51	4.57	0.62	4.77	0.45
IT1	The diary filling function meets the usage requirements (click on the screen after the university is accompanied by the classroom)	4.55	0.73	4.47	0.71	4.61	0.63
IT2	Textbook upload function meets the needs of use	4.56	0.76	4.49	0.67	4.61	0.68
IT3	JoinNet platform operation is stable	**3.79**	1.08	**3.96**	0.97	4.37	0.83
IT4	Computer classroom network connection is stable	**3.99**	0.85	4.10	0.85	4.27	0.86
IT5	Computer classroom hardware equipment is stable	4.22	0.82	4.19	0.83	4.38	0.74

4.2 ANOVA and Post Hoc Tests

From Table 3, W1 (I am willing and actively involved in education and training), S2 (The needs of the schoolchildren provided by the school team helped me), S3 (The preparation materials provided by the school team helped me), S9 (The class teacher does counsel the teaching problems and tracking), and IT3 (JoinNet platform operation is stable) are significant variables. In the post hoc test of Scheffe method, W1 in the first semester was significantly higher compared to the second semester. The scores in S2, S3 and S9 were significantly better in the third semester compared to the second semester. About the IT support (IT3), it was significantly different in the third semester among three semesters.

Table 3. ANOVA and post hoc tests of semesters and behavior assessment

		Sum of squares	d.f.	Mean sum of squares	F	Sig.	Post hoc test
W1	Between groups	3.194	2	1.597	3.582	0.029*	1 > 2
	Within groups	121.716	273	0.446			
	Total	124.909	275				
S2	Between groups	3.983	2	1.991	4.727	0.010*	3 > 2
	Within groups	115.013	273	0.421			
	Total	118.996	275				
S3	Between groups	5.296	2	2.648	5.586	0.004*	3 > 2
	Within groups	129.410	273	0.474			
	Total	134.707	275				
S9	Between groups	2.034	2	1.017	3.674	0.027*	3 > 2
	Within groups	75.543	273	0.277			
	Total	77.576	275				
IT3	Between groups	16.820	2	8.410	9.207	0.000*	3 > 1; 3 > 2
	Within groups	249.365	273	0.913			
	Total	266.185	275				

Note: *p-value is significantly smaller than 0.05

4.3   Visualization

Total scores of 28 items in self-evaluation, administration assessment and information technology support were transferred into Tableau software and trends for three semesters were observed using dashboards shown in Fig. 2. The lowest mean among

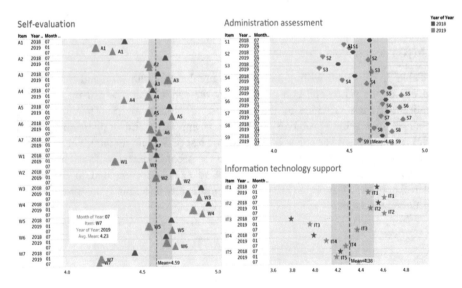

Fig. 2. Dashboard of behavior assessment

three indexes was for IT support (mean = 4.38). However, it was seen to be growing and improving gradually step by step. One problem was about self-evaluation related to the undergraduates' willingness to continue to participate in several student programs in W7. Although the average score was still above 4.0, the trend decreased from the first semester to the third semester.

5 Discussion

5.1 TTQS

In the (Key Performance Indicators) KPI of the e-learning partner project, there were 23 items including the number of primary school students, the number of undergraduates, the frequency and time of e-learning partner in computer classes, total time of training courses, number of training video upload, environment set of learning side, the learners' need survey, management system set of tutoring, preparing the teaching materials and set of undergraduates' counselling system, diary reply and sampling system, the selection of excellent teaching materials, real meeting activity, monthly reports upload, monthly reports-sampling and observation, monthly reports-excellent teaching videos upload, monthly reports-warm stories, questionnaires survey, e-learning platform training, participate in national e-learning partner project conferences and seed teachers training, participating in the real observation of tutors in partner universities, participating in the workshop of assistants, participate in national e-learning partner project conferences and sharing seminars of excellent models. From the previous plan, design, process do, review and post outcome, the TTQS can be carried out to detect the teaching quality of e-learning partner project well. In the result, the questionnaire of undergraduates proved very important and necessary in the case university during design, do and review stages. In recent three semesters, the behavior assessments were filled and discussed only in the outcome stage. For the goal of MOE, the quality management of teaching and training played heavy roles. Therefore, the suggestion can be offered to the host of e-learning partner project to improve in the next semester.

5.2 Retention

Through visualization graphs, the dashboard showcased the challenges for the decision maker. In this research, the hosts told us the retention rate of undergraduates was low in the case university during the three semesters. It's similar to the result of information visualization. Retention rates were used for the purpose of institutional accountability and institutional management [24, 25]. According to the behavior assessment, the scores of willingness to continue to participate in several student programs (W7) was decreasing year by year. There were many factors which affected the retention of teaching and interaction with primary school students. In the further study, the mid-term requirement questionnaire of undergraduates could be designed and executed. Then the cost of human resources training could be declined.

6 Conclusions

Good quality management in education can improve the performance of training. In the present study, TTQS was carried out to explore the effect of the e-learning partner project with 18 indexes. Through the diagnosis among self-evaluation, administration assessment and information technology support of undergraduates during three semesters at the case university, we detected the hidden problem and offered efficient visualization results to the hosts of e-learning partner projects. The factor of retention is necessary to be discussed in the future. Questionnaire design and implementation in the mid-term stage was also important. From the suggestions and improvement, decreasing the gap of undergraduates' expectations, raising institutional performance and promoting sustainability in e-learning partner project can be achieved.

References

1. The Ministry of Education: E-learning partner plan (2016). https://etutor.moe.gov.tw/zh/about/begin
2. Shuai, J.J., Jhou, J.H.: The key successful factors for after school's online tutoring. Minghsin J. **40**(1), 141–161 (2014)
3. Lee, Y.B., Ho, R.G.: The use of internet to reduce the digital divide: case study of e-learning partner. Taiwan Educ. Rev. **670**, 2–11 (2011)
4. Lin, L.J., Lin, H.Y., Lu, T.H.: A construction of social engagement index for university students in online tutoring services. Res. Educ. Commun. Technol. **108**, 17–29 (2014)
5. ChanLin, L.J., Lin, H.Y., Lu, T.H.: Online after-school learning for bridging the digital divide. In: Proceedings of 12th IEEE International Conference on Advanced Learning Technologies (ICALT 2012), pp. 436–437, Rome, Italy (2012)
6. Chen, Q., Kao, T.C.: Applying the flipped classroom instructional model to rural online tutoring program in upper elementary mathematics. Contemp. Educ. Res. Q. **27**(2), 01–37 (2019)
7. Elakovich, D.M.: Does a student's use of self-regulation change in the flipped classroom? (Unpublished doctoral dissertation). Montana State University, USA (2018)
8. Lo, C.K., Hew, K.F.: Using, "first principles of instruction" to design secondary school mathematics flipped classroom: the findings of two exploratory studies. Educ. Technol. Soc. **20**(1), 222–236 (2017)
9. Lu, C.H., Lin, H.Y.: The implementation mode and strategy of distance education counseling in primary and secondary schools in rural areas-take FuJen University's long-distance course for example. In: Conference on Computer and Network Technology in Education (2008)
10. Lin, C.H., Li, W.U.: A study of relationship between teaching efficacy and teaching satisfaction in the university students joining distance learning companion project. J. Cheng Shiu Univ. **31**, 43–64 (2018)
11. Liu, W.T., Lyau, N.M., Lin, C.H.: Construction of a remedial instruction effectiveness evaluation model for online after-school tutoring: an integrated perspective of students' learning motivation. J. Educ. Res. **49**(2), 65–94 (2015)
12. Lin, S.G.: Comparative study of TTQS and ISO training quality system. TTQS E-J. **9** (2016). https://ttqs.wda.gov.tw/EDocs/2016009/%E7%9F%A5%E8%AD%98%E8%AC%9B%E5%A0%822.html

13. Hsieh, S.C., Lin, J.S., Lee, H.C.: Analysis on literature review of competency. Int. Rev. Bus. Econ. **2**, 25–50 (2012)
14. Lo, C.L., Tsai, C.Y., Lan, C.C., Lin, M.H.: Implementing the Taiwan TrainQuali System (TTQS): a case study of a pharmaceutical company. World Trans. Eng. Technol. Educ. **19**, 233–238 (2011)
15. Shi, M.F.: Application of TTQS to manage the internship quality of vocational school students. Eval. Bimon. **37**, 45–48 (2012)
16. Hoelscher, J., Mortimer, A.: Using Tableau to visualize data and drive decision-making. J. Account. Educ. **44**, 49–59 (2018)
17. Ko, I., Chang, H.: Interactive visualization of healthcare data using Tableau. Healthc. Inform. Res. **23**(4), 349–354 (2017)
18. Riley, M., Harelson, D., Monte, M., Chong, M., O'Hern, M., Smith, M., White, K.P.: Criminal justice system data analysis and visualization. In: Systems and Information Engineering Design Symposium (SIEDS), pp. 295–299. IEEE (April 2015)
19. Haara, A., Pykäläinen, J., Tolvanen, A., Kurttila, M.: Use of interactive data visualization in multi-objective forest planning. J. Environ. Manag. **210**, 71–86 (2018)
20. Keim, D., Huamin, Q., Kwan-Liu, M.: Big-data visualization. IEEE Comput. Graph. Appl. **33**(4), 20–21 (2013)
21. Kung, S.-Y.: Visualization of big data. In: IEEE 14th International Conference on Cognitive Informatics & Cognitive Computing (ICCI* CC), pp. 447–448 (2015)
22. Medeiros, C.P., Alencar, M.H., Almeida, A.T.: Hydrogen pipelines: enhancing information visualization and statistical tests for global sensitivity analysis when evaluating multidimensional risks to support decision-making. Int. J. Hydrog. Energy **41**(47), 22192–22205 (2016)
23. Bureau of Employment and Vocational Training: 2011 TTQS plan: Operating standards, Taipei (2011)
24. Gold, L., Albert, L.: Graduation rates as a measure of college accountability. Am. Acad. **2**(1), 89–106 (2006)
25. Hovdhaugen, E., Frolich, N., Aamodt, P.O.: Informing institutional management: institutional strategies and student retention. Eur. J. Educ. **48**(1), 165–177 (2013)

The Application of Traveling Gravity Model on International Tourism

Chi-Yueh Hsu[1], Bo-Jun He[1], and Jian-Fa Li[2($\boxtimes$)]

[1] Department of Leisure Services Management,
Chaoyang University of Technology, Taichung City, Taiwan
cyhsu@gm.cyut.edu.tw, gp222444777@gmail.com
[2] Department of Finance, Chaoyang University of Technology,
Taichung City, Taiwan
jfli@cyut.edu.tw

Abstract. As a kind of tertiary industry, tourism plays an invisible role of best diplomacy. Tourism is not only a policy with instant effect, but also becomes a crucial economic resource. Most nations put emphasis on tourism development and how to appeal international tourists is one of the tasks. This study used gravity model for tourism demand, derived from Newton's law of gravitation, as empirical model to examine across 9 countries from 2003 to 2016. To investigate factors that influence international tourists visiting Taiwan, the longitudinal data was collected and estimated by classical linear regression model, fixed effects regression model, and random effect regression model. According to the study, it indicates that the GDP of outbound country has significant negative effect to Taiwan tourism, fare from Taiwan to the foreign country has significant negative impact, the number of outbound travelers has significant positive effect, and hardware facilities, number of hotel rooms has significant positive impact. Differed from previous studies, this study shows that the GDP of foreign country has negative impact on Taiwan tourism.

Keywords: Outbound tourists · Gravity model for tourism demand · Fixed effects model · Random effects model

1 Introduction

1.1 Background and Motivation

As a kind of no-chimney industry, tourism is play an invisible role of best diplomacy. Liu (2012), Tourism is also "a kind of self-expression. Séraphin et al. (2018) point out that tourism is to some degree important to regional economic development that not just stimulates tourist areas, attractions, and recreation and entertainment related but bridges the gap between urban and the rural. Tourism not only is a development policy with instant effect, but also becomes a significant economic resource. Li et al. (2018) investigated the essential of tourism which are accommodation, food, and quality. Facilities like infrastructure and transportation are indispensable consideration as developing tourism. Not only are environmental and cultural features themselves more attractive, but also various industries and facilities combined are needed if an area or a country develops tourism. So far, the number of outbound tourists (especially in Asia-Pacific area) has increased rapidly. United Nations World Tourism Organization (UNWTO) predicts that the average growth of international tourist comes to 3.3% in 2010–2030 that it will be 1.4 billion people in 2020 and 1.8 billion in 2030. It is obvious that tourism is not just a new trend but also a key factor of global economic development (UNWTO 2018).

Because of great potential, Taiwanese tourism would be both the mainstay of domestic demand and the best way to enhance international communication (UNWTO 2016). World Travel & Tourism Council (WTTC) predicts that total value of tourism will reach NT$1.2 trillion which will create 833 thousand job opportunities (WTTC 2016). UNWTO also gives another prediction that the number of tourists in Asia will grow by 102% by next 10 years.

In terms of Taiwan tourism development, in fact, 14,795 tourists had visited Taiwan in 1956 when Taiwan was just open. More than 1 million tourists visited in 1976. Over 3 million tourists visited Taiwan in 2005, composed of Japanese tourists at 33% which is over 1 million people and Americans visitors at 12% (Taiwan Visitors Association 2006; Taiwan Tourism Bureau 2010). Total tourism income in 2009 was 11.8 billion dollars which is 2.92% of Taiwanese GDP, including 5.9 billion dollars foreign exchange earnings (Taiwan Tourism Bureau 2009). Total value of tourism grew to 25.67 billion dollars in 2016 which is 4.84% of Taiwanese GDP, including 13.37 billion dollars (Taiwan Tourism Bureau 2016). Besides, China, Japan, Singapore, US, and Hong Kang were top 5 mainstays origins of tourism revenue in 2008–2016 (Taiwan Tourism Bureau 2017). Moreover, Chinese tourists become the biggest source that have brought great economic effects in 2008–2017 (Taiwan Tourism Bureau 2017).

In order to appeal to tourists from mainland China, related regulations have been amended. For example, 2011 tourism policy comprehensively open to Chinese free individual travelling not only enhanced Taiwan attraction to Chinese tourists but stimulated number of Chinese tourists that went up to almost 4 million in 2014 of 9,910,204 and became the biggest outbound tourists market. Hong et al. (2016) pointed out that Chinese tourists have greater and greater impact on Taiwan economics, especially on tourism industry. Number of international tourists visiting Taiwan grew to 10,739,601 in 2017, and next year Taiwan government strongly put forward tourism sustainable development concept called "Tourism 2020" which put emphasis on environmentally, socially, and economically sustainable development (Taiwan Tourism Bureau 2018). These acts did bring about great influence in Taiwan tourism industry (Taiwan Tourism Bureau 2017).

Chinese tourists formally joint tourism battle in 2008 and have grown by leaps and bounds. Mei and Chang (2010) empirically showed the evidence that the positive economic benefit the policy comprehensively open to Chinese tourists brought about was much more than those brought by tourists from Japan, Hong Kang and Macao or other countries. However, number of Chinese tourists dramatically dropped during 2014–2016 due to Taiwan regime change and freezing cross-strait relation (Taiwan Tourism Bureau 2017). As a result, the government carried out new tourism policy toward Southeast Asian countries (Taiwan Tourism Bureau 2017).

The study uses the gravity model for tourism demand to explore the factors that affect outbound tourists visiting Taiwan (Zipf 1946; Stewart 1948). Morely et al. (2014) proposed using bilateral gravity model to discuss the distance between two countries. Wu et al. (2009) empirically showed that round-trip ticket fare has negative influence while total population, GDP per capita, and number of outbound travelers all have positive impact. Guo (2007) gave a complete explanation of gravity model application on tourism and it came out that similar economic development and certain distance affects tourism development most after examination on tourist given market. Ruijuan et al. (2007) used gravity model to establish model analyzing area tourism factors.

Despite previous studies on the application of gravity model on nation's tourism. There is few study employing the gravity model to investigate the tourism demand in Taiwan. Therefore, In the present study, we aim to explore the determinants of the international tourists visiting Taiwan by employing the gravity model of tourism demand.

2   Empirical Model

In the present study, we employ the gravity model for tourism demand adopted Ma and Lee (1999) and add other variables to examine the factors of affecting Taiwan tourism development through the international tourism.

2.1   Hypothesis

The hypothesis of this study is stated as follows:

H1.1 the number of outbound tourists has positive relationship with average of inbound tourists.

H1.2 GDP of outbound country has positive relationship with average of inbound tourists.

H1.3 Number of hotel rooms has positive relationship with average of inbound tourists.

H1.4 fare from Taiwan to a country has positive relationship with average of inbound tourists.

H2.1 the number of outbound travelers has positive relationship with total expenditure of inbound tourists.

H2.2 GDP of outbound country has positive relationship with total expenditure of inbound tourists.

H2.3 number of hotel rooms has positive relationship with total expenditure of inbound tourists.

H2.4 fare from inbound country to outbound country has positive relationship with total expenditure of inbound tourists.

H3.1 the number of outbound travelers has positive relationship with total inbound tourists.

H3.2 the GDP of foreign country has positive impact on inbound tourists.

H3.3 number of hotel room has positive relationship with inbound tourists.

H3.4 fare from inbound country to outbound country has positive impact on inbound tourists.

2.2 Regression Model

The earliest gravity model for tourism demand based on Newton's law of gravity is illustrated as (Stewart 1948)

$$I = \frac{P1P2}{D} \tag{1}$$

where I stands for attraction index, P1 and P2 are population for cities 1 and 2, respectively. D denotes the distance between 2 cities. This model was amended and became the common and classical model as below (Crampon 1966):

$$T_{ij} = G\left(P_i A_j \big/ D_{ij}^b\right) \tag{2}$$

where T_{ij} represents time of visitors traveling from area i. and area j; P_i is the measure of population, capital, or travel preference of origins where visitors came from i;

A_j is measure of attraction index or capacity of the destination j; D_{ij} is the distance between origins where visitors came from country I and country j; G and b is slope coefficient.

Ma and Lee (1999) used population of particular origins, GDP per capita, round-trip ticket fare, total travel expenditure per capita, and travel price to establish inbound travel gravitational model

$$X_{it} = g_i\left(x_{it-1}^{\mu i} y_{it-1}^{\alpha i} z_{it-1}^{\beta i} P_{it-k}^{\theta i}\right) \big/ W it^{\nu i} (w_{it} = u_{it} + v_{it}) \tag{3}$$

where X_{it} denotes tourism demand represents number of tourists visiting the country i in year t; y_{it} is population of origin; P_{it} denotes number of inbound tourists; z_{it} is GDP per capita, u_{it} is round-trip ticket fare, v_{it} represent total travel expenditure per capita, g_i, μ_i, α_i, β_i, ν_i, and θ_i are slope coefficients.

This research used variables of Ma and Lee (1999) model as analyzed foundation and amended to predict factors and influence of 9 main origins visiting Taiwan. The new model this research used is:

$$\begin{aligned}
\mathbf{TBL_{it}} &= \mathbf{g_i}\left(\mathbf{ROO}_{it^{\mu i}} + \mathbf{TIC}_{it^{Vi}} + \mathbf{GDP}_{it^{\beta i}} + \mathbf{OUT}_{it^{pi}}\right) \\
\mathbf{TOT_{it}} &= \mathbf{g_i}\left(\mathbf{ROO}_{it^{\mu i}} + \mathbf{TIC}_{it^{Vi}} + \mathbf{GDP}_{it^{\beta i}} + \mathbf{OUT}_{it^{pi}}\right) \\
\mathbf{CON_{it}} &= \mathbf{g_i}\left(\mathbf{ROO}_{it^{\mu i}} + \mathbf{TIC}_{it^{Vi}} + \mathbf{GDP}_{it^{\beta i}} + \mathbf{OUT}_{it^{pi}}\right)
\end{aligned} \tag{4}$$

where i represent the particular country, t denotes year; TBL_{it} is ratio of inbound tourist from outbound country i to total Taiwan inbound tourist in year t; TOT_{it} is total inbound tourists from outbound country i in year t; CON_{it} is total expenditure per capita from outbound country i in year t; OUT_{it} stands for tourists from outbound country i in year t. origin; GDP_{it} is GDP at country i in year t. origin; ROO_{it} is number of hotel rooms from country i in year t. origin; TIC_{it} stands for fare from Taiwan to outbound country i in year t; g_i, β_i, p_i, v_i, and μi are slope coefficient.

3   Empirical Results

3.1   Data Source

The present study took 9 countries of international tourists as subjects, including China, South Korea, US, Japan, Singapore, Hong Kang and Macao, New Zealand, and Australia. Due to the limitation of longitudinal data collection, study time setting of this study was 2003 to 2016. Although this research takes tourism industry, one of Taiwan important economic resource, as subject, based on Taiwan tourism bureau research survey, are only taken as objects to carry out longitudinal data collection of variables. The number of tourists visiting Taiwan was tourism yearbooks from Tourism Bureau, MOTC. The population was from Knoema-World and regional statistics, national data, maps, rankings. The longitudinal data of GDP of outbound country was public data and global stock market economic data from Google. The number of outbound travellers was from China statistical yearbooks, Knoema, South Korean tourism organization, and analyzation of Japanese outbound tourism market. The longitudinal data of each fare from Taiwan to each country, which represents the variable of distance, was annual average fare from Star travel Agency statistics. Total expenditure of each country tourists was from tourist expenditure and trend survey conducted by Tourism Bureau, MOTC. The number of hotel rooms, which represents variables of hardware facilities, was from Tourism Bureau, MOTC monthly report of operation of home stay, hotel, and tourism hotel.

3.2   Results from the Regression Analysis

The estimation is based on classical linear regression model, fixed effects regression model, and random effect regression model to explore the relationship between factors that affect international tourists visiting Taiwan and Taiwan tourism development, and discussed the result into three sections: TBL as average of inbound tourists from outbound country, CON as the expenditure of inbound tourists, and TOT as total inbound tourists. The empirical result is the following:

Table 1 reports each country number of outbound travelers, GDP of outbound country, number of hotel rooms, and fare from Taiwan to each country were independent variables while average of total tourists from each country was dependent variable analyzed by classical linear regression analysis. There was significant difference between average number of each country tourists and each country number of outbound travelers, GDP of outbound country, and fare from Taiwan to each country separately ($p < 0.01$, $p < 0.1$, adjusted $R^2 = .24$), and t statistic were 3.39, −1.71, and −2.97 separately. This result showed that each country number of outbound travelers has positive relation with average number of tourists from each country while fare from Taiwan to each country has negative relationship with average number of each country tourist, which was consistent with Wu et al. (2009) that the more outbound travelers is, the more average number of each country tourists increases while the higher fare from Taiwan to each country and the longer distance is, average number of each country tourists decreases. In term of GDP of outbound country, the result was not consistent to Wu et al. (2009) that increasing GDP of outbound country brings about tourism income effect. It was a negative relation that tourism choices and expectation of resorts increase that Taiwan is not the first choice anymore when the GDP of outbound country increases.

Table 1. Estimation based on classical linear regression model (dependent variable: average of inbound tourists)

Variable	Coefficient	Standard error	t statistic
(Intercept)	9.9975e − 02***	3.6035e − 02	2.7744
(OUT)	8.6528e − 06***	2.5526e − 06	3.3898
(GDP)	−5.5793e − 07 *	3.2606e − 07	−1.7111
(ROO)	1.1207e − 07	7.7137e − 08	1.4528
(TIC)	−2.3874e − 06***	8.0336e − 07	−2.9718

"***", "**", and "*" denote significance at the 1%, 5%, and
10% levels, respectively.
Total Sum of Squares: 1.0528
Residual Sum of Squares: 0.76761
R-Squared: 0.2709
Adjusted R-Squared: 0.23954
F-statistic: 8.63855, p-value: 5.6635e − 06

As shown in Table 2, each country number of outbound travelers, GDP of outbound country, number of hotel rooms, and fare from Taiwan to each country as independent variables while average of outbound tourists as dependable variable analyzed by fixed effect regression analysis. There was significant difference between total tourists from each country and each country number of outbound travelers and fare from Taiwan to each country (p < 0.01, p < 0.1, adjusted R^2 = 0.17) and t statistic were 4.90 and −1.67 separately. The result showed that each country number of outbound travelers has positive relation to average of total tourists from each country and fare from Taiwan to each country has negative relation to average of total tourists from each country, which is consistent with those in Wu et al. (2009) that the more each country number of outbound travelers, the more average of total tourists from each country increases while the higher fare from Taiwan to each country and the longer distance is, average number of total tourists from each country decreases.

Table 3 illustrates outbound travelers, GDP, the number of hotel rooms, and fare from Taiwan to outbound country as independent variables while average number of total international tourists as dependable variable analyzed by random effect regression model. There was significant difference between total tourists from each country and

Table 2. Estimation based on fixed effect regression model (dependent variable: average of inbound tourists)

Variable	Coefficient	Standard error	t statistic
(OUT)	1.3090e − 05***	2.6707e − 06	4.9012
(GDP)	−7.6987e − 07	5.6259e − 07	−1.3685
(ROO)	1.0712e − 07	7.0451e − 08	1.5205
(TIC)	−2.2349e − 06*	1.3393e − 06	−1.6687

"***", "**", and "*" denote significance at the 1%, 5%, and 10% levels, respectively.
Total Sum of Squares: 0.45014
Residual Sum of Squares: 0.33373
R-Square: 0.25861
Adjusted R-Squared: 0.17339
F-statistic: 7.58676, p-value: 2.7238e − 05

Table 3. Estimation based on random effect regression model (dependent variable: average number of inbound tourists)

Variable	Coefficient	Standard error	t statistic
(Intercept)	8.9943e − 02*	4.7502e − 02	1.8934
(OUT)	1.2803e − 05***	2.5816e − 06	4.9592
(GDP)	−7.3128e − 07	5.0938e − 07	−1.4356
(ROO)	1.0481e − 07	6.6122e − 08	1.5851
(TIC)	−2.1905e − 06*	1.2167e − 06	−1.8003

"***", "**", and "*" denote significance at the 1%, 5%, and 10% levels, respectively.
Total Sum of Squares: 0.46688
Residual Sum of Squares: 0.34653
R-Squared: 0.25778
Adjusted R-Squared: 0.22586
F-statistic: 8.07491, p-value: 1.2396e − 05

each country number of outbound travelers and fare from Taiwan to each country ($p < 0.01$, $p < 0.1$, adjusted $R^2 = 0.23$) and t statistic were 4.96 and −1.8 separately. The result showed that each country number of outbound travelers has positive relation to average of outbound tourists and fare from Taiwan to outbound country has negative relation to average of inbound tourists, which is consistent with Table 2 under fixed effect regression analysis.

In Table 4, there was significant difference ($p < 0.01$, $p < 0.05$, adjusted $R^2 = 0.41$) and t statistic were 3.59, −3.28, 5.74, and −2.53 separately. It showed the evidence that number of hotel rooms has positive relation to total expenditure of inbound tourists from each country and fare from Taiwan to each country has negative relation to average of inbound tourists, which is consistent with Wu et al. (2009) that

Table 4. Estimation based on classical linear regression model (dependent variable: total expenditure of inbound tourists)

Variable	Coefficient	Standard error	t statistic
(Intercept)	−8.9993e + 07	6.0922e + 07	−1.4772
(OUT)	1.5495e + 04***	4.3155e + 03	3.5907
(GDP)	−1.8062e + 03**	5.5125e + 02	−3.2765
(ROO)	7.4744e + 02***	1.3041e + 02	5.7316
(TIC)	−3.4417e + 03***	1.3582e + 03	−2.5340

"***", "**", and "*" denote significance at the 1%, 5%, and
10% levels, respectively.
Total Sum of Squares: 3.8491e + 18
Residual Sum of Squares: 2.194e + 18
R-Squared: 0.43
Adjusted R-Squared: 0.40548
F-statistic: 17.5391, p-value: 9.3425e − 11

the more number of outbound travelers is, the more expenditure of inbound tourists increase, and the hardware facilities and number of hotel rooms appeal tourists to visit Taiwan and stimulate consumption while the higher fare from Taiwan to outbound country and the longer distance is, the more unwilling tourists are on consumption. In term of each country GDP, the result didn't agree with Wu et al. (2009) that increasing GDP of outbound country brings about tourism income effect.

In Table 5, each country number of outbound travelers, the GDP of outbound country, number of hotel rooms, and fare from Taiwan to each country as independent variables while total expenditure of outbound tourists as dependable variable analyzed by fixed effect regression analysis. There was significant difference between total expenditure of outbound tourists from each country and each country number of outbound travelers, the GDP of outbound country, and number of hotel rooms ($p < 0.01$, $p < 0.1$, adjusted $R^2 = 0.44$) and t statistic were 5.54, −1.75, and 5.44 separately. The result showed that each country number of outbound travelers and number of hotel rooms has positive relationship to total expenditure of tourists from each country, which is consistent with Wu, Huang, and Lee's research (2009) that the more number of outbound travelers is, the more expenditure of tourists from each country increase, and the hardware facilities and number of hotel rooms not just meeting accommodation demand but combining with local resource appeal tourists to visit Taiwan and stimulate consumption. In term of the GDP of outbound country, the result didn't correspond to Wu et al. (2009) that increasing GDP brings about tourism income effect.

Table 6 reports number of outbound travelers, the GDP of outbound country, number of hotel rooms, and fare from Taiwan to each country as independent variables while total expenditure of outbound tourists as dependable variable analyzed by random effect regression model. There was significant difference between total expenditure of international tourists and number of outbound travelers, the GDP of outbound country, and number of hotel rooms ($p < 0.01$, $p < 0.1$, adjusted $R^2 = 0.27$) and t

Table 5. Estimation based on fixed effect model (dependent variable: total expenditure of inbound tourists)

Variable	Coefficient	Standard error	t statistic
(OUT)	26326.83***	4748.38	5.5444
(GDP)	−1746.98*	1000.26	−1.7465
(ROO)	681.95***	125.26	5.4442
(TIC)	−2377.44	2381.31	−0.9984

"***", "**", and "*" denote significance at the 1%, 5%, and 10% levels, respectively.
Total Sum of Squares: 2.0996e + 18
Residual Sum of Squares: 1.055e + 18
R-Squared: 0.49755
Adjusted R-Squared: 0.43979
F-statistic: 21.5376, p-value: 2.2522e − 12

Table 6. Estimation based on random effect model (dependent variable: total expenditure of inbound tourists)

Variable	Coefficient	Standard error	t statistic
(Intercept)	−1.2686e + 08	7.8469e + 07	−1.6166
(OUT)	2.5484e + 04***	4.5766e + 03	5.5683
(GDP)	−1.7736e + 03*	8.8927e + 02	−1.9944
(ROO)	6.8916e + 02***	1.1654e + 02	5.9134
(TIC)	−2.4923e + 03	2.1258e + 03	−1.1724

"***", "**", and "*" denote significance at the 1%, 5%, and 10% levels, respectively.
Total Sum of Squares: 2.1618e + 18
Residual Sum of Squares: 1.1004e + 18
R-Squared: 0.49096
Adjusted R-Squared: 0.46907
F-statistic: 22.4243, p-value: 5.5084e − 13

statistic were 5.57, −1.99, and 5.91 separately. The result showed that each country number of outbound travelers and number of hotel rooms has positive relation to total expenditure of outbound tourists from each country while GDP of outbound country has negative relation, which correspond to Table 5 under fixed effect regression model.

As shown Table 7, each country number of outbound travelers, GDP of outbound country, number of hotel rooms, and fare from Taiwan to each country as independent variables while total international tourists as dependable variable analyzed by classical linear regression analysis. There was significant difference ($p < 0.01$, adjusted $R^2 = 0.46$) and t statistic were 4.70, −2.89, 6.18, and −2.72 separately. The result showed that number of outbound travelers and number of hotel rooms has positive relation to total expenditure of tourists from each country while fare from Taiwan to

Table 7. Estimation based on the classical linear regression model (dependent variable: total inbound tourists)

Variable	Coefficient	Standard error	t statistic
(Intercept)	−4.6228e + 05*	2.3705e + 05	−1.9502
(OUT)	7.8867e + 01***	1.6791e + 01	4.6968
(GDP)	−6.1946e + 00***	2.1449e + 00	−2.8881
(ROO)	3.1361e + 00***	5.0742e − 01	6.1805
(TIC)	−1.4386e + 01***	5.2847e + 00	−2.7222

"***", "**", and "*" denote significance at the 1%, 5%, and 10% levels, respectively.
Total Sum of Squares: 6.4595e + 13
Residual Sum of Squares: 3.3217e + 13
R-Squared: 0.48576
Adjusted R-Squared: 0.46365
F-statistic: 21.9628, p-value: 8.7443e − 13

Table 8. The estimation based on fixed effect model (dependent variable: total inbound tourists)

Variable	Coefficient	Standard error	t statistic
(OUT)	113.36382***	19.53668	5.8026
(GDP)	−7.67484*	4.11546	−1.8649
(ROO)	3.06519***	0.51537	5.9476
(TIC)	−11.73640	9.79764	−1.1979

"***", "**", and "*" denote significance at the 1%, 5%, and 10% levels, respectively.
Total Sum of Squares: 3.8156e + 13
Residual Sum of Squares: 1.7859e + 13
R-Squared: 0.53195
Adjusted R-Squared: 0.47816
F-statistic: 24.7197, p-value: 1.0973e − 13

each country has negative relation, which is consistent with those in Wu et al. (2009) that the more number of outbound travelers is, the more expenditure of tourists from each country increases, and increasing hardware facilities and number of hotel rooms appeal tourists to visit Taiwan and stimulate consumption. On the other hand, the higher fare from Taiwan to each country and the longer distance is, the more unwilling tourists are on consumption. GDP of outbound country for each country has negative relation to total tourists from each country that increasing GDP of outbound country brings about tourism income effect.

In Table 8, there was significant difference between total tourists from each country and each country number of outbound travelers, GDP of outbound country, and number of hotel rooms ($p < 0.01$, $p < 0.1$ adjusted $R^2 = 0.48$) and t value were 5.80, −1.86, and 5.95 separately. The result showed that each country number of outbound

Table 9. Estimation based on random effect regression model (dependent variable: total inbound tourists)

Variable	Coefficient	Standard error	t statistic
(Intercept)	−5.5806e + 05*	3.0432e + 05	−1.8338
(OUT)	1.1003e + 02***	1.8712e + 01	5.8801
(GDP)	−7.3414e + 00**	3.5830e + 00	−2.0490
(ROO)	3.0531e + 00***	4.7409e − 01	6.4399
(TIC)	−1.1664e + 01	8.5722e + 00	−1.3607

"***", "**", and "*" denote significance at the 1%, 5%, and 10% levels, respectively.
Total Sum of Squares: 3.9306e + 13
Residual Sum of Squares: 1.8592e + 13
R-Squared: 0.52699
Adjusted R-Squared: 0.50664
F-statistic: 25.9032, p-value: 1.9415e − 14

travelers and number of hotel rooms has positive relation to total expenditure of tourists from each country, which is consistent with Wu et al. (2009), while each country GDP of outbound country has negative relation to total tourists from each country that increasing GDP of outbound country brings about tourism income effect.

Table 9 reports that there was significant difference between total tourists from each country and each country number of outbound travelers, GDP of outbound country, and number of hotel rooms (p < 0.01, p < 0.05 adjusted R^2 = .51) and t value were 5.88, −2.05, and 6.44 separately. The result showed that number of outbound travelers and number of hotel rooms has positive relation to total expenditure of inbound tourists while the GDP of outbound country has negative relation to total inbound tourists, which correspond to Table 4.2.8 under fixed effect regression analysis.

Results of Hypothesis Testing
Dependent variables (average of total outbound tourists, total expenditure of outbound tourists, total outbound tourists) all have negative relationship to independent variables the GDP of outbound country and fare from Taiwan to outbound country. On the other hand, dependent variables all have positive relation to independent variable, number of outbound travelers. There was significantly positive relationship in H2.3 and H3.3 between dependent variables and the independent variable, number of hotel rooms, while there was no significant difference in H1.3.

	Hypothesis	Verification
H1.1	The number of outbound tourists has positive relationship with average of inbound tourists	Positive
H1.2	GDP of outbound country has positive relationship with average of inbound tourists	Negative
H1.3	Number of hotel rooms has positive relationship with average of inbound tourists	Not significant

(*continued*)

(continued)

	Hypothesis	Verification
H1.4	Fare from Taiwan to a country has positive relationship with average of inbound tourists	Negative
H2.1	The number of outbound travelers has positive relationship with total expenditure of inbound tourists	Positive
H2.2	GDP of outbound country has positive relationship with total expenditure of inbound tourists	Negative
H2.3	Number of hotel rooms has positive relationship with total expenditure of inbound tourists	Positive
H2.4	Fare from inbound country to outbound country has positive relationship with total expenditure of inbound tourists	Negative
H3.1	The number of outbound travelers has positive relationship with inbound tourists	Positive
H3.2	The GDP of foreign country has positive impact on inbound tourists	Negative
H3.3	Number of hotel room has positive relationship with inbound tourists	Positive
H3.4	Fare from inbound country to outbound country has positive impact on inbound tourists	Negative

(1) Number of outbound travelers and number of hotel rooms had positive influence on Taiwan tourism development, which indicated that increasing number of outbound travelers and hardware facilities help Taiwan tourism development growth. Thus, meeting tourists' accommodation demand helps stimulate the will and consumption of international tourists visiting Taiwan.

(2) Fare from Taiwan to outbound country had negative effect that the higher fare and the longer distance is, the lower average of total tourists from foreign country decreases. Goeldner and Brent (2006) suggested that attraction plays an important role in resort competition, such as scenery, shopping, entertainment, culture, recreation facilities and event. It needs to put more emphasis on promoting attraction above and maintain tourism quality when developing Taiwan tourism.

(3) GDP of outbound country had negative impact. It implies that increasing GDP of outbound country brings about tourism income effect that the tourism choices and expectation of resorts increase, and tourists are more willing to increase expenditure. Liu and Shi (2009) indicated that tourism attraction has positive influence on tourists that the higher attraction is, the more willing tourist is to travel. This research suggests that Taiwan not just promote more resort attraction as each country GDP increases but maintain tourism quality to make an excellent impression on international tourists and to become first tourism choice.

4 Conclusion and Suggestions

4.1 Conclusion

We employ the gravity model of tourism demand to investigate the determinants of tourism through the longitudinal data of international tourists across 9 countries including China, Japan, U.S., South Korea, Singapore, Hong Kang and Macao, New Zealand, and Australia from 2003 to 2016. The empirical results show that the fare from Taiwan to each country has negative effects on Taiwan tourism, GDP of outbound country has negative impact, number of outbound travelers from each country has positive affection, and number of hotel rooms has positive effects. We find the evidence that it brings about tourism income effect on Taiwanese economy as foreign country's GDP of outbound country increase. However, Taiwan could uncertainly be the first choice as outbound country's income increase the tourism choices and expectation of resorts. The big issue here how to attract the foreign tourists through upgrading the tourism quality.

4.2 Suggestions

The number of inbound travelers and number of hotel rooms could promote Taiwan tourism. The future market development of tourism not just goes on with accommodation but also needs to combine with other industries to spread the unique of Taiwan tourism. As for fare from Taiwan to other countries, putting forward tourism project is feasible or keeping air ticket price constant and cooperating with other industry promotes the quality and the value of Taiwan tourism. Taiwan tourism industry needs more attraction, and it is necessary to promote tourism further and to upgrade the quality to impress the world as the GDP of outbound country increases.

References

Goeldner, C.R., Ritchie, J.R.B.: Tourism Principle Practice, Philosophies, 11th edn. Wiley, Hoboken (2006)

Hong, C.-Y., Hsu, C.-J., Huang, C.-H., Chen, L.-P., Li, J.-F.: Economic effects from change in tourism policy on an island economy. Glob. J. Bus. Res. **10**(3), 35–48 (2016)

Kou, Z., Cai, H.: Understanding bike sharing travel patterns: an analysis of trip data from eight cities. Phys. A **515**, 785–797 (2018)

Li, J., Xu, L., Tang, L., Wang, S., Li, L.: Big data in tourism research: a literature review. Tour. Manag. **68**, 301–323 (2018)

Liu, Y.L., Shih, Y.L.: A study of visitor's activity attraction, satisfaction and revisiting willingness in dragon boat festival event LU-KANG. J. Leis. Recreat. Ind. Manag. **2**(1), 28–49 (2009)

Liu, H.M.: The study of aborigine tourism management and development strategy a case study Ren-ai town, Unpublished master thesis, Department of Public Policy and Administration, College of Humanities, National Chi Nan University, Nantou (2012)

Ma, Y., Li, T.: China's Inbound Tourism Research. Science Press, Beijing (1999)

Molin, E., Mokhtarian, P., Kroesen, M.: Multimodal travel groups and attitudes: a latent class cluster analysis of Dutch travelers. Transp. Res. Part A Policy Pract. **83**, 14–29 (2016)

Morely, C., Rossello, J., Gallego, M.S.: Gravity models for tourism demand: theory and use. Ann. Tour. Res. **48**, 1–10 (2014)

Seraphin, H., Gowreesunkar, V., Roselé, C.P., Jamont, J.D.Y., Korstanje, M.: Tourism planning and innovation: the Caribbean under the spotlight. J. Destin. Mark. Manag. **9**, 384–388 (2018)

Séraphin, H., Gowreesunkar, V., Roselé-Chim, P., Duplan, Y.J.J., Korstanje, M.: Tourism planning and innovation: the Caribbean under the spotlight. J. Destin. Mark. Manag. **9**, 384–388 (2018)

Mei, W.S., Chang, W.P.: The economic effects of Chinese tourists on Taiwan economy. Prospect Q. **11**(3), 133–175 (2010)

Stewart, J.Q.: Demographic gravitation: evidence and applications. Sociometry **11**(1/2), 31–58 (1948)

Ruijuan, S., Ren, L., Wang, H.: The construction and empirical analysis of the gravity model of regional tourism trade taking the domestic tourist market in Nanjing as an example. World Sci. Technol. Res. Dev. **6**(29), 61–64 (2007)

Taiwan Visitors Association: 1956–2006 Taiwan visitors association 50th memorial special magazine. Taiwan Visitors Association, Taipei (2006)

UNWTO (2016). http://www2.unwto.org/

UNWTO (2018). http://www2.unwto.org/

Wu, K., Huang, F., Li, J.: A study on mainland people to travel in Taiwan based on the gravitation model. Reform. Strategy **25**(187), 143–146 (2009)

Zipf, G.K.: The P1P2/D hypothesis: on the intercity movement of persons. Am. Sociol. Rev. **11**(6), 677–686 (1946)

Prediction of Breast Cancer Recurrence Using Ensemble Machine Learning Classifiers

M. S. Dawngliani[1]([✉]) [iD], N. Chandrasekaran[1,2],
Samuel Lalmuanawma[3], and H. Thangkhanhau[4]

[1] Martin Luther Christian University, Shillong, Meghalaya, India
dawngliani@gmail.com
[2] CDAC, Pune, India
[3] Department of Management, Mizoram University, Aizawl, India
[4] Department of Computer Science, GZRSC, Aizawl, Mizoram, India

Abstract. Breast Cancer is the most common type of cancer prevalent among female cancer patients, while it is also the second most dreaded disease, causing cancer deaths among women. This study proposes new criteria for the prediction of survival of breast cancer patients, based on the analysis performed using four ensemble machine learning techniques, which include, AdaBoost M1, Bagging, Voting, and Stacking. For this study, we have used a breast cancer dataset consisting of 23 attributes and containing 575 samples obtained from Mizoram State Cancer Institute of Aizawl, Mizoram, India. We have employed ensemble machine learning classifiers to predict the recurrence of breast cancer within a period of three years evaluated based on the comparison of their performance. We have used 10 fold cross-validation technique and ROC curve to arrive at the results. From the dataset, attributes are ranked according to their contribution towards the prediction.

Keywords: Adaboost M1 · Bagging · Data mining · Ensemble method · Stacking · Voting

1 Introduction

Of the two types of breast cancer, the invasive type of cancer is more dangerous as it spreads often to attack the nearby tissues. Due to this reason, this type of cancer can damage other organs including the liver and lungs. As its name implies, the non-invasive type of breast cancer remains with the breast tissue region, where it originates [1]. That is why it is identified as benign, but yet is called pre-cancerous, because of the possibility that the cancer can potentially spread outside the breast tissue region at a later time to become invasive type of breast cancer [2].

In Mizoram, breast cancer is the most common cancer affecting the female population, and hardly any studies had been done till date to predict the chances of survival of breast cancer patients, especially using advanced data mining techniques. Most of the studies pertaining to this topic have been performed using medical data acquired from developed countries and also a few from developing countries, but not necessarily from this region of interest. For the current study, we have made use of the medical data collected from within Mizoram. This data has been analyzed using various data mining classification techniques so as to extract meaningful patterns for the prediction of survival period and also to determine the major factors which may affect the results. The medical database that is found in the hospitals, are mostly in the form of files (i.e. in paper form) and in the computer (but not in organized form). The records contain massive amounts of information, which include patients' personal information, data reports obtained from laboratories, etc. It is pertinent to point out that this particular database is growing at a fast pace. With the help of data mining algorithms, useful patterns of information can be found within the data, which will be utilized for further research and evaluation of reports [3]. For the purpose of computation, the disease of a patient is classified as 'recurred' if it is observed to occur at some subsequent time and 'non-recurred' otherwise.

2   Breast Cancer Statistics

According to the World Cancer Research Foundation, breast cancer is the second most common cancer in women worldwide, with over 2 million new cases diagnosed in 2018 alone [4]. This represents about 12% of all new cancer cases and 25% of all cancers in women. According to U.S. National Cancer Institute's Surveillance, Epidemiology, and End Results (SEER) Cancer Statistics Factsheets (2015) [5], the number of new cases of female breast cancer was estimated to be 124.8 for every 100,000 women per year, while the number of deaths was determined to be 21.9 per 100,000 women per year. 2010–12 data has concluded that approximately 12.3% of women will be diagnosed with female breast cancer at some point during their lifetime.

That means worldwide, it is the most common form of cancer in females, which is affecting approximately 12.3% of all women at some stage of their life. 97% of women will survive for 5 years or more. It is also reported that one in 22 women in India suffer from breast cancer [6], while in the USA, the disease affects 2.75 times more women as one in 8 will be affected by this disease [7].

The number of breast cancer victims registered in the Mizoram State Cancer Institute is increasing each year in Mizoram and have shot up by more than 250% in the last 7 years alone. Table 1 lists the breast cancer statistics collected from Mizoram State Cancer Institute during the 2009–2016 period.

Table 1. Number of breast cancer victims registered in Mizoram State Cancer Institute, India.

Year	Male	Female	Total
2009	0	40	40
2010	0	51	51
2011	0	56	56
2012	0	66	66
2013	0	67	67
2014	2	97	99
2015	0	94	94
2016	2	100	102
Total			**575**

3 Related Studies of Machine Learning on Breast Cancer Dataset

Till date, several studies have been conducted employing ensemble machine learning methods to facilitate the diagnosis and prognosis of breast cancer. Both the SEER database and Wisconsin dataset have been used to measure the effectiveness of different data mining techniques. The results have varied depending upon the data set, feature selection and the ensemble method used.

Safiyari and Javidan [8] have analyzed the lung cancer dataset from SEER to predict the probability of survival in the first 5-years of the onset. The dataset was fairly large as it contained 149 attributes and 643,924 samples. After preprocessing, it was reduced to 24 attributes and 48,864 samples. Five ensemble methods were used for the evaluation and out of which AdaBoost outperformed other algorithms in both accuracy and AUC metrics. The performance of almost all of the base learners (except the J48 algorithm in accuracy and Naïve Bayes algorithm in AUC) increased by the application of ensemble techniques.

Kumar et al. [9] used an ensemble machine learning technique called Voting Classifier on a breast cancer dataset. Three combinations they chose involved, voting with SMO, voting with both SMO and Naïve bayes and voting with SMO, Naïve Bayes, and J48. They concluded that the combination of Naïve Bayes, SMO and J48 tree produced the highest accuracy.

Sidodia [10] used three ensemble learning techniques including bagging, boosting, and random forests to arrive at a classifier model for classifying a given headline into the clickbait or non-clickbait. The performances of learners were evaluated using accuracy, precision, recall, and F-measure. It was observed that the random forest classifier detects click baits better than the other classifiers with an accuracy of 91.16%, a total precision, recall, and F-measure of 91%.

Mohebian et al. [11] have reported developing a Hybrid Computer-aided-diagnosis System for Prediction of Breast Cancer Recurrence (HPBCR) using Optimized Ensemble Learning. Among 579 patients who participated in the study, 19.3% had a recurrence of cancer within five years after diagnosis. The authors employed a hybrid of three algorithms, viz., decision tree, SVM and MLP to conclude that they were able to obtain the best results using the hybrid approach.

Tarek et al. [12], in their work had developed and presented a new ensemble system for Cancer classification based on gene expression profiles. This approach had resulted in the development of a simple system that outperforms the ensemble system study suggested by Okun [13]. It also successfully addressed the three drawbacks, namely, enhancing result accuracy, covering more cancer types, and mitigating the effect of over-fitting. This work used the kNN classifier as a base member of the ensemble.

Lavanya and Usha Rani [14] studied the two-hybrid approaches, viz., bagging (CART decision tree classifier with feature selection) and boosting ensemble method. They had performed this study using three Wisconsin breast cancer dataset. The authors achieved accuracy on bagging and boosting for Breast Cancer dataset of 74.47% and 65.03% and for Breast Cancer Wisconsin (original) dataset of 97.85% and 95.56%. Breast Cancer Wisconsin (diagnostic) dataset produced an accuracy of

95.96% and 95.43%. Through this study, they concluded that the bagging ensemble method was preferable for the diagnosis of breast cancer data than boosting.

Abed et al. [15] had suggested a hybrid classification algorithm that is based on the Genetic Algorithm (GA) and k Nearest Neighbor algorithm (kNN). This algorithm was tested by applying Wisconsin Breast Cancer dataset from UCI Repository of Machine Learning databases using different datasets including Wisconsin Breast Cancer Database (WBCD) and Wisconsin Diagnosis Breast Cancer (WDBC). The accuracy of the algorithm was compared with different classifier algorithms, vis-a-vis, using the same database. The evaluation results of the algorithm proposed have achieved 99% accuracy.

Avula and Asha [16] proposed a Hybrid Machine Learning algorithm using two supervised algorithms, viz., Naïve Bayes and JRIP. The methodology adopted in their paper for proposing a new hybrid machine learning algorithm was implemented using R programming language and WEKA software tool. Further, a comparative study was made between the individual algorithms and the proposed hybrid algorithm in order to prove the improvement in prediction accuracy on medical datasets. The proposed algorithm showed enhanced performance compared to the individual classifiers.

Dawngliani et al. [17] compared four classifiers with ensemble classifier to evaluate the percentage accuracy for predicting breast cancer recurrence. They used WEKA to analyze their dataset and AUC and cross-validation to validate their results. The ROC value of one of the ensemble methods, bagging was determined to be 0.7959, which was the highest compared to all other classifiers.

4 Methodology

4.1 Data Mining Concept

Data mining is the process of discovering patterns in large datasets involving methods, which are common to machine learning, statistics, and database systems [18]. Its overall objective is to extract information from a data set employing intelligent methods and to transform the same into a comprehensible structure for further use [19]. According to Eapen [20], the four different methods of learning used in data mining are Classification Learning, Numeric Prediction, Association Learning and Clustering. The salient features are described in the above paper.

4.2 Tools Used

A model is to be proposed and tested for its accuracy using WEKA. WEKA is a machine learning tool for data pre-processing, classification, regression, clustering, association rules, and visualization [21]. This package has been developed by the University of Waikato in New Zealand. It provides numerous machines learning functionality and has been found to be easy to use. The default input file format is arff file, but with the help of Microsoft Excel, one can easily transform the data tables into comma-delimited format (.csv) file in order to import them into WEKA. Once data is imported, one can easily transform the data into the desired format as also perform tasks like replace the missing value, remove the redundant value, etc. [22]. Numerous algorithms are available for carrying out the required analysis. In this study, the Breast Cancer dataset was analyzed using J48 data mining classifiers.

4.3 Data Pre-processing

Completeness, accuracy and consistency of the data are factors that define data quality. Data preprocessing is an important step in the data mining process to satisfy data quality requirements. Data preprocessing is a data mining technique that involves transforming raw data into an understandable format [23]. Therefore, for the data to be analyzed, the collected attributes having numeric value should be coded in groups ranging from 1 to 10. Since machine learning models are based on mathematical equations, we need to encode the categorical value in numbers [24]. The ranges of the attributes are as shown in Table 2.

Table 2. Acronyms and ranges of the dataset attribute

Sl. No	Attributes	Abbr	Range	Sl. No	Attributes	Abbr	Range
1	Age	AG	1–10	Comorbid Condition			
2	Sex	SX	1–2	15	Tuberculosis	TB	1–2
3	Laterality	LT	1–3	16	Hypertension	HP	1–2
4	BMI	BMI	1–5	17	Diabetes	DB	1–2
5	Morphology	MP	1–10	18	Heart disease	HD	1–2
6	Socioeconomic status	SS	1–7	19	Asthma	AS	1–2
7	Axillary lymph node	ALN	1–6	20	allergy	AL	1–2
8	Skin involvement	SI	1–2	21	Hepatitis	HPT	1–2
9	Stage Grouping	SG	1–9	22	Others	OT	1–2
Habitual data				23	Aids	AI	1–2
10	Cigarette	CG	1–2	24	Recurred	RC	1–2
11	Tobacco	TBC	1–2				
12	Alcohol	ALC	1–2				
13	Pan masala	PM	1–2				
14	Betelnut	BN	1–2				

4.4 Attribute Selection

Attribute selection, is extensively used and hence is an area of active research interest pertaining to pattern recognition, statistics, and data mining in the medical domain.

Feature selection invariably involves reducing the number of attributes to improve the accuracy of the outcome. The attributes are reduced by removing irrelevant and redundant attributes, which do not have much importance in determining the outcome [25]. The main goal is to find an optimal feature-subset (one that maximizes the prediction accuracy) [26]. The feature selection evaluators that we use are Infogain, GainRatio, and ChiSquared. They evaluate the worth of an attribute by measuring the information gain with respect to the class [27]. The evaluators check each subset using the ranker method and rank them according to their significance and relevance as shown in Table 3.

Table 3. Result of the rank of the attribute using evaluators

Rank	Info gain		Gain Ratio		Chi-Squared	
	Value	Attribute	Value	Attribute	Value	Attribute
1	0.1454	SG	0.2076	SG	68.2867	SG
2	0.0448	ALN	0.0693	ALN	20.5535	ALN
3	0.0355	MP	0.0549	PM	14.0160	MP
4	0.0130	PM	0.0492	AL	5.4488	SI
5	0.0119	SI	0.0384	HD	4.6993	LT
6	0.0105	SS	0.0343	TB	3.8391	SS
7	0.0095	LT	0.0224	SI	2.9234	PM
8	0.0082	AL	0.0156	MP	1.8336	AL
9	0.0023	HD	0.0088	LT	0.7104	BN
10	0.0019	HP	0.0047	SS	0.7047	HP
11	0.0018	BN	0.0042	HP	0.5146	HD
12	0.0012	TB	0.0021	BN	0.4142	DB
13	0.0010	AS	0.0021	AS	0.3825	AS
14	0.0010	DB	0.0020	OT	0.3823	TBC
15	0.0010	TBC	0.0018	DB	0.3200	OT
16	0.0009	OT	0.0014	ALC	0.2564	TB
17	0.0005	CG	0.0012	HPT	0.1941	CG
18	0.0003	ALC	0.0012	TBC	0.1098	ALC
19	0.0001	AI	0.0009	AI	0.0541	HPT
20	0.0001	HPT	0.0006	CG	0.0513	AI
21	0.0000	SX	0.0000	SX	0.0004	SX
22	0.0000	BMI	0.0000	BMI	0.0000	BMI
23	0.0000	AG	0.0000	AG	0.0000	AG

4.5   Ensemble Method

In this paper, we are using four ensemble machine learning algorithms. Ensemble algorithms are a powerful class of machine learning algorithms that combine the predictions from several base models in order to produce one optimal predictive model [28]. The base models have sought to produce one strong classifier for which the classification error can be made arbitrarily small. In ensemble learning, instead of creating one strong classifier, a huge set of weak classifiers are created and then we combine their outputs into one final decision [29]. According to Condorcet's theorem (1785), under proper conditions we can expect that the ensemble model can attain an error rate that is arbitrarily close to zero. This theorem has long been applied in political science and implies that collective wisdom of the people (even if it is weak) is superior to strong individual opinions. Another important point to note is that creating a lot of weak classifiers tends to become a much easier task than to create one strong classifier. Figure 1 below illustrates the important steps involved in creating the ensemble method [30]:

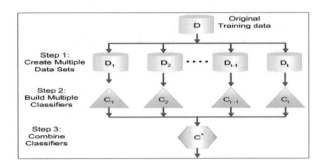

Fig. 1. Ensemble method

The ensemble classifiers we have used are Boosting, Bagging, Voting and Stacking, which are discussed below:

Boosting: Boosting is an ensemble machine learning algorithm typically used for tackling classification problems. We have used AdaBoostM1, a boosting model, which has successfully been implemented to increase the accuracy of the model. AdaBoost uses short decision tree models, normally referred to as decision stumps, each with a single decision point. Each instance in the training dataset is weighed and the weights are updated based on the overall accuracy of the model and whether an instance was classified correctly or not. Subsequent models are trained and added until a minimum accuracy is achieved or no further improvements are possible. Each model is weighted based on its skill and these weights are used when combining the predictions from all of the models on new data [30].
    The AdaBoost.M1 algorithm [31]:

```
1 weight_of_objects[1..n] = 1n;
2 ensemble = NULL;
3 weight_of_objects = NULL;
4 for number_of_classifiers do
5 //Construction of the classifier with a weighted vote
6 data_set_classifier = choose_objects(data_set, weight_of_objects);
7 new_classifier = build_classifier(data_set_classifier);
8 new_classifier.determine_the_weight_of_voting();
9 ensemble.add(new_classifier);
10 //Calculate the new weight of objects
11 for number_of_objects do
12 κ = classifies_object(current_object, ensemble);
```

```
13 weight_of_objects[current_object] =
weight_of_objects[current_object] •κ;
14 endFor
15 endFor
16 result = ensemble;
```

Bagging: The full form of bagging is Bootstrap Aggregation. It is an ensemble algorithm that can be used for classification as well as regression. Bagging is a statistical estimation technique where a statistical quantity like a mean is estimated from multiple random samples of the dataset (with replacement). It is a technique that is best used with models that have low bias and high variance, meaning that the predictions they make are highly dependent on the specific data from which they were trained. Multiple random samples of the training dataset are drawn with replacement and used to train multiple machine learning models or algorithms. All data subsets have exactly the same cardinality as the initial learning set. Each model is then used to make a prediction and the results are averaged to give a more robust prediction. The most used algorithm for bagging that fits this requirement of high variance is decision trees [32].

Assuming that the dataset consisted of 'n' elements, exactly 'n' elements are selected for every pseudo-sample—whereby every case in the learning sample is selected with exactly the same probability equal to 1n. Every data set is used as a basis for constructing an independent decision tree classifier. Each of the generated classifiers is granted exactly a single vote, and the set under consideration is assigned to the class which has obtained the greatest number of votes.

Algorithm: The bagging algorithm [31]:

```
1 ensemble = NULL;
2 for number_of_classifiers do
3 //Construction of the classifier
4 data_set_classifier = choose_objects(data_set); //bootstrap aggregat-
ing
5 new_classifier = build_classifier(data_set_classifier);
6 ensemble.add(new_classifier);
7 endFor
8 result = ensemble;
```

Voting: Voting is the simplest ensemble algorithm, and is often very effective for solving classification or regression problems. Voting works by creating two or more sub-models, with each one of sub-models making predictions. The sub-models are combined in some way that, by taking the mean or the mode of the predictions. Each sub-model can vote to determine the eventual outcome. In this study, we have done voting with four classifiers such as J48, Naive Bayes, MLP and SMO. In majority voting, the predicted class label for a particular sample is the class label that represents the majority (mode) of the class labels predicted by each individual classifier.

For example, if the prediction for a given sample is classifier 1 -> class 1, classifier 2 -> class 1 and classifier 3 -> class 2, the Voting classifier would classify the sample as "class 1" based on the majority class label [33] (Fig. 2).

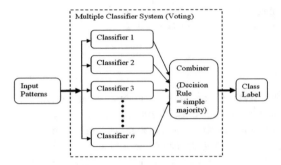

Fig. 2. Voting classifier [34].

Stacking: Stacked Generalization or Stacking is a simple extension to Voting that can also be used to solve classification and regression problems. In addition to selecting multiple sub-models, stacking allows specifying another model to learn how best to combine the predictions from the sub-models. The latter is also known as meta-classifier. Because a meta-classifier is used to best combine the predictions of sub-models, this technique is sometimes called blending, which implies that predictions can be blended together [18]. In this study, we have done stacking with four classifiers such as J48, Naive Bayes, MLP and SMO. For our work, J48 has been used as a meta classifier (Fig. 3).

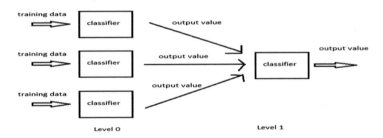

Fig. 3. Stacking classifier [35].

5   Experimental Result

From the dataset, 80% of the patients who were diagnosed with breast cancer recover from the disease, while 20% of the patients have a positive recurrence. Most of the patients who are diagnosed with breast cancer are in the age group between 25–60 years. No patient who is less than 20 years of age or more than 90 is diagnosed with breast cancer. And also, the majority of patients are diagnosed with invasive ductal carcinoma. Table 4 shows the result of the performance evaluation using WEKA.

In our study, we use a 10 fold cross-validation technique to determine the accuracy of the prediction. The boosting algorithm namely AdaboostM1 has the highest prediction accuracy of 82.807% followed by Bagging with an accuracy of 82.45 61%. The stacking method has the least accuracy of 81.4035%. The Tp rate, precision, recall, F-measure are also the highest in AdaboostM1, while the Vote method has the highest ROC Area. The performance is also measured in terms of error statistics as shown in Table 5. The Kappa Statistics is highest in AdaboostM1, which signifies that the attribute measurement is best compared to other ensemble methods. Also, the error rate, which is measured in terms of mean absolute error, root mean square error, relative absolute error, and root-relative square error, is relatively low in Adaboost M1.

Table 4. The performance measure of different ensemble machine learning

Performance	Ensemble Method			
	AdaboostM1	Bagging	Stacking	Vote
Accuracy	82.807	82.4561	81.4035	81.7544
Tp Rate	0.828	0.825	0.814	0.818
FP Rate	0.416	0.456	0.433	0.534
Precision	0.819	0.811	0.806	0.796
Recall	0.828	0.825	0.814	0.818
F Measure	0.823	0.815	0.809	0.798
ROC Area	0.758	0.794	0.732	0.796

Table 5. Error statistics of different ensemble machine learning methods

Error Statistics	Ensemble Method			
	AdaboostM1	Bagging	Stacking	Vote
Kappa Statistics	0.4371	0.4054	0.3994	0.3347
Mean Absolute Error	0.2304	0.2442	0.2609	0.2225
Root Mean Square Error	0.3599	0.3549	0.3838	0.3623
Relative Absolute Error	70.7647	75.0199	80.1303	68.3498
Root Relative Square Error	89.3871	88.1342	95.3283	89.9824

The following figure shows the ROC curve of the four ensemble methods that we have employed in our studies. The area under the ROC curve (AUC) of AdaboostM1 equals 0.7583, see Fig. 4(a), while the AUC determined by Bagging equals 0.7939 as in Fig. 4(b). Figure 4(c) shows the ROC curve of Stacking where the AUC equals 0.7317 and in Fig. 4(d), ROC curve of Vote indicates that AUC = 0.7959; The areas under the curves are large for all the ensemble methods, which means that the ensemble algorithms are the most efficient for prediction. The ensemble method with the largest area is the Vote method which has an area of 0.796.

242   M. S. Dawngliani et al.

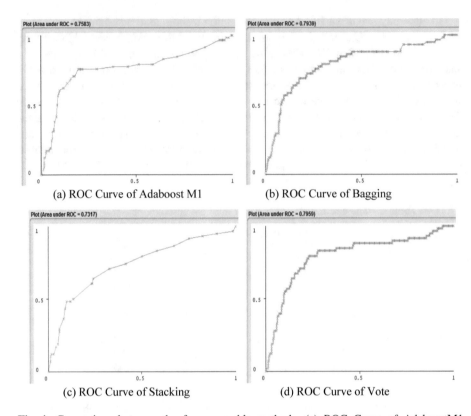

(a) ROC Curve of Adaboost M1

(b) ROC Curve of Bagging

(c) ROC Curve of Stacking

(d) ROC Curve of Vote

Fig. 4. Comparison between the four ensemble methods. (a) ROC Curve of AdaboostM1 (b) ROC Curve of Bagging (c) ROC Curve of Stacking and (d) ROC Curve of Vote

6   Conclusion

When a person is first diagnosed with breast cancer, there is an uncertainty and fear that the cancer may come back after some time. A cancer recurrence happens for 20% of the entire breast cancer patient. The recurrence may occur in the same place (local recurrence) or in some other organ (distant recurrence). This may lead to fatal, so it is of great concern for the patient and their family. From the three evaluators used, the stage grouping and the number of axillary lymph node affected has the two topmost ranks which means that higher the cancer stage and lymph node affected, higher is recurrence rate.

Data mining plays an important role in extracting information contained in the medical database. It is clear from the analysis that the stage of cancer plays a vital role in cancer survival. So, early screening is very important to increase the chance of survival. Resorting to preprocessing and attribute selection play an important role in improving the data quality. From our study, we can conclude that the four ensemble machine learning methods are very efficient when employed to predict the probability of breast cancer recurrence. Though their performances do not differ much, the performance accuracy of AdaboostM1 seems to be best. The performance has been evaluated using both the accuracy and computing the area under the ROC curve.

We would like to thank Dr. Thangchungnunga, Director of Mizoram State Cancer Institute (MSCI), Aizawl, for supporting this research work by giving us permission to collect breast cancer datasets from MSCI. The authors express their sincere gratitude to MLCU for facilitating and extending all possible help to complete this particular research work. Also, an ethical committee of MLCU has given all necessary clearance and that no reference to any personal records of any person who has undergone or undergoing treatment in the Mizoram State Cancer Institute (MSCI), Aizawl has been accessed and no name of the patient or his or her personal medical record has been cited.

References

1. NCI. https://www.cancer.gov/types/breast. Accessed March 2019
2. Medical News Today. https://www.medicalnewstoday.com/articles/37136.php
3. Witten, I.H., Frank, E., Hall, M.A.: Data Mining: Practical Machine Learning Tools and Techniques, 3rd edn. Morgan Kaufmann, San Francisco (2011)
4. World Cancer Research Fund and American Institute for Cancer Research. https://www.wcrf.org/dietandcancer/cancer-trends/breast-cancer-statistics
5. SEER. https://seer.cancer.gov/statfacts/html/breast.html
6. Indian against cancer. http://cancerindia.org.in/cancer-statistics/
7. National Breast Cancer Foundation, INC (2016). https://www.nationalbreastcancer.org/breast-cancer-facts
8. Safiyari, A., Javidan, R.: Predicting lung cancer survivability using ensemble learning methods. In: 2017 Intelligent System Conference IntelliSys 2017, vol. 2018, no. September, pp. 684–688 (2018)
9. Kumar, U.K., Nikhil, M.B.S., Sumangali, K.: Prediction of breast cancer using voting classifier technique. In: 2017 IEEE International Conference Smart Technology Management Computer Communication Controlling Energy Mater. ICSTM 2017 - Proceedings, no. August, pp. 108–114 (2017)
10. Sisodia, D.S.: Ensemble learning approach for clickbait detection using article headline features. Inf. Sci. **22**(2019), 31–44 (2019)
11. Mohebian, M.R., Marateb, H.R., Mansourian, M., Angel, M., Mokarian, F.: A hybrid computer-aided-diagnosis system for prediction of breast cancer recurrence (HPBCR) using optimized ensemble learning. Comput. Struct. Biotechnol. J. **15**, 75–85 (2017)
12. Tarek, S., Elwahab, R.A., Shoman, M.: Gene expression-based cancer classification. Egypt. Inf. J. (2016)
13. Okun, O.: Feature Selection and Ensemble Methods for Bioinformatics: Algorithmic Classification and Implementations (2011). https://doi.org/10.4018/978-1-60960-557-5
14. Lavanya, D., Usha Rani, K.: Ensemble decision making system for breast cancer data. Int. J. Comput. Appl. **51**(17), 19–23 (2012)
15. Abed, B.M., et al.: A hybrid classification algorithm approach for breast cancer diagnosis. In: IEACon 2016 – 2016 IEEE Industry Electronics and Applications Conference, pp. 269–274 (2017)
16. Avula, A., Asha, A.: Improving prediction accuracy using hybrid machine learning algorithm on medical datasets. IJSER **9**(10), 1461–1467 (2018)

17. Dawngliani, M.S., Chandrasekaran, N., Lalmuanawma, S.: A comparative study between data mining classification and ensemble techniques for predicting survivability of breast cancer patients. Int. J. Comput. Sci. Mob. Comput. **8**(9) (2019)
18. KDD home page. https://www.kdd.org/curriculum/view/introduction
19. Data Mining Britannica 2010. https://www.britannica.com/technology/data-mining
20. Eapen, A.G.; Application of Data mining in Medical Applications. University of Waterloo (2004)
21. Hall, M., Frank, E., Holmes, G., Witten, I.H., Cunningham, S.J.: Weka: practical machine learning tools and techniques. In: Workshop on Emerging Knowledge Engineering and Connectionist-Based Information Systems (2007)
22. Aksenova, S.S.: Machine learning with WEKA. Mach. Learn. **11**(1), 1–37 (2006)
23. What Steps should one take while doing Data Preprocessing? https://hackernoon.com/what-steps-should-one-take-while-doing-data-preprocessing-502c993e1caa
24. Data Pre Processing Techniques You Should Know – Towards Data Science (2018). https://towardsdatascience.com/data-pre-processing-techniques-you-should-know-8954662716d6
25. Khalid, S., Khalil, T., Nasreen, S.: A survey of feature selection and feature extraction techniques in machine learning. In: Proceedings of 2014 Science and Information Conference SAI 2014, pp. 372–378 (2014)
26. Feature Selection and Feature Extraction in Machine Learning: An Overview (2018). https://medium.com/@mehulved1503/feature-selection-and-feature-extraction-in-machine-learning-an-overview-57891c595e96
27. Machine learning mastery (2013). https://machinelearningmastery.com/perform-feature-selection-machine-learning-data-weka/
28. Ensemble Methods in Machine Learning: What are They and Why Use Them? https://towardsdatascience.com/ensemble-methods-in-machine-learning-what-are-they-and-why-use-them-68ec3f9fef5f
29. Ensemble Learning. http://www.inf.u-szeged.hu/~tothl/ML/10.%20Ensemble%20learning.ppt
30. Rokach, L.: Ensemble-based classifiers. Artif. Intell. Rev. **33**(1–2), 1–39 (2010)
31. Kozak, J.: Ensemble methods. Stud. Comput. Intell. **781**, 107–118 (2019)
32. Opitz, D., Maclin, R.: Popular ensemble methods: an empirical study. J. Artif. Intell. Res. **11**, 169–198 (1999)
33. Ensemble methods (2019). https://www.toptal.com/machine-learning/ensemble-methods-machine-learning
34. Bostock, J.: Automated cardiac rhythm diagnosis for electrophysiological studies, an enhanced classifier approach (2014). https://openaccess.city.ac.uk/id/eprint/12186/
35. Multiclassifiers; Ensembles and Hybrids; Bagging, Boosting, and Stacking - PRIMO (2019). http://primo.ai/index.php?title=Multiclassifiers;_Ensembles_and_Hybrids;_Bagging,_Boosting,_and_Stacking

An Impact of Capital Gains Tax for Securities on Taiwan Stock Market by Overreaction, Lock-in Effect and Market Microstructures

Ming-Min Lo[✉], Jian-Fa Li, and Kuo-Ching Chiou

Department of Finance, Chaoyang University of Technology, Taichung, Taiwan
Mingminlo@gm.cyut.edu.tw, {jfli,kcchiou}@cyut.edu.tw

Abstract. Capital gains tax on securities in Taiwan was resubmitted in Mar 2012 and reimplement from Jan 2013, but this taxation suffered the huge public pressure in the society and caused the decline of stock trading volume and finally abolished on the end of 2015. This study intends to combine behavioral finance and tax neutrality principle base on Lewin field theory to investigate whether investors' psychology, investors' behavior and the outer environment could be interact affected by taxation in Taiwan stock market. The empirical results have shown as follows: (1) Investor psychological impact-overreaction (GJR model): The taxation announcements (bad news) would affect investors into irrational bias and due to decrease original profit for investor would amplify the negative asymmetrical volatility in return rate of stock. (2) Investor behavior impact- Lock-in effect (CLRM and DID model): in a long term, the trading volume of Hong Kong stock market has been back to normal after the extension period to dilute other interference effect. However, the stock trading volume in Taiwan still has a 19.64% decrease comparing to that in Hong Kong. Such result indicates due to the taxation substitution effects, investors would shift their investment to other low-tax or free-tax goods. (3) Environment impact-adjustment of market microstructure, due to stock trading volume is depression by 25.38% after taxation announcement and implement, individual investor structure which has higher stock demand elasticity in Taiwan have been reduced by 12.99%. The microstructure adjustment and trading ecology of stock market could be interact affected by taxation lock-in effect and overreaction effect, which is worthy of the reference of the authorities.

Keywords: Capital gains tax for securities · Overreaction · Lock-in effect

1 Introduction

The capital gains tax on securities have been levied (hereafter SCG tax) three times in Taiwan. However, they have led to failure in the end. The first one was from March 1973 to the end of 1975. The second one was the most famous announcement in September 1988, causing Taiwan stock market index to fall during 19 days and the volume of transactions fell by nearly 90%. The third announced in March 2012 reintroduced the tax from 2013, but it also faced significant controversy. The volume of stock market has been depressed after the policy announcement. Even though the

© Springer Nature Switzerland AG 2020
L. C. Jain et al. (Eds.): SICBS 2019, AISC 1145, pp. 245–257, 2020.
https://doi.org/10.1007/978-3-030-46828-6_21

government continues to offer a number of revitalization programs, it is not able to achieve an effective upgrade in stock trading volume. Due to the turnover rate in the Taiwan stock market is still not rejuvenate though three and a half years and suffer the huge public pressure in the society, the capital gains tax for securities in Taiwan were finally abolished in the end of 2015.

This present paper aims to combine behavioral finance and tax neutrality principle based on Lewin field theory to investigate whether investors' behavior could be affected personal psychological and the interaction of outer environment by capital gains tax for securities in Taiwan stock market. The empirical evidence might serve as a reference for developing future government taxations. The research goals of this study are: (1) Does stock investors have an irrational cognitive bias for SCG tax event, resulting in the volatility of long term stock return rate asymmetry in over reaction? (2) Does the SCG tax affect investors' behavior to cause the decline of stock transactions, resulting in long term lock-in effect of stock volume? (3) Does the SCG tax cause the change in stock market microstructures by adjustment of the investor structure in the Taiwan stock market?

2 Literature Review

Lewin (1936) proposed the field theory, in which he considered that the individual's behavior (B) was primarily a function of the inherent Personal psychological conditions (P) and the external environment (E). The model is described as follows:

$$B = F\ (P, E) = F\ (P \sim P_1, P_2 ... Pn\ ;\ E \sim E_1, E_2 ..... E_n) \qquad (1)$$

Where $P \sim P_1, P_2 ....\ P_n$ represents the various psychological and physiological factors that make up an individual's internal conditions; E (Environment) represents the extrinsic environment which the individual is located, $E \sim E_1, E_2 ....\ E_n$ represent elements such as politics and the economy that make up the external environment.

Yin (2005) put forward the model of the psychological behavior mechanism in the capital market subject with field theory and pointed out the capital market tax mechanism would release the income effect and substitution effect of taxation through the investor's psychological expectation mechanism and would affect capital market. Previous studies have pointed out due to that SCG tax would decrease investors' original profit and would inhibit investors' confidence in trading levied tax assets, create a lock-in effect of stock trading volume, and easily have a negative effect on the stock market (Somers 1948; Stiglitz 1983; Sahm 2008; Dai et al. 2008). In behavioral finance theory, Kahneman and Tverskey (1979) mentioned that there are many emotions of conservatism or impulsivity, because people are more exposed to external circumstances or their own personality in decision-making. One of the assumptions of behavioral finance is that human beings are mostly limited rational in perspective theory. Common irrational behavior of individual comes from the investor's cognitive bias, often based on the rule of thumb or intuition for decision-making. Kahneman and Tversky (1974) pointed out that people would tend to overreaction effect by the similar representativeness event. Previous studies have pointed out investors' overconfidence and under confidence would affect the decision (Griffin et al. 1992). Feldstein and Yitzhaki (1978) argued that investment decisions of individual are highly sensitive to tax changes, because tax is a cost and would decline the gains from stock transactions. Somers (1948) mentioned investors would have lock-in effects on securities. The key is the investor's tax payment in order to avoid, which might lead to a reduction in the willingness to buy and sell financial commodity or to affect the investor's behavior of buying and selling. Stiglitz (1983) also obtained the conclusion that the capital gains tax is not conducive to the development of the securities market. Dai et al. (2008) explains the capitalization effect and pointed out under the effect of capitalization, investors would be affected by tax increase to reduce the purchase of stocks (reduce demand); Sahm (2008) pointed out that capital gains tax is based on realized gains tax, which would make investors tend to retain profitable parts and sell losses to regulate the taxed income, which tends to create a lock in effect, and will lead to distortions in stock liquidity and investment decisions. Even though some study pointed out that some kind of capital gains tax effects like locked-in effect could not represent the reversals in long-term return of U.S. stock market (Bhootra 2013). But most study like George and Hwang (2007) also argued that lock-in effect would let investors to change transcation decision to delay selling profitable stocks in order to avoid capital gains tax and cause U. S. stock price long-term reversal. Falsetta et al. (2013) indicated that a tax reduce or increase would affect the taxpayers to increase or decrease the risky investment. Diaz et al. (2016) pointed out the taxation of capital gains with be possible change individual taxpayer's behavior in Spanish, the taxpayers will decrease investment will and have lock-in effect in investment. Maleki et al. (2016) also indicated in some Europe Union countries, tax system on capital formation might reduce investors' motivation in the existing and future, and could affect finance and economy to develop and growth. Hayashida and Ono (2016) proposed that tax reforms would affect stock return volatility in Japan.

3 Empirical Model

3.1 GJR Models

We employed the model proposed by Glosten et al. (1993) (hereafter GJR model) to explore whether the implementation of the capital gains tax for securities (bad news) in March 2012 could have an impact on investors' psychology into irrational emotion and cause the overreaction in Taiwan stock market. The return rate of stock is set to be an explained variable. GJR model could be restated as follows:

$$y_t = \alpha x_t + \varepsilon_t \tag{2}$$

$$h_t = \sigma_t^2 = \alpha_0 + \sum_{j=1}^{p} \beta_j h_{t-j} + \sum_{i=1}^{q} \alpha_i \varepsilon_{t-i}^2 + \sum_{i=1}^{q} \gamma_i s_{t-i}^- \varepsilon_{t-i}^2 \tag{3}$$

$$\varepsilon_t | \Omega_{t-1} \sim N(0, h_t) \tag{4}$$

$s_{t-1}^- = 1$ if $\varepsilon_{t-i} < 0$ $s_{t-1}^- = 0$ if $\varepsilon_{t-i} > 0$

$\alpha_0 > 0$, $\alpha_i > 0$, $i = 1,2,\ldots,q$ $\beta_j > 0$, $j = 1,2,\ldots,p$

where y_t is time series data; x_t denotes the conditional mean of y_t; Ω_{t-1} stands for the collection of information collected from t-1; h_t is Conditional heterogeneous variation of y_t; γ_i represents asymmetry effect.

Where the long term observation period is set to from year 2007 to 2015 (the impact of the overreaction observation period after the taxation declaration from Mar 2012). Considering that the stock price trading information usually has the cluster volatility effect, it is based on the GJR model of GARCH/TARCH as the research method. GJR model illustrates virtual variables to represent the different effects of positive and negative shocks and by increasing the weight of the impact caused by bad news to investigate asymmetry of stock price fluctuations by leverage Glosten et al. (1993).

3.2   CLRM Model and DID Model

In order to investigate whether the capital gains tax on securities could affect investors to reduce the stock trading volume and result in a long term lock-in effect in stock market, the trading volume of stock is set to be an explained variable. The explanatory variables included in stock rate of return, stock amplitude rate, and 30-day average stock rate of return. The long term lock- in effects are observed during the reintroduce and implementation periods of capital gains tax in Taiwan from Mar 28, 2012 to Dec 31, 2015.

Classical Linear Regression Model (CLRM Model)
In order to examine investors' behavior in terms of long term lock-in effect on stock trading volume, the multiple regression model is as follows:

$$Y_t = \alpha + \beta_1 \times Time_t + \sum_{i=1}^{n} \gamma_i X_{it} + \varepsilon_t \tag{5}$$

Where Y_t denotes the stock trading volume in the stock market; α is a constant term; $Time_t$ is a dummy variable to denote whether the period of SCG tax had been announced (before the announcement = 0; after the announcement = 1); β_1 is the time effect test coefficient; ε_t is a residual; and. X_{it} represents the other explanatory variables including the stock return rate of each market (R), the stock amplitude rate of each market (W), and the 30-day average stock return rate of each market (R30).

Difference in Differences Model (DID Model)
Thanks to the explanory variables in the classical regression model (CLRM model), it could be limited by sample selection or have key factors fail to control, which could incline to lead to overestimation. Ashenfelter and Card (1985) proposed the difference in difference Model (DID model) to deal with overestimation. We divided the sample into experimental group and control group, subtracted the difference between the experiment group and the control group after the taxation in DID model, and obtained the taxation effect by the difference between the experiment group and the control group before and after the taxation. In order to exclude the international influence

factors, in the selection of the control group, since Taiwan and Hong Kong stock market are more relevant in Asia economic area, we set Taiwan stocks as the experimental group and Hong Kong stocks as a counterpart group, for the two stock market volume for further analysis.

Referring to Lo (2015), the DID model in Taiwan stock market could restated as:

$$Y_{it} = \alpha + \beta_1 \times T_t + \beta_2 \times N_i + \beta_{12} \times T_t \times N_i + \sum_{i=1}^{n} \gamma_i X_{it} + \varepsilon_t \qquad (6)$$

where Y_t denotes the two stock trading volume in the market for observe lock-in effect; Tt is a dummy variable in the period indicating whether the SCG tax had been announced. In the DID model for contrast to classical linear regression model, two dummy variables are included in this model: N_i and $T_t \times N_i$. The term N_i is a dummy variable indicating the economy area effect (Taiwan Stock Market = 1; Hong Kong Stock Market = 0). $T_t \times N_i$ is referred as the cross-effect to multiply by time factor and economy area factor. A particularly critical aspect of the DID model is the β_{12} cross-effect coefficient. If the announcement of the capital gains tax for securities had a major influence in Taiwan stock market, the difference between by degree of decline in stock trading volume of Taiwan market should be more serious than that in Hong Kong stock market in the same period. The term X_{it} represents the other explanatory variables, including the stock return rate of each market (R), the stock amplitude rate of each market (W), and the 30-day average stock return rate of each market (R30).

3.3 Market Microstructures

In order to study the changes in Taiwan stock market trading information in the 45 months before and after the event, the first part checks the investor structure of the Taiwan stock market (e.g. individual investors, foreign institute), and the second part checks the micro-structural adjustment of the market, whether the trading information of Taiwan stocks such as stock price index, volume and turnover rate have an impact. Research hypothesis is under the principle of tax neutrality, the government should keep that theoretically SCG taxation implement should not affect the development of the capital market (e.g. stock market investor structure and trading ecology). Variable X_1 (45 months before the SCG tax event: July 2007 to March 2012), variable X_2 (45 months after the SCG tax event: April 2012 to December 2015). The t-test in-pair maternal average difference is checked, and the trading information of Taiwan stock market is checked with alpha-0.1 whether there is a significant change before and after the SCG tax event.

H_0: There was no significant change in the Taiwan stock market before and after the SCG tax

t test statistic is as follows:

$$|Z| = \frac{|\bar{X}_1 - \bar{X}_2|}{\sqrt{\frac{\sigma_1^2}{n_1} - \frac{\sigma_2^2}{n_2}}} \qquad (7)$$

4 Empirical Results

4.1 Data Sources and Descriptive Statistics

Data Source and Variable Design
Table 1 reports two main stock market descriptive statistics for Taiwan Capitalization weighted stock index of Taiwan Stock Exchange (TWSI) and Hong Kong Hang Seng stock index of Hong Kong Stock Exchange (HKSI) including stock trading volume (Y) of each market, stock return rate (R) of each market, stock amplitude rate (W) of each market, and 30-day stock return rate (R30) of each market which served as variables in the observation period from year 2007 to year 2015.

Table 1. Descriptive statistics

Taiwan and Hong Kong stock market information in observation period (Year 2007~2015)

Variable	Average	Median	Max	Min	S.D.	Numbers
Stock volume(Y)-TW	104650	97574	322003	23837	34175	2023
Return rate (R)-TW	0.0001	0.0007	0.0674	-0.0878	0.0136	2023
Amplitude (W)-TW	0.0132	0.0102	0.0723	0.0026	0.0094	2023
30-day rate of returns (R30)-TW	0.0001	0.0004	0.0103	-0.0132	0.0026	2023
Stock volume(Y)-HK	26280	23596	102603	6872	11117	2023
Return rate (R)-HK	0.0002	0.0001	0.1435	-0.1367	0.0176	2023
Amplitude (W)-HK	0.0149	0.0120	0.1633	0.0032	0.0108	2023
30-day rate of return (R30)-HK	0.0002	0.0002	0.0132	-0.0207	0.0036	2023

Source and Remarks: The financial database (Datastream), except for the volume of transactions, the remaining units are %.

4.2 Investors' Personal Psychology (P) – Overreaction Effect

In Table 2, we employed GJR model to explore whether investors' personal emotion would be affected by SCG tax event and get into long term irrational overreaction effect in Taiwan stock market. The return rate of stock is a dependent variable. In the present study of the long-term observation period set to year 2007 to 2015, we use TAX variable as a dummy variable set to 1 to observe the impact of SCG tax after the announcement from March 2012 to the implementation of the end of 2015 period). The empirical evidence (referring to Table 2) shows that in implementation period of the SCG tax event (TAX) for the return rate of Taiwan stock market has a long-term negative effect.

As for the leverage effect, we employed the GJR model using virtual variables (RESID(-1)^2*(RESID(-1) < 0) to indicate whether SCG tax event would have the asymmetry of stock return rate fluctuations. If bad news would amplify the positive leverage. The empirical results illustrate the variables of variance Equation were significant. Among them, the virtual variables (RESID (-1) ^2*(RESID (-1) < 0) indicates stock return rate have a positive leverage significant. The fluctuation asymmetry by weight of the impact of aggravating shocks show up the stock return rate affected by SCG tax event (bad new) have long term overreaction effect is significant during implementation period of the SCG tax event.

Table 2. The results of GJR model- Overreaction effect

Long-term overreaction effect of SCG tax event on the stock return rate in Taiwan
Long-Term (Year 2007 ~ 2015) (Taxation period: Mar 2012 to Dec 2015)

Variable	Coefficient	Probability		
C	0.003518	0.0000		
R(-1)-TW	-0.016884	0.3286		
R30-TW	0.269354***	0.0004		
W-TW	-0.282796***	0.0000		
Y-TW	0.006178***	0.0000		
TAX	-0.001471***	0.0000		
R-HK	0.432127***	0.0000		
R_CH	-0.019658*	0.0562		
R_US	0.151270***	0.0000		
	Variance equation			
C	7.00E-07	0.0121		
RESID(-1)^2	0.052648***	0.0000		
RESID(-1)^2*(RESID(-1) < 0)	0.102515***	0.0000		
GARCH(-1)	0.900362***	0.0000		
R-squared	0.463202	Mean variable		0.000118
Adjusted R-squared	0.461068	S.D. variable		0.013618
S.E. of regression	0.009997	Akaike information		-6.665749
Sum squared residual	0.201181	Schwarz criterion		-6.629669
Log likelihood	6752.072	Hannan-Quinn		-6.652509
Durbin-Watson stat	2.026097			

Note: "***", "**", and "*" denote significance at the 1%, 5%, and 10% levels, respectively.

4.3 Investors' Behavior (B) - Lock-in Effect

Long-term Lock-in Effect from CLRM Model

As shown in Table 3, we use classical linear regression model to explore the lock-in effect on Taiwan stock market transcation volume during SCG tax event announcement and implement period, and Hong Kong stock market as be a comparison group. We set as Time effect is a dummy explanatory variable indicating whether the SCG tax had been announced (after the announcement = 1) to affect investors' behavior in stock trading volume. The empirical results show up the stock trading volume in Taiwan stock market had declined by as much as 23.1% after tax announcement, even though it had pass though almost three and half years from Mar 2012 to Dec 2015.

Table 3. Estimates of regression model in Lock-in Effect (Long-Term)

Object	Taiwan and Hong Kong stock trading volume by CLRM model					
Period	Long-Term (Year 2007 ~ 2015) (Taxation period: Mar 2012 to Dec 2015)					
Item	Taiwan stock Market (Experimental group)			Hong Kong stock market (Comparison group)		
	Trading volume affect-TW(Y)			Trading volume affect-HK(Y)		
Variable	Coefficient	Std. Error	Prob.	Coefficient	Std. Error	Prob.
Intercept (C)	11.519	881.937	0.0000	9.866	596.821	0.0000
Time effect	-0.231***	-19.367	0.0000	0.022	1.455	0.1458
Day return rate	0.393	0.969	0.3323	-0.037	-0.087	0.9305
Amplitude rate	5.972***	8.877	0.0000	15.146***	20.129	0.0000
30 day return rate	57.134***	25.807	0.0000	16.098***	7.274	0.0000
R-squared	0.4009			0.1768		

Note: "***", "**", and "*" denote significance at the 1%, 5%, and 10% levels, respectively.

Long-term Lock-in Effect from DID Model

For the further to observe the long-term lock-in effect of decreased stock trading volume in Taiwan stock market whether caused by SCG tax event not affected by other factors. We apply Difference in Differences model and use Taiwan stock market as experimental group and Hong Kong stock market as control group to be compare their differences for dilutes the impact the global and international economic and political overall factors to them. The most important variable by DID model is the verification statistics of cross-effect variable, which dummy variable is multiply by Time factor and Economy area factor that referred to as the cross-effect factors. Time dummy variable is temporal factor (set the event period after the SCG tax announcement = 1, from Mar 2012 to Dec 2015) and Economy dummy variable is economy area factor (Taiwan Stock Exchange = 1; Hong Kong Stock Exchange = 0).

Table 4. Estimates of DID model in Lock-in Effect (Long-Term)

Object	Taiwan and Hong Kong stock trading volume by DID model			
Period	Long-Term (Year 2007 ~ 2015) (Taxation period: Mar 2012 to 2015)			
Variable	Coefficient	Std. Error	t-Statistic	Prob.
Intercept(C)	9.936	768.103	0.0000	9.936
Time effect	-0.001	-0.071	0.9437	-0.001
Economy effect	1.508***	120.531	0.0000	1.508***
Cross effect	-0.196***	-10.152	0.0000	-0.196***
Day return rate	0.301	0.967	0.3335	0.301
Amplitude rate	10.923***	20.607	0.0000	10.923***
30 day return rate	29.997***	18.380	0.0000	29.997***
R-squared	0.8491			0.8491

Note: "***", "**", and "*" denote significance at the 1%, 5%, and 10% levels, respectively.

As shown in Table 4, although the observation of the time effect variable has a small negative relationship and has not been significant. But the most important cross-effect variable by using DID model, which showed a significantly negative relationship. This results show after other effects of international interference factors like European debt crisis were diluted, the stock trading volume in Taiwan stock market still showed up a 19.6% decline than in Hong Kong stock market after a long time taxation period from almost half and three years. It shows that the SCG tax event has a negative and significant long-term Lock-in effect on the stock trading volume of Taiwan market.

4.4   Stock Market Environment (E) - Market Microstructures

The Adjustment of Investors' Structure in Taiwan Stock Market
To explore the adjustment of stock market structure (outside environment) could be affected by SCG tax announcement, we investigate the differences of stock market environment by t-test model between before and after the SCG tax announcement from March 2012. As shown in Table 5, we could find out the proportion of individual investor was 62% and the foreign institute was just 22.6% at the SCG tax announcement in year 2012. However, when SCG tax was implemented until year 2015, the proportion of individual investors falling to 53.3% and the proportion of foreign institute investors rising to 28.4%. In Table 6, we observe the proportion of investors' structure in Taiwan stock market by 45 months before and after the SCG tax announcement in March 2012.

The most important part of market structure is individual investors, but the proportion of individual investors is from original 66.27% to fall in 57.66% (down 12.99%) before and after the SCG tax announcement. As shown in Table 6, the empirical results of t-test model for adjustment of individual structure in Taiwan stock market is a significant reduction. Due to the previously empirical results in Lock-in effect of decreased the stock trading volume in Taiwan. Such result indicates due to the taxation substitution effects, investors will shift their investment to other low-tax or free-tax goods. Therefore, it could be caused the adjustment of investor's structure in Taiwan.

Table 5. Investors' structure of Taiwan stock market before and after the announcement

The proportion of investors' structure in Taiwan stock market				
Period	Year 2008, 2012, 2015 (Announcement day: 2012/3/28)			
Year	Individual	Domestic institute	Foreign individual	Foreign institute
2008 (Before)	61.7%	14.0%	2.3%	22.1%
2012	62.0%	15.4%	0.1%	22.6%
2015 (After)	53.3%	18.3%	0.0%	28.4%

Source: Taiwan Stock Exchange.

Table 6. Estimates of t-test for adjustment of structure in Taiwan stock market

The proportion of investors structure (45 months before and after the SCG tax event)			
Item	Individual investors	Domestic institute investors	Foreign institute investors
45 months before the event	66.27%	13.70%	19.75%
45-months after the event	57.66%	16.88%	25.43%
Variance of change	-12.99%	+23.22%	+28.73%
T-value (two-tailed check)	2.038	5.519	1.448
p-value	$p < 0.05\%**$	$p < 0.01\%***$	

Note: "***", "**", and "*" denote significance at the 1%, 5%, and 10% levels, respectively.

The Adjustment of Trading Information in Taiwan Stock Market
In order to observe the adjustment of trading information and stock transaction ecology (outside environment) could be affected by SCG tax announcement, we investigate the differences of stock market environment by t-test between before and after the SCG tax announcement from March 2012. As shown in Table 7 and Table 8 were the trading information in Taiwan stock market, we could find out that compare with year 2008 before tax announcement, after SCG tax announcement and implement at the end of year 2015, although the stock price index improved due to the stability of the economy. But the other trading information such like average annual trading volume drop down, average monthly turnover rate from 11.51% fall to 6.2%, and the number of branches of securities brokers declined. In Table 6, we observe the trading information by t-test model in Taiwan stock market by 45 months before and after the SCG tax announcement in March 2012. Although average annual stock price index from the original average of 7280.71 points to 8426.60 points (up 15.74%). But the most important stock trading volume is affected by SCG tax event, average daily stock

volume level of NT$106,134 million significantly decreased to NT$79,196 million (a decrease of 25.38%) before and after 45 months SCG tax event. The average monthly turnover rate, which represented the strength of retail trading activity, from the original average of 11.68% decreased to 6.67% (a decrease of 42.9%). The microstructure adjustment of stock market and trading ecology could be inter affected by taxation lock-in effect and overreaction.

Table 7. Stock trading information before and after the announcement

Stock trading information of Taiwan stock market

Period | Year 2008, 2012, 2015 (Announced in March 2012)

Item	Average annual stock price index (Unit: point)	Average annual trading volume (NT$ million)	Average monthly volume turnover (%)	Numbers of securities brokers (firms)
2008	7,024.06	26,115,408	11.51	978
2012	7,481.34	20,238,166	7.08	998
2015	8,959.35	20,191,486	6.20	929

Source: Taiwan Stock Exchange.

Table 8. Estimates of t-test model for adjustment of Taiwan stock trading information

Taiwan stock trading information before and after taxation

Item	Average stock price index	Average trading volume (NT$ million)	Average monthly turnover rate (%)
45 months before taxation	7280.71	106134.91	11.68
45-month after taxation	8426.60	79196.05	6.67
Variance of change	+15.74%	-25.38%	-42.9%
t-value (two-tailed check)	3.538	4.444	1.243
p-value	P < 0.01%***	P < 0.01%***	

Note: "***", "**", and "*" denote significance at the 1%, 5%, and 10% levels, respectively.

5 Concluding Remarks

Capital gains tax on securities was resubmitted in Mar 2012 and implement from 2013 in Taiwan, but this taxation suffered the huge public pressure in the society and caused the decline of stock trading volume in Taiwan stock market. Therefore, this taxation was finally abolished on the end of 2015. This study intends to combine behavioral finance and tax neutrality principle base on Lewin field theory to investigate whether investors' personal psychology, investors' behavior and the outer environment of stock market could be interact affected by capital gains tax for securities in Taiwan stock market. The empirical results have shown as follows: First, the impact of investor psychology (P) - investor overreaction (GJR model): empirical results show that the SCG tax events to investors have a long term overreaction effect for negative impact on return rate of Taiwan stock market, bad news will increase the asymmetry volatility of return rate in stock market. Taiwan stock investors for SCG taxation event always existed irrational cognitive biases (such as representative bias) even though SCG taxation had offered three times in pass.

Second, for investor behavior impact (B)- Lock-in effect (CLRM model and DID model): the empirical results show that even though SCG taxation have been announcement and implement for half and three years from Mar 2012 to year 2015, the trading volume of Taiwan stock market had still decreased as much as 23.1% in CLRM model. We apply DID model to use Hong Kong stock market as control group to be compare their differences for dilutes the impact the global and international economic and political overall factors. The cross-effect variable by using DID model, which showed a significantly negative relationship. This results show after the effects of international interference factors like European debt crisis were diluted, the stock trading volume in Taiwan stock market still showed up a 19.6% decline than in Hong Kong stock market. It shows that the SCG tax event has a negative and significant long-term Lock-in effect on the stock trading volume in Taiwan market.

Thirdly, for environment impact (E)-adjustment of stock market microstructure: due to stock trading volume caused by SCG taxation was depression by 25.38% after taxation announcement and implement though 45 months refer as had a lock-in effect, individual investor structure in Taiwan stock market have been significantly reduced by 12.99%. From the point of view in stock's investor demand elasticity to discuss individual investors' behavior, in general the demand flexibility of individual investors' stock investments is relatively higher than other foreign institutes. Due to originally the individual investors' percentage in Taiwan stock market is more than 66%, SCG taxation would affect the adjustment of individual investor structure in Taiwan.

From the point of view in long-term over-reaction in negative impact and asymmetry of stock return rate, lock-in effect in stock trading volume and adjustment of stock structure, they showed up that the stock investors' sense of tax deprivation psychological reaction. And Taiwan SCG taxation for stock investment will decrease investors' original profit and result in change investors' investment behavior to choose other low-tax or free-tax good to cause the lock-in effect of declined Taiwan stock investment and adjustment of investors' structure for stock market. The empirical results show up that SCG taxation in stock market could affect the investors' psychology and behavior mechanism with field theory and in further to affect the trading information of the stock market and distort the resource allocation in the capital market for adjustment of stock market microstructures, which is worthy of the reference and attention of the competent authorities.

References

Ashenfelter, O., Card, D.: Using the longitudinal structure of earnings to estimate the effect of training programs. Rev. Econ. Stat. **67**(4), 648–660 (1985)

Bhootra, A.: On the role of intangible information and capital gains taxes in long-term return reversals. Finan. Manag. **42**, 537–573 (2013)

Dai, Z., Maydew, E., Shackelford, D.A., Zhang, H.H.: Capital gains taxes and asset prices: capitalization or lock-in? J. Finan. **63**(2), 709–742 (2008)

Diaz Caro, C., Crespo Cebada, E.: Taxation of capital gains and lock-in effect in the spanish dual income tax. Eur. J. Manag. Bus. Econ. **25**(1), 15–21 (2016)

Feldstein, M., Yitzhaki, S.: The effect of the capital gains tax on the selling and switching of common stocks. J. Publ. Econ. **9**(1), 17–36 (1978)

Griffin, D., Tversky, A.: The weighing of evidence and the determinants of confidence. Cogn. Psychol. **24**, 411–435 (1992)

Glosten, L.R., Jagannathan, R., Runkle, D.: On the relation between the expected value and the volatility of the nominal excess return on stocks. J. Finan. **48**, 1779–1801 (1993)

George, T.J., Hwang, C.Y.: Long-term return reversals: overreaction or taxes? J. Finan. **62**(6), 2865–2896 (2007)

Hayashida, M., Ono, H.: Tax reforms and stock return volatility: the case of Japan. J. Asian Econ. **45**, 1–14 (2016)

Falsetta, D., Rupert, T.J., Wright, A.M.: The effect of the timing and direction of capital gain tax changes on investment in risky assets. Accou. Rev. **88**(2), 499–520 (2013)

Kahneman, D., Tversky, A.: Subjective probability: a judgment of representativeness. Concept Prob. Psychol. Exp. **8**, 25–48 (1974)

Kahneman, D., Tversky, A.: Prospect theory: an analysis of decision under risk. Econometrica **47**, 263–291 (1979)

Lewin, K.: Principles of Topological Psychology. Translated by Fritz Heider and Grace M. Heider. McGraw Hill, New York (1936)

Maleki, B., Sameti, M., Sameti, M., Ranjbar, H.: The effect of capital gain tax on capital formation, financial development and economic growth, case study of selected Europe Union countries. J. Glob. Pharma Technol. **8**(12), 493–503 (2016)

Lo, M.M.: A study on lock-in effect of capital gains tax for securities in Taiwan stock market— an application of DID model. Mod. Econ. **6**, 954–964 (2015)

Somers, M.: An economic analysis of the capital gains tax. Natl. Tax J. **1**, 226–232 (1948)

Stiglitz, J.E.: Some aspects of the taxation of capital gains. J. Publ. Econ. **21**, 257–294 (1983)

Sahm, M.: Methods of capital gains taxation and the impact on asset prices and welfare. Natl Tax J. **61**(4), 743–768 (2008)

Yin, Y.P.: Theoretical analysis on the capital market tax system. Financ. Econ. **209**, 68–72 (2005)

Analysis of the Influence of Strengthening Visual Message Oriented Design on Way-Finding Problems

Chi-Jui Tsai[1,2] and Kuo-An Wang[2,3(✉)]

[1] Department of Advertising and Strategic Marketing,
College of Communication, Ming Chuan University, Taipei City, Taiwan
tsaichi@mail.mcu.edu.tw
[2] Department of Industrial Education and Technology,
National Changhua Normal University, Changhua City, Taiwan
[3] Department of Management Information Systems,
Central Taiwan University of Science and Technology, Taichung City, Taiwan
gawang@ctust.edu.tw

Abstract. People receive various kinds of information through visual senses in their lives. Vision is the most important source of human body's acceptance of external information and then causes various activities. Therefore, this study investigated people's information generation from the aspects of message processing, message perception and information identification. This study will investigate whether the subject's acceptance of the message will affect his decision-making behavior. The research objects are college students, and the school is the field. In the experiment, there are two paths for the students to arrive at the destination. One of the paths is selected to interfere with the information processing and information marking. According to the five-element theory of environmental information identification, set the location of the visual message and record the student's action path to the classroom by RFID sensors every week. We use the sensors record to determine whether the student will be affected by message processing and informational mark, and change his decision on path selection. From the difference in the path, it is possible to study whether the student being tested will be affected by the message processing and information indication, and make his path (way-finding) decision. In four times of perceptual data, the number of choices has continued to grow positively for path designed with message processing and informational indications. Conversely, the path that has not been processed by the message shows negative growth. The results of this study will help us understand the post-information processing. If appropriate message processing design is used and information mark are set up in appropriate locations, then these message processing can influence the decision-making choices of the students participating in the experiment. This experience can be used as a reference method for message processing and message setting in the environment, so visually oriented design can be used to make the flow guidance more effective.

Keywords: Message processing · Information marking system · Way-finding

© Springer Nature Switzerland AG 2020
L. C. Jain et al. (Eds.): SICBS 2019, AISC 1145, pp. 258–269, 2020.
https://doi.org/10.1007/978-3-030-46828-6_22

1 Introduction

People in the environment are walking from a certain point to a certain point, when walking and selecting a path, people will be affected by the information and information processing set by the designer. This study will focus on the five cognitive elements of message processing and information orientation, information message design, visual message design, and environmental identification. The respondents are mainly based on the students and school environment, and the students in the survey can take either of the two paths to reach their destination. This study wants to interfere with one of the paths and treat the path with post-additional information. Then we assess whether students will be affected by these additional information messages to change the path of choice. If the survey and experiment in this study can interfere with the student's choice, then the post-additional information and message processing in this study can be further extended to a larger environment, as a reference for information-oriented and crowd guide.

2 Message Processing and Information Orientation

2.1 Elements of Message Recognition and Identification

How to apply visual messages and information to setup the design in the environment, we need to explore the relationship between environmental perception and information cognition. This study explores the phenomenon of human beings dealing with this information from the setting of information vision [1]. The so-called path can be thought of as a corridor or a road that can be used as a person or a car. The edge is a linear component that cannot be regarded as a road. It is a dividing line between two regions, like a wall, a coast, etc. Area refers to the large area of urban lands. These lands or areas have obvious commonality or use characteristics in architecture, landscape and signage; nodes are the intersections of typical channels or concentrated places with characteristics. Landmarks have special features, which are real objects and are often considered to be an outward reference point, often evident in distance [2].

Generally speaking, in the process of environmental message orientation and "way-finding", the user establishes a Spatial Conception through spatial cognition to determine the way-finding strategy, and according to the direction of each predetermined reference point. Moving forward, the sequence formed by each reference point becomes the "path". The selection of the path is determined by the space under its "conceptualization", and the confirmation of the reference point in the path is made to reach the destination to which it is going. The content can be roughly divided into three stages, namely, the spatial cognition establishment stage, the decision stage and the execution stage. In the spatial cognition phase, Passni [3] considers the recognition of spatial orientation as the most important factor in establishing the concept of space. The general term for "orientation" is the combination of "direction" and "location". Among them, the ability to judge the orientation and the ability to judge the direction are closely related. It is a kind of ability to maintain the direction of going ahead. In the process of moving forward, the relative position is further distinguished according to its relative relationship. In short, it means that "people can make a rough connection between the current position and the direction they want to go." Therefore, in the process of way-finding decision-making, the cognition of "orientation" in the concept of space is extremely important.

Under the setting of environmental messages and visual information, this orientation and way-finding mode is also affected by some environmental conditions, such as information features on the plane, the number of decision points, planning on the moving line, and so on. In the study of O'Neill [4], it is pointed out that when the floor plan configuration increases, the way-finding error does not necessarily increase, but when the planes of different buildings are compared, the buildings with higher plane complexity are compared. In less complex building planes, the way-finding error rate will increase. The decision point is the reference point in the way-finding process. When the decision point is too much, the uncertainty of the user's judgment is increased, the choice is easy to be troubled, and the difficulty of finding the path is increased [5]. The number of decision points depends on the complexity of the plane. The planning of moving lines tends to focus more on function and convenience in the design stage. Moderate and direct convenience in space helps to reduce way-finding problems [6]. However, when the number of intersections in the route increases, the number of "decision points" increases, but it is easy to cause trouble. If the moving line is too complicated, the planning of "post-additional information" in the environment becomes more and more important [7].

2.2   Message Processing and Information Marking System

The marking system is one of the sources that provide a source of information for the environment. It is a kind of "post-additional information". "Marking" is a tool for indicating or conveying a message through a way that can be perceived by one's senses. The form of the mark is usually presented in the form of color, pattern, signal, text, etc., and through the overall labeling plan, systematic application and configured in the appropriate space and location to help solve the problem of way-finding [8]. The most straightforward way to find a way-finding problem is to find the location of the marking system. It is expected that the marking system will provide an answer to the way-finding decision process.

Although the marking system is one of the many links to solve the problem of way-finding, the function of the marking system has also been listed as an important research item in the study of the topic of "way-finding". Arthur's discussion of marking systems in visual communication suggests that the marking system is often associated with the user's way-finding decisions in the built environment. He also stated that the mark is a requirement for people to solve problems, such as helping users identify their direction, location and destination, to provide reference for decision making [9].

There is currently no specification for mark design in our country. Usually the design of the marking system is determined by the different needs of the case. According to the definition of "Taiwan Public Logo Design Manual", the so-called mark design is a combination of words, patterns and colors. Provide visual design which had functions such as identification, guidance, explanation, warning, etc., with clear shapes and patterns. According to the function of the mark and the content of the message, the mark are classified into six types [10] (Table 1).

Table 1. Mark types and definition

Type	Definition
Identificational mark	Identificational mark represents the label of the object itself, indicating the name of the object, called the name mark
Directional mark	Directional mark has indicators that direct the user to a specific target or direction, mostly presented in lines, line markers, and arrows, guiding the sequential continuity in the environment
Orientational mark	Orientational mark presents the relative relationship, overall condition and related facilities in the environment or building with floor plan or map. Generally speaking, it appears in the space entrance, traffic and other places to provide an "overview" of spatial awareness
Informational mark	Informational mark describes the content of the thing, the method of operation, related specifications, activity content and notice
Regulatory mark	Regulatory mark is used to remind, prohibit or manage the rules and guidelines for the use of behavior, and to maintain the function of safety and order
Ornamental mark	Ornamental mark modifies or emphasizes the individual elements of the environment and has the landscaping features such as archways and wall decorations

2.3   Message Processing and Related Information Technology

This study hopes to record the data of each stage of way-finding behavior by recording the students in the process of going to the destination. To understand the factors that influence the way-finding, and the interaction between the subjects and the environmental information. As a reference for future environmental information planning and design, it can also be used as the basis for planning and formulating strategies for way-finding and other related issues [11]. The definition of related nouns is as follows:

Way-Finding Behavior and Guided Design: It is a behavioral process that seeks to answer spatial questions. It includes behavioral responses when people are lost, environmental sensing and cognitive status, pathfinding decisions, and action plans. The plan was put into action at the appropriate location until a series of responses before the arrival of the destination [12, 13].

Spatial Conception: Through the information obtained from environmental information, such as spatial orientation, shape and spatial function, people can distinguish orientation and direction. According to the spatial structure under "conceptualization", the relative relationship between the current position and other related environments is formed in the brain with structured overall spatial knowledge, which is the product of the individual through spatial cognition and learning [14].

Environment Information: Environmental information is the meaning and message of the space conveyed by the environment. The overall spatial perception formed by the spatial objects such as stairs, walls, windows, or the size and scale of space is called "native information". In addition, messages conveyed via facilities or auxiliary equipment, such as the name of the space, the way of using, and the distribution of functions, are "post-additional information" [15]. In the way-finding process, the information content provided by "Native Information" helps people to directly identify and judge the location, and "post-addition information" can give users assistance of space information that can't be provided by the environment [16].

The overall hardware and software deployment, described as follows:

In order to achieve the goal of reading the node signal, this study uses RFID wireless identification technology. This architecture is mainly composed of readers and tags and back-end applications. The reader emits radio waves, touches the RFID tag in the sensing range, and the current generated by the electromagnetic induction supplies the RFID tag to operate and emits an electromagnetic wave response sensor. The information is linked to the back-end database to identify, track, and confirm the status of the subject through the node.

Overall, there are the following main components to achieve the goal of reading the tag signal through the node.

Passive Tag:

1. Wafer: used to store and process data, modulate and demodulate radio signals.
2. Antenna: used to receive the signal sent by the reader and send the requested data back to the reader.

Reader: The two-way radio wave transceiver sends a signal to the tag and interprets its response, and in combination with the application system, receives the command from the host side, and transmits the data stored in the tag to the host in a wired or wireless manner.

Back-End Database: For the details of the tag, when the reader receives the UHF tag signal, it will connect to the backend application database via the wireless network. After obtaining the details of the UHF tag, it is used by other applications (Fig. 1).

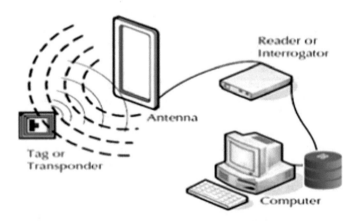

Fig. 1. Hardware and software deployment of RFID system.

When designing this overall hardware layout, the four key points to focus on are as follows

1. Master the requirements, complete reader construction and antenna configuration
2. Use Manual pasting tag
3. Avoid noise interference in the environment around the nodes
4. Repeated testing of readers and tags

2.4   Design of Visual Messages

The design of visual messages can be set by means of communication and creativity, or it can be designed with digital information. It breaks through the feeling of visual flatness and static form, and shows the attraction of visual shape and dynamic display. This study lists some of the ways in which information and visual messages can be used in design [17, 18].

Change the Visually Viewing Angle: Ordinarily people used to watch things with eyes parallel or up and down. Consequently, designers apply the visual elements to ceiling, floor, wall, etc. that can change the custom of watching, break through the limit of watching by standing on the ground, and create the visually magic illusion that walls and floors involve another space (Fig. 2(a) & (b)).

(a) Change the visually viewing angle-1

(b)Change the visually viewing angle-2

(c) Mirroring and mirroring replace-1

(d) Mirroring and mirroring replace-2

(e) Flipped design of swerve and corner

(f) Reverse and differentiation design of crossing and penetration

Fig. 2. Design of visual messages

Mirroring and Mirroring Replace: the generation of mirroring can increase the spacial depth sense. With the visual creation design words, through the objects' transmission of reflection and mirroring, forming more symmetrical relation about attachment and penetration spatially and visually (Fig. 2(c) & (d)).

Flipped Design of Swerve and Corner: turning and corner is the transition of angle in action. Normally, environmental visual design would not be used at the turning and corner. Therefore, building corner surprise in urban, is not only presenting delightful visual design as well as promoting the application of the corner, could catch people's attention. It is no longer as the corner people pass by, hence the flipped design for the visual custom (Fig. 2(e)).

Reverse and Differentiation Design of Crossing and Penetration: in a way of jumping off the spatial reality, constructing design penetrating from walls and ground, reverse and differentiation break through people's cognitive rules and visual-Inertial in ordinary life experience. It brings the visual impact, impels psychological reaction and reach the surprising visual effect by producing variation, visual phenomenon and spatial illusion against the normality (Fig. 2(f)).

2.5   Design of Nodes Data Collection

This study uses the campus environment to identify five elements (Paths, Edges, Districts, Nodes and Landmarks), combined with RFID sensing technology to collect students' walking line sample data. This survey is based on the students who take the database system course once a week. When the students move from the school gate to the classroom, the students can reach the classroom by two paths. At the beginning of the experiment. The two paths do not arrange information and visual information. At this time, we will first conduct a survey on the path of student walking. The two paths do not arrange information and visual information. At the beginning of the second week, we continue conducting a survey of the student's walking path, and then make a visual message and information arrangement for one of the paths.

Experimental design steps for this study are as follows:

Step1: The selected Path one (P1) is a flat road. After advancing to the general staircase and then climbing up to the fourth floor, it travels along the corridor path to the destination classroom. Path 2 (P2) travels from the flat road to the elevator entrance. After taking the elevator to the fourth floor, it goes in the opposite direction to the path (P1) and travels along the corridor to the destination classroom.
Step2: The two paths do not process any information and visual information. First, collect the travel route data of the tested students to the destination classroom.
Step3: Visual information processing and information labeling are performed on the important nodes of Path One (P1). According to these visual principles, application and design are applied, so that it is possible to effectively guide students to select Path One (P1) to reach the target classroom. Path 2 (P2) remains as it is and does nothing.

In this study, 50 students who taking the course "Database System" were requested to use the RFID tag to sense the reader in the classroom as a basis for attending the course. The students surveyed are not aware of participating in the survey to avoid interfering with the results. They thought that their bringing RFID tag just to be a attending record.

The way in which this research survey recorded is as follows:

Path 1 (P1) and Path 2 (P2) were used the same sensing points at the start and end points. Each path has four sensing points. In addition to the starting and ending sharing sensors, the other two sensing points. Path 1 sets a sensing point (P1a) at the entrance of the stairs on the first floor. Nodes also have a sensing point (P1b) when the stairs reach the fourth floor. Path 2 (P2) is the elevator entrance on the first floor (P2a) and the elevator exit on the fourth floor (P2b). Set a little sensing point. Therefore, students will leave a record when they travel through the sensing point. If a student walks a certain path to the classroom, then all four sensing points will have a record, and we can know which path the student chooses to reach the destination classroom (Fig. 3).

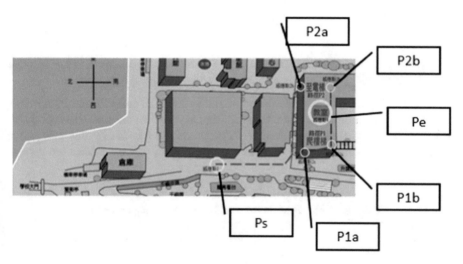

Fig. 3. Design of nodes (RFID tag reader location) and path diagram. *Path1: Ps(start)-P1a-P1b-Pe(end). *Path2: Ps(start)-P2a-P2b-Pe(end)

2.6   Assessment and Data Analysis

Based on the student's usual path to the classroom, we set four RFID tag sensing points on the important nodes of the two paths. The starting and ending points of the two paths are similar. After the paths are separated, there are two sensing points for identification. This is an important node for recording the student's travel route. According to the principle of information processing and environmental messages identification, the node is an important reference position in the environment identification information. The following table sample explains the choice of 50 student travel paths (Table 2):

For the record interpretation about path is:

1. If the node P1a or P1b has an inductive record, it means that the student is taking the Path 1 route.
2. If the node P2a or P2b has an inductive record, it means that the student is taking the Path 2 route.

Table 2. Descriptive records of travel paths at (week2)

Ps	P1a	P1b	P2a	P2b	Pe
1	1	1	0	0	1
1	1	1	0	0	1
1	0	0	1	1	1
1	1	1	0	0	0
0	0	0	1	1	1
1	1	1	0	0	1
0	0	0	1	1	1

1: an inductive record. 0: a non-inductive record

3. If there is no record of two paths a or b, it means that the student is taking other paths to the classroom and is not listed as valid data for the experiment.

In Table 3, the first week's sensing data shows that both paths P1 and P2 have not yet increased the information and message design to guide. At this time, the data shows that there are more students in Path 2 (P2) who choose to walk by this path to the destination classroom. Because there are elevators on this route, students are used to taking this path.

Table 3. Path record statistics of four weeks during experiment

Week	Path 1	Path 2
1	12	35
2	20	28
3	33	16
4	38	10

Starting in the second week, we added information and label design to Path1 as a guide. Attracting the visual attention and attention of pedestrians, while Path 2 (P2) does not do any message processing and labeling. Comparing this week's data with the first week's data, it shows that Path One (P1) adds 8 students to choose this path to reach the destination classroom.

In the third week of the sensory data, path one (P1) also has the previously set information message and label design to attract the pedestrian's visual attention. Path 2 (P2) does not do any message processing as usual. The data obtained is compared with the data of the second week, showing that Path one (P1) increases 13 students choose this path to reach the destination classroom. On the contrary, the number of students in Path two (P2) is reduced by 12.

The fourth week was continuously observed, and the experimental design was not changed. The data obtained is compared with the data of the third week, showing that Path one (P1) increases the number of students who choose this path to reach the destination classroom. On the contrary, the number of students in Path two (P2) is reduced by 6.

Comparing the two paths P1 and P2, the change in data during the four weeks shows that the number of students who choose the path P1 as the route to the classroom is increasing. The number of people who choose path P2 is continuously decreasing (Fig. 4).

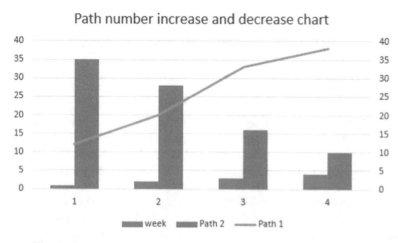

Fig. 4. Increase or decrease in the number of people in different paths

We also conducted a visual sensibility questionnaire survey of the subjects after the experiment. The questionnaire is designed on four levels. The four levels are as follows: visual viewing angle, color impression, symbol and appearance, and spatial perspective. The average of the scores of these four levels is the value of visual sensibility. Through the data of the questionnaires obtained from these four levels, statistical analysis is carried out to assess the situation of the information messages received by the subjects. And gender is used as a variable to compare different genders of subjects, whether they have differences in visual sensibility, to explore the characteristics of men and women on the topic of way-finding.

From the basic descriptive statistics analysis (Table 4), we found that women have higher visual sensibility than men in the two nodes P1a and P1b where we have completed information messages design. To verify this phenomenon in more detail, we performed a t-test on the sample and found that women were indeed higher in visual sensibility than men and reached a significant level (Table 4).

Table 4. Descriptive statistics analysis of visual sensibility during different gender

Gender	Nodes	Minimum	Maximum	Average	Standard deviation	t-test
Female (N = 18)	P1a	2	5	4.2894	.12474	0.412[*]
	P1b	2	5	4.1313	.11707	0.443[*]
	P2a	1	4	2.2175	.34759	
	P2b	1	4	2.6663	.56287	
Male (N = 32)	P1a	2	5	3.8906	.30813	
	P1b	2	5	3.6503	.27582	
	P2a	1	4	2.4233	.50469	
	P2b	1	5	2.4197	.53004	

* $p < .05$

3 Conclusion

Usually, people's recognition and location reference to the environment will be influenced by five factors to establish their own position in the environment. The five elements are Path, Edges, Region, Nodes and Land Mark. The results of the research survey showed that when the first week was not added to the post-additional information processing, most of the students chose the path P2 to reach the upstairs classroom through the elevator. However, starting from the second week, we add the message processing after the path P1 is placed, and the path P2 remains unchanged. At this time, due to the increase of the influence of new information and messages at path p1, the number of people maintained growing from the second week to the fourth week. The number of students who choose this path P1 to reach the destination classroom continues to increase, which shows that students are affected by the setting information and message orientation, and change the route they are used to, and change to the p1 path climbing from the stairs to the classroom. The processing of these post-addition information can be applied from the research to the fixed environment or urban environment. When more than one path can reach the destination, you can guide or disperse the crowd by setting up information and message processing. This is also the main purpose of this study. It is necessary to conduct an experimental investigation from a campus environment to verify that the information and message processing, if properly set, may change people's choices.

Therefore, it is valuable to present the influence of post-information processing and design through this research. It can be applied to larger or more complex environments to solve the problem when the original line is not good or there are multiple paths.

In addition, in the post-experimental questionnaire survey, we also found that female students in the overall visual sensibility of the four indicators were significantly higher than male students. This result also gives us a vision. Perhaps in many commercial or artistic marketing, female customers should pay more attention to their visual experience, such as strengthening packaging design or creating an attractive atmosphere. This is very beneficial to the recognition and support of female customers. Therefore, for women-oriented sales strategies, through certain designs and arrangements to enhance their visual perception, it will be able to effectively influence their action decisions.

Ethical Approval. All procedures performed in studies involving human participants were in accordance with the ethical standards of the institutional and/or national research committee and with the 1964 Helsinki declaration and its later amendments or comparable ethical standards. The survey data were anonymized. This study was also approved by the Department of Management Information Systems, Central Taiwan University of Science and Technology ethics committee in advance.

References

1. James, W.: The Principle of Psychology, vol. 1 and 2. Henry Holt, New York (1890)
2. Lynch, K.: The Image of the City. MIT Press, Cambridge (1960)
3. Passini, R.: Wayfinding in Architecture. Van Nostrand Reinhold Company Inc, New York (1939)
4. O'Neill, M.: Effects of signage and floor plan configuration on wayfinding accuracy. Environ. Behav. **23**(5), 553–574 (1991)
5. Down, R., Stea, D.: Image and the Environment: Cognitive Mapping and Spatial Behavior. Aldine, Chicago (1973)
6. Mack, R.L., Burdett, J.M.: When novices elicit knowledge: question-asking in designing, evaluating and learning to use software. In: The Psychology of Expertise: Cognitive Research and Empirical AI, pp. 245–268. Springer, New York (1992)
7. Devlin, A.S., Bernstein, J.: Interactive way-finding: map style and effectiveness. J. Environ. Psychol. **17**, 99–110 (1997)
8. Arthur, P., Passini, R.: Wayfinding-People, Signs and Architecture. McGraw-Hill Ryerson Ltd, Whitby (1992)
9. Montello, D.R., Lovelace, K.L., Golledge, R.G., Self, C.M.: Sex-related differences and similarities in geographic and environmental spatial abilities. Ann. Assoc. Am. Geogr. **89**(3), 515–534 (1999)
10. Nielsen, J.: Usability Engineering. Academy Press Professional, New York (1993)
11. Kitchin, R., Blades, M.: The Cognition of Geographic Space. I.B. Tauris, London (2002)
12. Robinson, A.H., Morrison, J.L., Muehrcke, P.C., Kimerling, A.J., Guptill, S.C.: Elements of Cartography, 6th edn. Wiley, New York (1995)
13. Shemyakin, F.N.: General problems of orientation in space and space representations. In: Ananyev, B.G. (ed.) Psychological Science in the USSR. U.S. Office of Technical Reports, Arlington (1962)
14. Paivio, A.U.: Imagery and Verbal Processes. Holt, Rinehart and Winston, New York (1971)
15. Arthur, P., Passini, R.: Wayfinding, People, Sign and Architecture. McGraw-Hill, New York (1992)
16. Wang, R.-H.: A Study on Wayfinding and Construction of Spatial Conception in Underground Arcade. Chung Yuan Christian University, Department of Architecture Institute (2003)
17. Smiciklas, M.: The Power of Infographics: Using Pictures to Communicate and Connect with Your Audiences. QUE Publishing, Indianapolis (2012)
18. Steuer, J.: Defining virtual reality: dimensions determining telepresence. J. Commun. **42**(4), 73–93 (1992)

Face Forgery Detection Based on Deep Learning

Yih-Kai Lin[(✉)] and Ching-Yu Chang

Department of Computer Science, National Pingtung University,
No. 4-18 Minsheng Road, Pingtung City 90003, Taiwan
yklin@mail.nptu.edu.tw, cbe104028@nptu.edu.tw

Abstract. In this paper, we propose two deep learning approaches for face forgery detection. The first approach uses neural networks to detect fake faces in individual images. Three methods of this approach are studied. The second approach uses long short-term memory recurrent neural networks to detect fake videos by checking inconsistencies between successive frames. The experiments show that accuracy of forgery detecting is around 90%.

Keywords: Forensics · Fake faces · LSTM

1 Introduction

Using face forgery techniques [5, 7, 13], the faces of the original video are replaced by other faces to change the facial identity with similar facial expression. Or we do not change the facial identity but modify the facial expressions. Since facial features play an important role in our society, face forgery becomes a critical issue when the powerful neural networks are used to replace the faces in images or videos. With today's neural network models, fake videos with face forgery are difficult to detect by human vision. In this paper, we propose four digital media forensic methods to automatically detect face forgery. Because the convolutional neural networks (CNNs) have a good result for pattern recognition [12], it is a high possibility that convolutional neural networks can identify feeble patterns of digital fake images which cannot be found by human vision. Thus, we use convolutional neural networks to detect face forgery. Also, because the face forgery technique swaps the faces in videos by manipulating each individual frame of the videos, it is possible that there exists a small inconsistency between successive frames which can be recognized by neural networks but cannot be detected by human vision. We use long short-term memory (LSTM) recurrent neural networks [10] to detect these abnormal patterns between video frames. Our proposed scheme is shown in Fig. 1.

Y.-K. Lin—Supported by the Ministry of Science and Technology of Taiwan under contracts MOST-108-2221-E-153-005.

L. C. Jain et al. (Eds.): SICBS 2019, AISC 1145, pp. 270–278, 2020.
https://doi.org/10.1007/978-3-030-46828-6_23

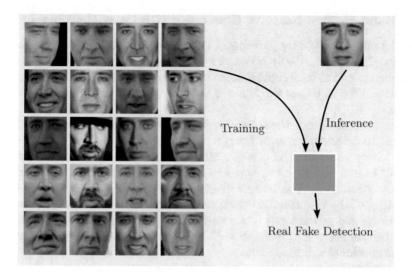

Fig. 1. The architecture for detecting the face forgery.

2 Related Work

Thies et al. [14] use a facial reenactment system to construct a 3D model from source video streams and use image-based rendering techniques to generate target videos with altered facial movements. FaceSwap [7] uses a computer graphics-based approach to make fake faces. They first detect facial landmarks of source faces then map these landmarks to a 3D template model. Generative adversarial networks (GANs) [9] have good performance for generating images, so they are used to generate fake faces [1]. Karras et al. [11] use GANs to synthesize fake faces. Upchurch et al. [15] use GANs to alter face attributes like skin color, mustache, age, etc. DeepFakes [5] uses two autoencoders to swap faces of images. Their method first crops the face from source images. Then, the method creates a fake image by applying a trained encoder and decoder to the target face.

The CNN-based forgery detection has been proposed in recent literature [2, 4]. The abnormalities of video frames like duplicated frames, dropped frames, chroma-key compositions, and copy-move manipulations are studied to detect fake videos [3,6,8]. The above approaches suffer robustness issues, since image resizing and compression are used to erase manipulation traces from the video format. Thus, as our methods use abnormalities of the face area in videos, it is the most robust feature for forgery detection.

3 The Proposed Scheme

We propose two approaches for forgery detection: (A) hand-crafted features combined with deep learning and (B) checking of continuous frames by LSTM.

3.1   Hand-Crafted Features Combined with Deep Learning

Traditionally, local binary patterns (LBP for short) and histogram of oriented gradient (HOG) are usually adopted for face recognition. The success of both methods shows that they capture some important features of the face. Therefore, it is natural for us to use the features adopted in these methods to detect face forgery.

Our methods merge these hand-crafted features with convolutional neural networks to detect forgeries of facial images.

We study three methods below. Method 1. The images are transferred to the LBP format and then are sent into a convolutional neural network for classification. Method 2. The images are transferred to the LBP format and the histograms of the resulting LBP images are calculated. Then the histograms of LBP are treated as the input of a convolutional neural network. Method 3. The images are transferred to 8-direction HOG format and then treated as the input of a convolutional neural network.

In the detail of Method 1, the images are first converted to gray level images. Then we apply the LBP converter to the gray level images with the parameters which number of circularly symmetric neighbour set point 24 and radius of circle 3. The resulting LBP images are sent to our convolutional neural network. Our network has 3 convolutional layers followed by 2 fully connected layers. The full network is shown in Fig. 2.

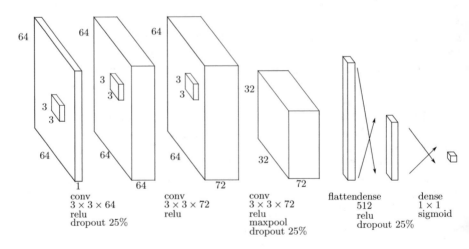

Fig. 2. The architecture for recognizing LBP features. The network has 3 convolutional layers followed by 2 fully connected layers.

In the detail of Method 3, the images are first transferred to LBP format, and the histogram width 256 is calculated from these LBPs. Then the histograms of every image are sent into the fully connected neural network as the input (Fig. 3).

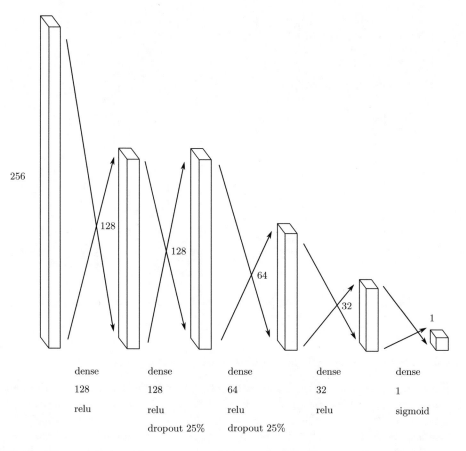

256

128

128

128

64

32

1

dense | dense | dense | dense | dense
128 | 128 | 64 | 32 | 1
relu | relu | relu | relu | sigmoid
| dropout 25% | dropout 25% | |

Fig. 3. The architecture for recognizing Histogram-LBP features. The network has 5 fully connected layers.

In the detail of Method 2, the images are first transferred to 8-direction, viewed as 8-channel, HOG features which are 2×2 cell and 1×1 cell block. Then the HOG features of every image are sent into the fully connected neural network as the input. The full network is shown in Fig. 4.

Method 1, Method 2, and Method 3 use the binary cross entropy lose function

$$-\frac{1}{N} \sum_{i=1}^{N} [y_i \log(P(y_i)) + (1 - y_i) \log(1 - P(y_i))]$$

where y_i is the target and $P(y_i)$ is the predicted probability for all N samples. The sigmoid activation function used in this paper is

$$f(x) = \frac{1}{1 + e^{-x}}.$$

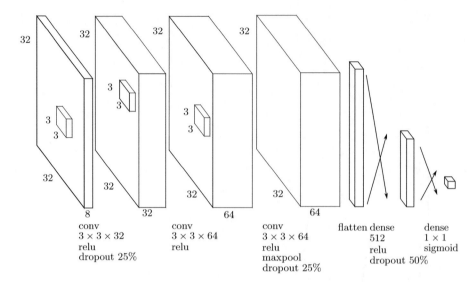

conv   conv   conv   flatten dense   dense
3 × 3 × 32   3 × 3 × 64   3 × 3 × 64   512   1 × 1
relu   relu   relu   relu   sigmoid
dropout 25%   maxpool   dropout 50%
                            dropout 25%

Fig. 4. The architecture for recognizing Hog features. The network has 3 convolutional layers followed by 2 fully connected layers.

3.2 The Analysis of Feature Continuity Between Frames

The previous mentioned three methods only detect individual frames split from videos. Now, we propose a method, denoted Method 4, that considers continuity between video frames. The method can be divided into three steps. First, the training set of the input video which contains continuous actions of the face is collected. For example, the video contains a side face turning to the front or a face swaying with the head. Then, the facial landmarks of each frame are fetched to form the feature sequences. These sequences are used to estimate gaze direction, shape of mouth and head pose of each frame by the neural network. Finally, the feature sequences of videos are fed into the recurrent neural network for analyzing the difference between the fake and real videos. The proposed method uses a 256-layer LSTM recurrent neural network. The feature sequences are divided into groups of 5 frames and each frame contains 30 facial landmarks. The 30 landmarks contain 3-axis right and left eye, 3-axis head pose direction, 3-axis shape of mouth location and one frame identification.

4   Experiment and Comparison

The proposed methods are implemented to evaluate the practical performance. The experiment sample images are two well-known persons Nicolas Cage and Donald Trump. Both persons are usually used for image processing research because they are easy to fetch from the Internet. Nicolas Cage's images are used as the real face, and Donald Trump's images are converted to Nicolas Cage's images to be the fake faces. All of the images are collected in the Trump−Cage dataset [5].

We present experiments of forgery detection in the Trump−Cage dataset. The dataset contains 313 real images of Cage's face and 306 fake images which are the original Trump's face but converted to Cage's face. From now on, the image sets are called real and fake respectively. Some random selected real images are shown in Fig. 5 and some random selected fake images are shown in the second row of Fig. 6. The first row of Fig. 6 shows the original images' correspondence to the fake images in the second row. That is, the images in the second row of Fig. 6 are drawn from fake and are converted from Trump's real images in the first row of Fig. 6 using the DeepFakes method [5]. Note that Trump's real images are not included in dataset real.

The image size of dataset fake and real is 64 × 64. The 526 images of real and fake are used as the training set and 93 images are used as the testing set. The testing set contains 46 real images (positive samples) and 47 fake images (negative samples).

Fig. 5. The random selected images from real which contain 313 Cage's real faces.

We use the Adam optimizer to train all three methods. Method 1 contains 31092561 trainable parameters, Method 2 contains 59777 trainable parameters, and Method 3 contains 5596577 trainable parameters. We set batch size 128 and epochs 30 to evaluate three models. In this paper, the metric used to measure the performance is

$$accuracy = \frac{TP + TN}{TP + TN + FP + FN}$$

Fig. 6. The random selected images from fake, which contains 306 Cage's fake faces, are shown in the second row. The corresponding original Trump's faces are shown in the first row.

where TP denoted true positive, TN denoted true negative, FP denoted false positive and FN denoted false negative. The results are shown in Table 1. Based on Table 1, Method 3 has the best accuracy 0.9784 among other methods. And Method 1 has the worst performance. The TP, TN, FP, and FN samples predicted by Method 3 are shown in Fig. 7. Since Method 3 uses the HOG features as the input of the neural network, the HOG images of Fig. 7(a) and (b) are shown in Fig. 8(a) and (b) respectively. It is difficult to decide which images are fake from checking Fig. 8 by human vision.

Table 1. Result using Trump−Cage data set

	Method 1	Method 2	Method 3
Accuracy	0.4516	0.9462	0.9784
TP	46	43	45
FP	47	2	1
TN	0	45	46
FN	0	3	1

To evaluate our continuity checking method, we collect the videos of Trump and Cage to form a data set Trump−CageVideo. Trump's videos are split into 3385 frames and Cage's videos are split into 2414 frames. Every 5 frames, called clips, are treated as a unit of input of LSTM. There are 232 clips used as the test set. We set the batch size to be 128 and epochs from 100 to 800 to evaluate the ability of continuity checking. The results are shown in Table 2.

Based on Table 2, the maximum accuracy 0.900 occurs at epochs 500. To compare with our previous Method 1, Method 2, and Method 3, the accuracy of continuity checking is lower than that of Method 2 and Method 3. This result suggests that checking for forgery frame by frame is a reasonable approach.

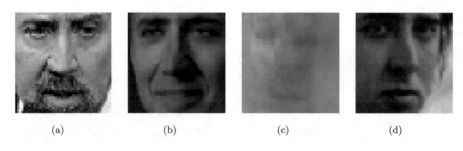

(a)   (b)   (c)   (d)

Fig. 7. Classification example of method 3: (a) a TP image (b) a TN image (c) a FP image (c) a FN image.

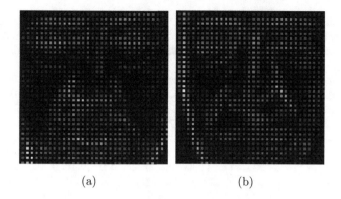

(a)   (b)

Fig. 8. The HOG images of Fig. 7(a) and (b).

Table 2. Result of continuity of using the `Trump−CageVideo` data set

Epochs	100	200	300	400	500	600	700	800
Accuracy	0.689	0.853	0.879	0.887	**0.900**	0.814	0.870	0.866

5 Conclusions

We have shown that the proposed two approaches can detect fake faces with high accuracy. If a given video is suspected of being a fake video which contains face A, then we can collect the known real images of face A and generate the fake faces with the help of well-known face forgery techniques and the real images of face A. Thus, we can use these real and fake image sets to train our proposed detection models and use the trained models to detect whether the given video is fake. Our future work will focus on generic face forgery detection without re-training the model for each identity.

Acknowledgements. We thank the anonymous referees for making valuable suggestions and comments which greatly improved the contents as well as the presentation of this paper.

References

1. Antipov, G., Baccouche, M., Dugelay, J.L.: Face aging with conditional generative adversarial networks. In: 2017 IEEE International Conference on Image Processing (ICIP), pp. 2089–2093. IEEE (2017)
2. Bayar, B., Stamm, M.C.: A deep learning approach to universal image manipulation detection using a new convolutional layer. In: Proceedings of the 4th ACM Workshop on Information Hiding and Multimedia Security, pp. 5–10. ACM (2016)
3. Bestagini, P., Milani, S., Tagliasacchi, M., Tubaro, S.: Local tampering detection in video sequences. In: 2013 IEEE 15th International Workshop on Multimedia Signal Processing (MMSP), pp. 488–493. IEEE (2013)

4. Cozzolino, D., Poggi, G., Verdoliva, L.: Recasting residual-based local descriptors as convolutional neural networks: an application to image forgery detection. In: Proceedings of the 5th ACM Workshop on Information Hiding and Multimedia Security, pp. 159–164. ACM (2017)
5. Deepfakes: GitHub (2018). https://github.com/deepfakes/faceswap
6. D'Amiano, L., Cozzolino, D., Poggi, G., Verdoliva, L.: A patchmatch-based dense-field algorithm for video copy-move detection and localization. IEEE Trans. Circ. Syst. Video Technol. **29**(3), 669–682 (2018)
7. Faceswap: GitHub (2018). https://github.com/MarekKowalski/Faceswap/
8. Gironi, A., Fontani, M., Bianchi, T., Piva, A., Barni, M.: A video forensic technique for detecting frame deletion and insertion. In: 2014 IEEE International Conference on Acoustics, Speech and Signal Processing (ICASSP), pp. 6226–6230. IEEE (2014)
9. Goodfellow, I., et al.: Generative adversarial nets. In: Advances in Neural Information Processing Systems, pp. 2672–2680 (2014)
10. Hochreiter, S., Schmidhuber, J.: Long short-term memory. Neural Comput. **9**(8), 1735–1780 (1997)
11. Karras, T., Aila, T., Laine, S., Lehtinen, J.: Progressive growing of GANs for improved quality, stability, and variation. arXiv preprint arXiv:1710.10196 (2017)
12. LeCun, Y., Bengio, Y., et al.: Convolutional networks for images, speech, and time series. In: The Handbook of Brain Theory and Neural Networks, vol. 3361, no. 10 (1995)
13. Suwajanakorn, S., Seitz, S.M., Kemelmacher-Shlizerman, I.: Synthesizing Obama: learning lip sync from audio. ACM Trans. Graph. (TOG) **36**(4), 95 (2017)
14. Thies, J., Zollhofer, M., Stamminger, M., Theobalt, C., Nießner, M.: Face2Face: real-time face capture and reenactment of RGB videos. In: Proceedings of the IEEE Conference on Computer Vision and Pattern Recognition, pp. 2387–2395 (2016)
15. Upchurch, P., et al.: Deep feature interpolation for image content changes. In: Proceedings of the IEEE Conference on Computer Vision and Pattern Recognition, pp. 7064–7073 (2017)

A Study of Text-Overlaid Image Compositions Information System About Enterprise Documents, Logo and Slogan

Shao-Shiun Chang[1], Kuo-An Wang[1,2(✉)], and Hsing-Hui Chen[1]

[1] Department of Industrial Education and Technology,
National Changhua Normal University, Changhua City, Taiwan
chess1@cc.ncue.edu.tw, d0531012@gm.ncue.edu.tw
[2] Department of Management Information Systems,
Central Taiwan University of Science and Technology,
Taichung City, Taiwan
gawang@ctust.edu.tw

Abstract. This study focus on the combination of enterprise logo, slogan and announcement documents. With the three factors, we developed a text-overlaid image Compositions information system for business. "Logo" became an English word in 1937. The logo has become an important feature of the corporate image. It is interpreted as meaning of "identity" or "trademark" which is used to identify a small visual design works. Those works can be used for the purpose of cooperate identity in many industries. Being based on the aesthetic view point, the study found a method to calculate block visual weight and setting text color by color difference formula. We used the aesthetic formula to complete weight balance degree examination and font size prediction to get the optimization of text-overlaid image compositions. The optimization was applied to simplify the process of enterprise announcement preparation. The other object is to integrate logo and slogan to appear a remarkable impression. The developed text-overlaid image Compositions system play a role of automatic typesetting to find a good way of designing the location of image and text. The user just pay attention to the content of documents and never minded about the layout of logo and slogan. The major help of the system is to promote the enterprise announcement documents designing efficiency. We provide this system for some SMEs to try to understand the practicality of system. Finally we carried out a questionnaire survey to test the user's attitude about the system from three perspectives, perceived usefulness, perceived ease of use and behavior intention. The collection data of users' opinion shows that the text-overlaid image Compositions information system generally achieve satisfied standard. Overall, users give this system a lot of positive comments. The information system is also considered to be very helpful for the process of making announcements in practice.

Keywords: Logo · Slogan · Text-overlaid image compositions

L. C. Jain et al. (Eds.): SICBS 2019, AISC 1145, pp. 279–289, 2020.
https://doi.org/10.1007/978-3-030-46828-6_24

1   Introduction

When people begin to have a large demand for graphics editing, various drawing and image processing software will be produced. But this is a problem for users who are not familiar with the design. They may not know how to arrange, design, and adjust the best placement of text with picture to create a comfortable visual effect. Therefore, we estimate that some companies should have a lot of troubles and improvement in the production of documents containing pictures. We hope to make good use of the convenience of the document-making information system so that employees in the company can focus on their own business. With regard to the design and production of documents, using information systems to produce aesthetically pleasing documents containing pictures will save much labor costs. The use of computers and Internet has increased the demand for digital services. There are also a handful of companies or common users who use image processing software. But to integrate a picture and text, most users are not easy to arrange the appropriate font size, color, location, etc. The popular problem may be that the text arrangement and the color cannot be matched, that is, it is difficult to judge the position where the text and the image fit together, which are the problems often encountered in design [1]. Logo is a visual expression of brand strategy. Trademark, brand and corporate identity (CIS) design is used in all professional design fields of graphic design, web design and advertising design to convey the values represented by the company. In the past, the research on trademarks was rarely explored by combining the aesthetic effects of Logo and Slogan. This study will start from the perspective of how to combine logo with the representative Slogan. Systematization of the integration of graphic and text documents in the enterprise will be studied. What is to be presented is to combine the company's notice document with the trademark (logo) and slogan, so that the company's notice file will be automatically generated. Under the premise of balancing aesthetics and visual balance, the effect of graphic and text integration is improved, and the concrete expression is conveyed. The message can enhance the image of the company too.

2   Text-Overlaid Image Compositions Information System

2.1   Corporate Identity and Logo

2.1.1   The Concept of Corporate Identification

One of the key elements of a company's success is Corporate Identity (CI) which can be seen everywhere in everyone's life [2]. For example, the yellow logo [M] mentioned, what is immediately thought of in the heart? The answer is the effect of corporate identity. It is to enable consumers to create a good corporate image and to understand enterprises in the shortest period of time, such as Starbucks and McDonald's, which have established a good impression and fame in the minds of consumers, and of course further strengthen the competitiveness of enterprises in the market. It can ensure its leading position in market share.

2.1.2   Corporate Identification Components

In general, Corporate identity (CI) consists of the following three elements.

1. Mind Identity (MI)
   Concept recognition refers to the company's business philosophy, promotion strategy, spiritual slogan to the outside world, internal spiritual slogan, employee motto, etc. For example, McDonald's created the familiar 'I'm Lovin' it.
2. Behavior Identification (BI)
   Behavior Identification refers to all activities undertaken by an enterprise to enhance its image, strengthen its recognition ability, and increase marketing profit. Its contents include organization, management, employee education, external market research, product development, promotion activities, and social welfare activities.
3. Visual Identity (Visual Identity, VI)
   A representative totem that conveys corporate image by using visual elements such as image text color. This is the logo that everyone familiar with, such as trademarks/signs/business cards [3].

2.1.3   The Importance of the Corporate Logo

It is the most common aspect of a brand that past impression affects people's attitude towards the brand [4]. The use of trademarks can distinguish the most important and essential functions and sources of different goods or services, and can guide consumers to identify brands in shopping and consumption. A unique symbol of a company, products or service would be the most valuable assets. Uniqueness is a key factor in the company's competitiveness.

Uniqueness can promote the quality of products or services by producers or operators. The quality of goods or services is the basis of trademark reputation. While guiding consumers to identify brand consumption, it also promotes operators to improve the quality of products or services in order to maintain their trademark reputation. The logo is unique and is considered a registered trademark representing the company and protected by law [5]. The registered trademark and the designated goods or services are mutually beneficial. The trademark reputation can reflect the quality, the quality is stable and the trademark reputation can be improved. The use of a registered trademark of goods or services can also increase the customer's sense of security.

Trademarks represent the carrier of memory. Well-designed good trademarks will be more easily remembered by the public, which will bring benefits to enterprises [6]. Therefore, trademarks are a weapon of commercial warfare. A commodity must be opened to the market and recognized by the consumers. In addition to ensuring the reliability of quality, it must also be advertised through the trademark focus to stimulate consumers' desire to purchase, so that consumers can understand the goods in the quickest way. Therefore, trademarks are the pioneers in opening up the market.

Trademarks are intangible wealth and a carrier of credibility. Based on reliable product quality, it will make its trademarks well-known and continue to appreciate, making the economic benefits of the company better and better. Even the name of the origin place of the product is not known, but the name of the brand is widely known. In this sense, a registered trademark is an invaluable asset and an intangible asset.

2.2   The Meaning and Importance of Advertising Slogans

The slogan endorses the spiritual connotation of the brand and the company, and has a strong ability to bring out the content of the copywriting. The use of words is more streamlined than the copy. The words are shorter and less frequently replaced. Many widely circulated slogans are also very easy to impress consumers' psychology [7]. If the slogan can't attract the attention of consumers in a short period of time, there is no way to drive consumers to want to know more about their meaning. The unique fonts, layouts, and freely dispatchable slogans must be given according to different themes, and the way to communicate with the public can be found. Therefore, the performance of the slogan is mainly concise, and its purpose is to make the public deeply impressed the subject of communication [8]. A good advertisement will impress everyone, and a simple advertising slogan often becomes the daily vocabulary of the public. Successful advertising slogans can make full use of the characteristics of the text. Communicate rich concepts in a simple sentence, making people resonate or smile. In general, the importance of advertising slogans can be understood by the following features:

1. Reflecting the background of the times

What modern people see in their eyes every day, what they hear in their ears, and what they wear are all advertisements, but we are often trapped in it but not consciously. The slogan can also express the pulsation of the times, which has a keen relationship with the times, and can lively grasp the trend of the times [8]. For advertisers, understanding this feature will help to design a more effective slogan [9].

2. Enterprise screening function

The slogan has the role of "enterprise screening", which can be used to interpret the organization's business philosophy and spirit. It can faithfully present the corporate character and represent the core value of the corporate brand and will not change easily.

3. Reflecting company policy

The slogan may be changed based on the company's annual plan. Some companies change their advertising slogans every few years to match the company's overall annual plan. So the slogan can represent the company's current strategy.

4. Raising the imagination of the product

When consumers think of the brand, they think of the product, because the brand has been prompted, and the brand has unlimited imagination space on the extension. Advertising slogans refer to sentences that are repeated in advertising campaigns [9]. The slogan of an activity must emphasize the theme of advertising campaigns, so that the public can remember certain facts and associate their brands. This effect can make the value of the slogan.

5. Forming a social trend

Advertising slogans are designed according to the company's strategy [10] which usually changes the slogan if the advertising strategy changes. Putting a touching slogan is an important job for every business. As the social atmosphere evolves, the slogan of advertising also reflects the trend of contemporary society. The long-lasting advertising slogan has become a trend in people's mantra and daily society.

2.3   Aesthetic Quantitative Balance Analysis

There are many factors that affect the quality of a work that contains pictures and text. Perhaps because of the more creative text layout, it may be due to the more design-oriented text effects, or it may be the balance, symmetry, coordination, density and other factors after the text is placed [7].

In our information system, our application is to place the slogan text block on the existing Logo file and calculate the location that best fits the slogan text block. This is a typographic application of corporate identity and text integration. Our research focuses on the use of balance. In terms of balance, the main calculation goal is to find the visual center. The balance is divided into two parts, vertical and horizontal. As shown in Fig. 1, if the upper and lower blocks of the picture are equal in weight and the position of the visual center is one-half of the height of the picture, then we say that it has reached equilibrium in the vertical. As shown in Fig. 2, the left and right blocks are equal in weight, and the position of the visual center is also one-half of the width of the picture. Then we say that it has reached equilibrium in the horizontal plane. The block area represents the weight of the block. The center point coordinates of the block are the center of gravity of the block. We use the principle of leverage to find the position of the visual center [10].

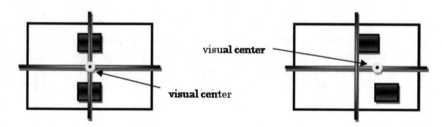

Fig. 1. Vertical balance

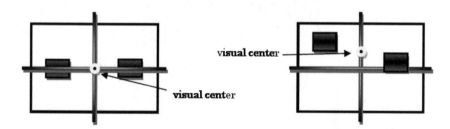

Fig. 2. Horizontal balance

Since the color of the logo will affect the color of the slogan text, it is necessary to make a comparison of the degree of color difference for further image analysis.

Therefore, finding the formula for calculating the color difference that is consistent with human cognition is a prerequisite for us to explore the quantification of aesthetics. After using the HSV color space to define the color of the image, use the HSV color difference formula proposed by Smith and Chang to calculate the color difference between (h1, s1, v1) and (h2, s2, v2) Degree, the formula is as follows: [11]

$$\Delta C = \frac{\sqrt{(v_1 - v_2)^2 + (s_1 \cos h_1 - s_2 \cos h_2)^2 + (s_1 \sin h_1 - s_2 \sin h_2)^2}}{\sqrt{5}}$$

2.4   Method and Results

The name of the system built in this study is "Enterprise Announcement Output System" (EAOS), which can quickly integrate the enterprise logo with the slogan and place it under the corporate notice file. The system can integrate the corporate logo, slogan and current notice into the automatic graphic according to the needs of the enterprise users. It brings considerable convenience to the editorial text, and the system operation interface is easy to understand and beneficial to customers. The user can freely adjust the position of the logo of the company and select the background color. According to these two conditions, the system will perform the operation under the principle of balance optimization to obtain the slogan x coordinate and y coordinate position to find the most appropriate location for corporate slogans. What is the setting of the text color? It is to calculate the color difference between the color appearing on the logo image and the background color selected by the user. The maximum color difference value is selected as the Slogan text color, which is to set the font color of the

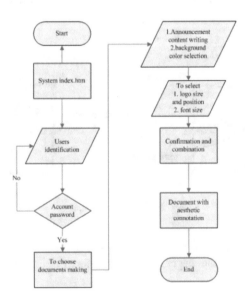

Fig. 3. "Enterprise Announcement Output System" (EAOS) processing flow chart

corporate slogan according to the principle of chromatic aberration. This system is developed in VB language and PHP language. We added the FLASH interactive web page feature to system. The notification file output made by the system finally can be previewed by the user or downloaded as a PDF file to print or transfer to another man. As shown in Fig. 3. "Enterprise Announcement Output System" (EAOS) processing flow chart was presented. Readers can understand what the process is to make a company notice automatically by the system. We showed some important system interface to demo actual operating procedure in Fig. 4. With the image and text description in Fig. 4, readers can realize the system function clearly. Finally, we show some samples of the notification documents completed by the system in Fig. 5.

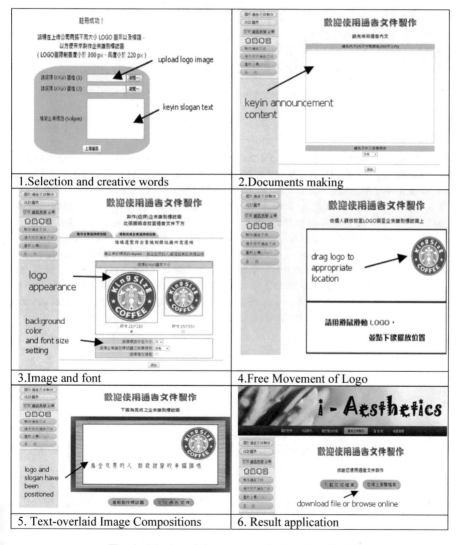

Fig. 4. Display of the system main functions (1–6)

Fig. 5. Enterprise announcement samples of system output

2.5   Evaluation

In this study, 38 employees from different companies having professional experience in administration, management, research, production and marketing were requested to use the "Enterprise Announcement Output System" (EAOS). Each employee was asked to fill a questionnaire (Table 4) based on the modified technology acceptance model (TAM) [12, 13] to evaluate perceived usefulness (PU), perceived ease of use (PEU), and behavior intention (BI) of the information system. All questions are graded on a 5-point Likert scale ranging from 1 to 5 points. The Cronbach Alpha of the questionnaire is 0.85 indicating its high reliability. As shown in Table 1, the scores of PU, PEU, and BI are all significantly greater than 3 ($p < 0.05$) indicating usefulness of the Enterprise Announcement Output System, ease of use of the website environment, and positive attitude toward using the function in EAOS.

Table 1. Descriptive statistics of modified TAM.

Technical Acceptance Model (N = 38)	Mean(SD)
A. Perceived Ease of Usefulness	3.65(0.93)
1. I found it is easy to operate Enterprise Announcement Output System (EAOS)	3.95(0.70)
2. I found EAOS is easy to do what I want it to do	3.75(0.73)
3. I found the user interface of EAOS is clear and understandable	3.74(0.77)
4. I found the interaction with EAOS is flexible	4.24(0.82)
B. Perceived Usefulness	3.93(0.79)
5. I agree EAOS can facilitate self-learning and accomplish tasks more quickly	3.91(0.66)
6. I agree EAOS can decrease learning time and increase productivity	3.72(0.65)
7. I agree EAOS can elevate using wiliness and effectiveness	3.41(0.64)
8. I agree EAOS can provide information for different industry groups	3.49(0.72)
9. I agree EAOS can promote notice quality in company	3.97(0.78)
10. I agree EAOS is useful for making announcement documents	3.44(0.81)
11. I agree EAOS is useful for integration of technology and aesthetics	3.67(0.75)
12. I agree EAOS is useful for the diversification of announcement documents	3.46(0.69)
C. Behavior Intention	3.57(0.94)
13. I intend to use EAOS as frequently as I need	3.96(0.72)
14. I will continue to use EAOS whenever possible in suitable circumstance	3.96(0.69)
15. I expect to use EAOS in other related activities and training in the future	3.81(0.69)

3 Conclusion

Under the requirements of efficiency improvement, enterprises pay attention to the speed and automation of processes. Because time is a key factor affecting profits.

Based on this concept, the functional objectives of this system are divided into two directions: In practice, the system can make the texts of all kinds of notification documents of the enterprise, and achieve basic automation to beautify and integrate documents. In marketing, the company concept (including the corporate identification logo and slogan) will be transmitted to employees and customers in the form of notification documents, which will gain good cohesiveness and publicity. The system is based on aesthetic theory and uses the aesthetic formula to apply the integration of graphics and texts in corporate notification documents. Complete the prototype of the system, and then make a second revision of the system according to the feedback from the user. In the process of researching the needs of users, this study continues to validate the color difference formula and aesthetic formula, hoping to make the output of the system closer to the model of human aesthetic experience. We also correct the misunderstandings that often occured on the system interface to make the system operation more friendly and convenient. In order to meet fast and convenient corporate goals. Information quality and system integration are two important factors which highly influences perceived usefulness and post adoption of an information system [13]. In our study, the system interface and function were studied, examined, and recorded by employees from different company. Those specialists greatly ensures quality of the information system.

In conclusion, the system has completed the pre-function, we first use the balance theory of color difference and block weight to do accurate calculations, find the coordinates for the positioning of pictures and text. At present, the basic architecture of graphic and text integration automation has been completed, and it is expected that future systems will be able to extend more advanced functions. Such as the background, text color system refinement, how to optimize the combination of logo graphics and corporate slogan fonts, statistical analysis of graphics coordinates. We hope that the graphic integration system will become more popular, and finally it will be integrated into the enterprise information system by the business operators and become a part of the company's important competitiveness.

Ethical Approval. "All procedures performed in studies involving human participants were in accordance with the ethical standards of the institutional and/or national research committee and with the 1964 Helsinki declaration and its later amendments or comparable ethical standards. The survey data were anonymized. This study was also approved by the Department of Management Information Systems, Central Taiwan University of Science and Technology ethics committee in advance."

References

1. Bauerly, M., Liu, Y.: Computational modeling and experimental investigation of effects of compositional elements on interface and design aesthetics. Int. J. Hum Comput Stud. **64**(8), 670–682 (2006)
2. Kim, J., Lee, J., Choi, D.: Designing emotionally evocative homepages. Int. J. Hum Comput Stud. **59**, 899–940 (2003)
3. Tractinsky, N., Lavie, T.: Assessing dimensions of perceived visual aesthetic of web sites. Int. J. Hum Comput Stud. **60**, 269–298 (2003)
4. Wilson, A., Chatterjee, A.: The assessment of preference for balance: introducing a new test. Empirical Stud. Arts **23**(2), 165–180 (2015)
5. Paschos, G.: Perceptually uniform color spaces for color texture analysis: an empirical evaluation. IEEE Trans. Image Process. **10**(6), 932–937 (2011)
6. Altaboli, A., Lin, Y.: Investigating effects of screen layout elements on interface and screen design aesthetics. Adv. Hum.-Comput. Interact. **2011**, 10 (2011). https://doi.org/10.1155/2011/659758
7. Bauerly, M., Liu, Y.: Computational modeling and experimental investigation of effects of compositional elements on interface and design aesthetics. Int. J. Hum Comput Stud. **64**, 670–682 (2006)
8. Bauerly, M., Liu, Y.: Effects of symmetry and number of compositional elements on interface and design aesthetics. Int. J. Hum.-Comput. Interact. **24**, 275–287 (2011)
9. Bi, L., Fan, X., Liu, Y.: Effects of symmetry and number of compositional elements on Chinese users' aesthetic ratings of interfaces: experimental and modeling investigations. Int. J. Hum.-Comput. Interact. **27**, 245–259 (2011)
10. Lai, C.Y., Chen, P.H., Shih, S.W., Liu, Y., Hong, J.S.: Computational models and experimental investigations of effects of balance and symmetry on the aesthetics of text-overlaid images. Int. J. Hum Comput Stud. **68**, 41–56 (2010)

11. Liu, Y.: Engineering aesthetics and aesthetic ergonomics: theoretical foundations and a dual-process research methodology. Ergonomics **46**, 1273–1292 (2003)
12. David, F.D.: Perceived usefulness, perceived ease of usefulness, and user acceptance of information technology. MIS Q. **13**, 319–340 (1989)
13. Saeed, K., Abdinnour-Helm, S.: Examining the effects of information system characteristics and perceived usefulness on post adoption usage of information systems. Inf. Manag. **45**, 376–386 (2008)

Multimodal Presentation Attack Detection Based on Mouth Motion and Speech Recognition

Chao-Lung Chou[✉]

Department of Computer Science and Information Engineering,
Chung Cheng Institute of Technology, National Defense University,
Taoyuan, Taiwan (R.O.C.)
chaolung.chou@gmail.com

Abstract. Face recognition systems have grown rapidly in a variety of applications recently, such as surveillance, access control, mobile payments, and forensic investigations. However, face recognition systems are highly likely to be deceived because imposters attempt to gain unauthorized access to the system by spoofing biometric data. In this paper, we proposed a multimodal presentation attack detection (PAD) method based on a challenge-response scenario. When the user speaks a word as required, the mouth movement is detected and the recognized speech is referenced to determine if the user is a real person or not. Two weighted score level fusion rules are adopted in the machine learning algorithm for training and testing. Experimental results show the proposed method is very effective in resisting photo-attack and video-attack targeting to face recognition systems.

Keywords: Multimodal · Presentation attack detection · Challenge-response · Fusion

1 Introduction

Biometric is the technology based on physical and behavioral characteristics of human, such as fingerprint, face, palm print, iris, palm vein, voice, signature, keystroke, gait and etc. [1]. Nowadays, biometric technology has been widely used in all areas of life, including surveillance, access control for homes or office buildings, digital device identity verification, mobile payments, border control, forensic investigations, and etc. Biometrics effectively improves the security and convenience of using traditional passwords and magnetic cards.

With the fast growing of biometric technology, the security threats against biometric systems are also increasing. The most common attack is that an attacker presents as a legitimate user by submitting fake biometrics and trying to cheat the system. This type of attack is also called as a presentation attack, spoof attack or direct attack. Among various biometric modalities, face is one of the most widely adopted modalities to perform presentation attack due to the popular of social networks media. It is easy to download face photos from by visiting social media or personal websites. Imposter can then use these photos to generate fake traits.

© Springer Nature Switzerland AG 2020
L. C. Jain et al. (Eds.): SICBS 2019, AISC 1145, pp. 290–298, 2020.
https://doi.org/10.1007/978-3-030-46828-6_25

Three types of spoofing attacks for face recognition systems are (1) photo attacks (2) video attacks and (3) 3D mask attacks [2]. Photo attacks is easy and simple to conduct in that usually printed out the photo of legitimate users and presents in front of the face recognition system sensor to perform the attack. Video attack, also known as replay-attack, is an advanced attack which instead of using static image in front of the sensor by displaying video through a smart phone or tablet in which shows behaviors similar to real faces. 3D mask attack is the most complex attack by wearing a 3D gel mask of a legitimate user, yet it is also most difficult and expensive attack because it needs special devices to manufacture the mask.

The countermeasure, namely presentation attack detection (PAD) or anti-spoofing, to deal with the problem of face spoofing attacks has been explored recently [3]. These existing face presentation attack detection algorithms can be broadly classified into hardware based and software based as Fig. 1 shows [4].

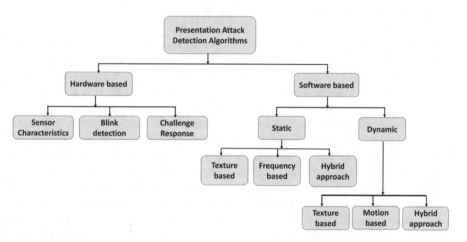

Fig. 1. Face presentation attack detection algorithms categories [4].

The hardware-based approaches require user cooperation and specialized hardware, and can be broadly classified into sensor characteristics, blink detection, and challenge response. The software-based approaches can be divided into static methods and dynamic methods in which static methods are designed to work on a single image, while dynamic methods are designed to focus on video containing time information.

Challenge-response can be viewed as a one-time password authentication mechanism. The authentication server randomly generates a challenge and requests user to response within a certain time limit. This scenario may cause user inconvenience but it is effective for resisting both photo-attack and video-attack [4].

Frischholz et al. proposed a method requires the subject to make a significant head movement response to detect video attacks [5]. Gao et al. developed a sound-based CAPTCHA verification code analysis method, using the gap between human speech and synthetic speech as a basis for discrimination, the tester needs to read the system randomly selected sentences Then, the system automatically analyzes whether the audio responded is synthetic or not [6]. Sluganovic et al. proposed an interactive visual stimulus technique to randomly displaying the light spots on the screen and tracking the eye movement of the subject [7].

In this paper, we proposed a multimodal based PAD method using mouth motion and speech recognition in a challenge-response scenario. The server randomly generates a word (challenge) and request user to say it (response). In conventional cases, a biometrics system will authentic the user as long as the recognized speech is correct. We found that once people speak, their mouths will open and close interchangeably. Therefore, we use this feature to detect the mouth movements when the user replies the requested word and verify that the user is a real person.

The rest of this paper is organized as follows: Sect. 2 reviews the multimodal biometric fusion algorithm. Section 3 presents the proposed PAD method. Section 4 demonstrates the experimental results and Sect. 5 gives the conclusion.

2 Fusion of Multimodal Biometric System

A biometric system uses single biometric modality usually is called the unimodal biometric system. Likewise, a biometric system uses multiple biometric traits for authentication is called multimodal biometric system. Multimodal biometric system is more reliable and robust in preventing spoofing attacks because of an impostor is hard to spoof multiple biometric traits of a genuine user simultaneously [8].

A multimodal biometric system can work in serial mode or parallel mode. In the serial mode, user may not be required to provide the other traits if one modality has sufficient confidence on the identity. In the parallel mode, multiple modalities are submitted to respective sensors simultaneously and the decision is based on some fusion rules.

In a multimodal biometric system, fusion can be accomplished either at sensor level, feature level, score level or decision level. Score level fusion is usually preferred in multimodal biometric systems because it is easy to combine the scores presented by different modalities [8]. There is various score fusion rules [9]. In this paper, we adopt weighted sum and weighted product fused score which are computed as (1) and (2).

$$s = \sum_{i=1}^{N} w_i s_i \tag{1}$$

$$S = \prod_{i=1}^{N} S_i^{w_i} \tag{2}$$

where $0 \leq w_i \leq 1$ is the weight of the matcher i.

The advantage of weighted summation and product fusion is that each feature can be weighted [10]. Weights can be chosen to reflect the distinctiveness and acceptability of each modality for different applications.

Chibelushi et al. [11] use a form of weighted summation fusion combined speech and still face profile images as (3) shows

$$f = w_1 s_1 + (1 - w_1) s_2 \tag{3}$$

where s_1 and s_2 are the scores from the speech and face profile features respectively, and w_1 denotes the weights.

Brunelli et al. [12] proposed the weighted product approach combined speech and geometric features obtained from static frontal face images as (4) shows

$$f = s_1^{w_1} + s_2^{(1-w_1)} \tag{4}$$

where s_1 and s_2 are the scores from the speech and face profile features respectively, and w_1 denotes the weights.

From [11] and [12], the results show that the multimodal biometric system using speech and facial image recognition combined with the weighted summation fusion or weighted product fusion method is significantly better than that using only single feature.

3 The Proposed Method

The proposed method is mainly composited with two modules. The mouth motion detection module uses facial landmark technique to detect mouth changes and the speech recognition module is to identify the respondent's response words. These two modules generate feature vectors and feed it into a weight fusion score function to obtain a final score to discriminate real person and spoof one.

3.1 Mouth Motion Detection

In 2016, Soukupová and Čech proposed the EAR blink detection method [13]. The EAR is defined as shown in (5). The EAR method uses six feature points around the human eye. When the eyes are open, the distance between [P_2, P_6] and [P_3, P_5] is farther, as shown in Fig. 2(A); and when the eyes are closed, the distance between [P_2, P_6] and [P_3, P_5] is significantly closer, as shown in Fig. 2(B).

$$EAR = \frac{\|P_2 - P_6\| + \|P_3 - P_5\|}{2\|P_1 - P_4\|} \tag{5}$$

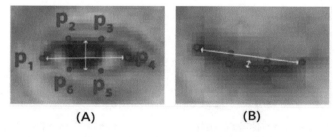

(A)   **(B)**

Fig. 2. Example of EAR (A) eyes open (B) eyes closed [13].

Inspired by the EAR method, it uses facial landmarks near the eyes to detect blinks. We use 8 of total 68 facial landmarks on the iBUG 300-W face landmark dataset as shown in Fig. 3 [14]. The 8 feature points around the mouth are selected and define the MAR (Mouth Aspect Ratio) parameter which as (6) shows.

$$MAR = H/W \qquad (6)$$

where $H = |P_{62} - P_{68}| + |P_{63} - P_{67}| + |P_{64} - P_{66}|$ and $W = |P_{61} - P_{65}|$.

When the mouth is opened and closed, the width and height of the mouth will change significantly. We define the absolute value function $\delta(MAR_i)$ to estimate the degree of mouth opening and closing at specific time as (7) shows.

$$\delta(MAR_i) = |MAR_{i+1} - MAR_i| \qquad (7)$$

where i denote the ith second.

Accordingly, the mouth motion feature f_M can be obtained as (8) shows.

$$f_M = \frac{\sigma^2}{\mu}, 0 \leq f_M \leq 1 \qquad (8)$$

where $\mu = \frac{1}{N}\sum_1^N \delta(MAR_i)$ denote the mean value and $\sigma^2 = \frac{1}{N}\sum_1^N \delta(MAR_i - \mu)$ denote the variance in continuous time N.

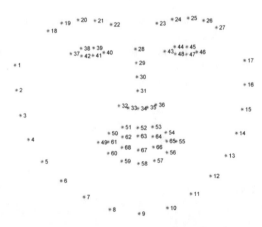

Fig. 3. Facial landmarks from the iBUG 300-W dataset [14].

3.2   Speech Recognition

Speech recognition is a technique used to convert human spoken words into computer text. The speech recognition module is used to immediately identify the respondent's response word and verify its similarity to the actual requested word.

Accordingly, the speech recognition feature f_S can be obtained by (9).

$$f_s = 2M/T, 0 \leq f_S \leq 1 \tag{9}$$

where T is the total number of the request and response characters, and M is the total number of identical characters in these two words. If two words matches completely, $f_S = 1$ and if it does not match at all, then $f_S = 0$.

3.3 The Complete Processes

The proposed method uses machine learning algorithms to train weighted score fusion through mouth motion and speech recognition features. The decision of real human or spoof is based on the trained model. The flowchart of the proposed method shows in Fig. 4.

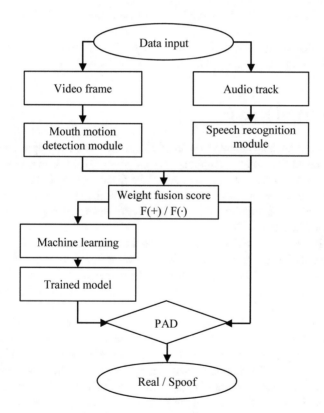

Fig. 4. Flowchart of the proposed method

Step1. All input video from client sensor will be separated into video frames and audio tracks simultaneously. Video frames and audio tracks are sent to the mouth motion detection module and the speech recognition module to obtain their feature values respectively.

Step2. The feature values f_M will be obtained by using (8) in the mouth motion detection module, and likewise the feature values f_S will be obtained by using (9) in the speech recognition module.

Step3. According to (3) and (4), we use $F(+)$ denote the weighted summation fusion and $F(\cdot)$ denote the weighted product fusion as (10) and (11) shows.

$$F(+) = w_1 f_M + (1 - w_1) f_s \quad (10)$$

$$F(\cdot) = f_M^{w_1} \times f_s^{(1-w_1)} \quad (11)$$

where and w_1 is the corresponding weight in the [0, 1] interval.

Step4. For supervised learning, the live clips are label as 1 and spoof ones are label as 0. The weight fusion scores are feeding into supervised machine learning algorithms to train the model. The trained model is further used for PAD testing.

Step5. The final stage is for PAD testing. The decision is based on the trained model to determine if the user is a real person or a spoof one.

4   Experimental Results

The experiment was implemented on 2.3 GHz, 8 GB RAM laptop computer with a 30 fps webcam and external microphone using Python 3.7, PyAudio, OpenCV2, Dlib, difflib, scikit-learn and Google translate API. The database used in the experiment was built in indoor environment with similar pose and illumination. The database collects total 662 videos which including 494 live clips and 168 spoof clips from 28 subjects. Each clip has duration of 6 to 8 s.

For the live clips, all subjects were requested to say a simple word includes in the predefined words bank. For the spoof clips, there are total three patterns simulating video-attack scenario as follows. (1) Staring at the camera and say nothing. (2) Moving the head back and forth from left to right and say nothing. (3) Answering words unrelated to the prompt.

There are three multimodal fusion scenarios in the experiment. The first one is simply summation of mouth motion and speech recognition without weighting. The second one is using weight summation fusion, and the third one is using weight product fusion. The weight w_1 for both weighted summation and product fusion is set to 0.5.

Three supervised machine learning algorithms, namely decision trees (DT), random forests (RF) and k-nearest neighbor (kNN) are used in the experiment. In the database, half of live and spoof clips from 28 subjects are randomly partition into training set, and the others are used for testing.

To evaluate the performance of the proposed method, we use accuracy (ACC) and half total error rate (HTER) criteria as presented in (12) and (13) respectively [15].

$$\text{Accuracy} = \frac{TP + TN}{TP + TN + FP + FN} \tag{12}$$

where TP (True Positive) means that the live human is correctly detected, TN (True Negative) represents the spoof one is correctly detected, FN (False Negative) represents the live human is wrongly detected as spoof one and FP (False Positive) represents the spoof one is wrongly detected as live human.

$$\text{HTER} = \frac{FAR + FRR}{2} \tag{13}$$

where $FAR = \frac{FP}{TN+FP}$ denote the false acceptance rate and $FRR = \frac{FN}{TP+FN}$ denote the false rejection rate.

Table 1 shows the performance evaluation of the proposed method using different multimodal fusion scenarios. From the table, the best ACC result is 94.86% with random forests classifier and the best HTER is 4.04% with decision tree, and both are using weighted summation fusion. According to Table 1, weighted summation fusion gives a better result in both ACC and HTER. Since the summation operations are much faster than production operations, the proposed method suggests using weighted summation fusion to achieve better performance and lower computing cost.

Table 1. Experimental results

Classifier	Summation		Weighted summation fusion		Weighted product fusion	
	ACC (%)	HTER (%)	ACC (%)	HTER (%)	ACC (%)	HTER (%)
Decision tree	93.05	5.84	94.56	4.04	91.84	5.47
Random forests	92.75	7.34	94.86	4.62	91.84	5.47
k-nearest neighbor	92.15	9.19	94.56	5.21	91.84	5.47
Average	92.65	7.45	94.66	4.62	91.84	5.47

5 Conclusions

In this paper, we proposed a multimodal PAD technology based on a challenge-response scenario. The facial landmark technology is used to detect mouth movement, and the speech recognition technology is used to recognize the response of the user. Two score level fusion rules combine the facial and speech biometrics are trained by three supervised machine learning algorithms. Experimental results show that the weighted summation fusions rule can achieve best result is 94.86% for ACC and 4.04% for HTER. It also demonstrated that the proposed method is very effective in resisting video-attack. Our future work is to focus on studying different complement modalities in different level of fusion rules to improve performance in accuracy, cost and reliability.

References

1. Jain, A.K., Ross, A., Prabhakar, S.: An introduction to biometric recognition. IEEE Trans. Circuits Syst. Video Technol. **14**, 4–20 (2004)
2. Akhtar, Z., Micheloni, C., Foresti, G.L.: Biometric liveness detection: challenges and research opportunities. IEEE Secur. Priv. **13**(5), 63–72 (2015)
3. Galbally, J., Marcel, S., Fierrez, J.: Biometric antispoofing methods: a survey in face recognition. IEEE Access **2**, 1530–1552 (2014)
4. Ramachandra, R., Busch, C.: Presentation attack detection methods for face recognition systems: a comprehensive survey. ACM Comput. Surv. **50**(1), Article no. 8 (2017)
5. Frischholz, R.W., Werner, A.: Avoiding replay-attacks in a face recognition system using head-pose estimation. In: IEEE International Workshop on Analysis and Modeling of Faces and Gestures, Nice, France, pp. 234–235 (2003)
6. Gao, H., Liu, H., Yao, D., Liu, X. Aickelin, U.: An audio CAPTCHA to distinguish humans from computers. In: Third International Symposium on Electronic Commerce and Security, Guangzhou, China, pp. 265–269 (2010)
7. Sluganovic, I., Roeschlin, M., Rasmussen, K.B., Martinovic, I.: Using reflexive eye movements for fast challenge-response authentication. In: ACM Conference on Computer and Communications Security, Vienna, Austria, pp. 1056–1067 (2016)
8. Ross, A., Nandakumar, K., Jain, A.K.: Handbook of Multibiometrics. Springer, New York (2006)
9. Zahid, A., Mohammed, R., Sandeep, K.: Multimodal biometric fusion: performance under spoof attacks. J. Intell. Syst. **20**(4), 353–372 (2011)
10. Sanderson, C., Paliwal, K.K.: Information fusion and person verification using speech and face information. Research Paper IDIAP-RR 02-33, Martigny, Switzerland, pp. 1-8 (2002)
11. Chibelushi, C.C., Deravi, F., Mason, J.S.: Voice and facial image integration for speaker recognition. In: IEEE International Symposium and Multimedia Technologies and Future Applications, Southampton, UK, pp. 157–160 (1993)
12. Brunelli, R., Falavigna, D., Poggio, T., Stringa, L.: Automatic person recognition using acoustic and geometric features. Mach. Vis. Appl. **8**, 317–325 (1995)
13. Soukupová, T., Čech, J.: Real-time eye blink detection using facial landmarks. In: 21st Computer Vision Winter Workshop (2016)
14. Intelligent behaviour understanding group (iBUG). https://ibug.doc.ic.ac.uk/resources/facial-point-annotations/
15. Chingovska, I., Andr, A., Marcel, S.: Biometrics systems under spoofing attack: an evaluation methodology and lessons learned. IEEE Signal Process. Mag. **32**(5), 20–30 (2015)

Unsupervised Learning: Using Clustering Algorithms to Detect Peer to Peer Botnet Flows

Andrea E. Medina Paredes and Hung-Min Sun[✉]

Department of Computer Science, National Tsing Hua University, Hsinchu, Taiwan
andrea_mp89@hotmail.com, hmsun@cs.nthu.edu.tw

Abstract. The war against botnet infection is fought every day by users that want to feel safe against any threat of compromise hosts. In this paper we are going to focus on the behavior of Peer 2 Peer (P2P) botnets, which along with hybrid botnets is a growing trend among attackers. The main approach will consist of a behavior comparison among features extracted from network flows, focusing only in the flows from P2P applications including P2P botnets.

Keywords: Clusters · Network flows · P2P botnets · Unsupervised learning

1 Introduction

Malicious software such as botnets has been around for quite a time already and it keeps improving, evolving and growing, as for the detection systems, they try to keep track of these new emerging botnets trends but some fail to provide a definite and accurate solution to this problem. For example, this past May "8chan", a website composed of user-created boards, reported a series of DDoS attacks coming from the "Hola!" network, a popular virtual private network use for viewing blocked videos and TV shows from other countries, which counts with a pool of 6 million IP addresses [1]. Even though the number of botnet assisted DDoS attacks and victims has declined in the first quarter of 2015 compare to the fourth quarter of 2014, it doesn't mean the trend is declining, now attackers are expanding their boundaries over other geographical areas and using all of their stole resources to target specific victims for extended amount of time [2].

Activities such as adding signatures to databases, protecting servers against hackers, the use anti-virus software to protect computers from getting infected, track C&C server activities and so many other actions are taken in consideration, but still cybercriminals find a way to go around the security measures. The use of supervised learning models is one of many approaches that can be use to deal with botnets, classifiers like support vector machines (SVM) have shown great accuracy separating botnet network flows from normal flows [3], other methods like decision tree algorithms have been put to the test as well, measuring how

L. C. Jain et al. (Eds.): SICBS 2019, AISC 1145, pp. 299–311, 2020.
https://doi.org/10.1007/978-3-030-46828-6_26

accurate the decision tree classifies the data [4]. The drawback of the previous mention methods is that most of them need labeled data in order to function and only yields better results when the botnet signature is already known. In hopes to contribute to these efforts, in this paper we propose the use of Unsupervised Machine Learning algorithms for the fight against botnet detection. A comparison among three clustering algorithms using network flows extracted from a set of features, will be carried out thorough out this paper. The rest of this paper, is organized with the following: Background and related papers, the method use, the experiment design with the respective observations and finally the analysis of the results.

2   Background

2.1   Botnet

A P2P network is a network architecture in which peers (computers in network) are connected and share their resources (content, storage, etc.) by direct exchange rather than going through a centralized server. In other words, each peer works as client as well as server at the same time. Today, one of the major uses of P2P networks is file-sharing, and there are various P2P file-sharing applications, such as eMule [5], Skype [6] and BitTorrent [7].

There are two major categories of P2P architectures:

1. Pure P2P network: All peers act as clients and server there is no central server and no central router (Shown in Fig. 1). The peers act as server when processing a file share query and client when requesting for a file. There is also the chance a normal peer, may also be refer to as "leaf peer", may become a "super peer" when they possess better computing abilities and network bandwidth. Acting as a local server for leaf peers.

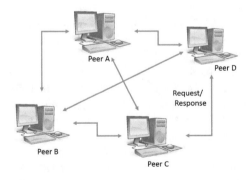

Fig. 1. This is how a pure peer to peer network looks like

2. Hybrid P2P Network: Has a central server that keeps information on direc-
tories and routing information (Shown in Fig. 2), it processes file search
queries and is in charge of coordinating downloads among peers. Peers are
responsible for hosting the information because the central server in this
situation doesn't store files, for letting the central server know what files
they want to share and for downloading its shareable resources to peers
that request it [8].

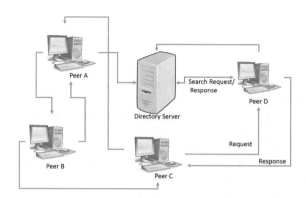

Fig. 2. A hybrid peer to peer network has a directory server just for control

2.2 Peer 2 Peer (P2P) Botnets

After attackers gradually realize the limitations faced using a traditional centralized botnet structure, shown in Fig. 3, a good number of P2P botnet structure
have emerge as an alternative. One of the upsides to this change is that in a
P2P structure, shown in Fig. 4, the communication will not be disrupted easily
if a few number of bots is missing, because the bots are connected to each other
and act as both parties, command and control (C&C) server and client. The life
cycle of a P2P botnet consist of the following four stages:

- Infection stage: The computer unintentionally contact malicious software and
  get infected. There are several ways to be infected, such as through drive-
  by downloads, malware installed by the end-user, or infected USB(Universal
  Serial Bus, USB) flash drive, etc.
- Rally stage: The bot join the P2P network by connecting to the computer in
  the peer list secretly.
- Waiting stage: The bot maintain P2P network connection and waits for the
  bot-master's command.
- Execution stage: The stage which it actually execute the command, such as
  a denial of service (DoS) attack, generate spam emails, etc. [9]

In order to avoid detection by Detection Systems and firewalls, these botnets
tend to maintain a stealthy communication pattern with the bot-master or other

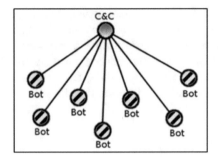

Fig. 3. How a centralized botnet structure looks like

bots. By generating little traffic these botnets are successful in fooling Detection Systems and Firewalls [10].

Zeus is a banking Trojan that has been a headache for online banking customers all over the world collecting personal information including passwords and log in credentials [11]. Waledac is also known as the Storm botnet [12] is a spamming botnet that can send up to 2000 malicious emails on a daily basis and once infected it modifies the Windows host file to redirect a number of popular search engines to malicious an IP address which keeps displaying advertisement pop-ups [13].

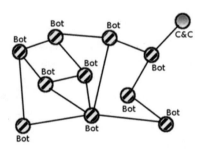

Fig. 4. How a decentralized botnet structure looks like

2.3 Cluster Analysis: Simple K-means, Farthest First and DB Scan

In this paper we choose to work with clustering algorithms to compare which is the best fit to classify P2P botnets traces using network flows extracted from the packets send within a network [14].

1. K-MEANS is a classical clustering technique that attempts to find a user-specified number of clusters (K) in the data. Its algorithm is quite naïve. First, it finds K centroids at the feature space. Second, it assign each data to their closest centroid. Third, it update the centroid by finding the centroid of

the data gathered by previous step. It repeat second and third steps until the position of these K centroids converge. K-means is simple and can be used for a wide variety of data types. However, it can only handle globular clusters and it is sensitive to the choice of K.

2. The Farthest First Traversal (FF-traversal) algorithm [15], which works somewhat like a reverse K-means, minimize the maximum distance of the farthest point to the closest centroid. This is a variant of k-means that places each cluster center in turn at the point furthest from the existing cluster centers, the point must lie within the data area. This method is fast and suitable for large-scale data mining datasets.

3. DBSCAN (Density-based spatial clustering of applications with noise, DBSCAN) is a density-based clustering algorithm, in which the number of clusters is automatically determined by the algorithm. Unfortunately in some cases DBSCAN may fail to cluster datasets with great density differences and can be expensive when the computation of nearest neighbors requires computing all pairwise proximities, as is usually the case for high-dimensional data [16].

2.4 Previous Work

Machine learning techniques have been use before to classify P2P traffic in real-time using a small number of packets, as few as the first five packets of a TCP connection and classifying those using decision tree algorithms which perform better than using K-means clustering [17]. The results from the use of Supported Vector Machines (SVM) prove worthy of use, extracting network flows in the forward and backward direction from the packets generated in the network, and a great accuracy to detect P2P botnet traffic flows [3].

Anomaly detection systems like BotMiner [18] try to identify botnets avoiding the use of protocol and botnet structures focusing on the principle that bots within the same botnet will exhibit similar communication and malicious activities patterns, but fails it purpose if evasion techniques are used. Some of the recent work has used supervised machine learning algorithms [4,19] to analyze features extracted and used as inputs to facilitate the learning process of the algorithms, Bayesian regularized artificial neural networks (BRANNs) [20] for real time P2P traffic classification and unsupervised [10] machine learning approaches combined with some supervised learning method proving that a hybrid system may be a viable answer to this problem.

3   Methodology

3.1 Approach Using Unsupervised Learning

Traffic can be classified by selecting its attributes which distinguishes their behavior, we want the unsupervised learning algorithm to find the patterns hidden among the P2P flows. To facilitate the algorithm detection a previous process to select the most relevant features will be carry out and then these input will be feed to the clustering algorithms in order to compare their overall performance creating clusters based on the characteristics of those features. Then the resulting cluster will be cross validated in order to ensure the legitimacy of the outputs. Figure 5 shows a flowchart with the overall process:

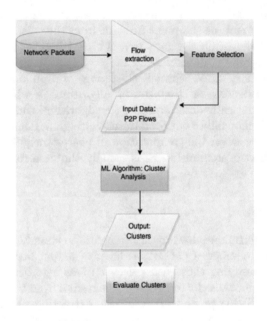

Fig. 5. Overall process of the approach using unsupervised learning

3.2   Feature Extraction from Flows

We need to use some appropriate 'features' as the inputs for Machine learning training models. For this paper, the samples from Normal P2P Traffic and Botnet P2P Traffic (as described in the previous section) collected are store in network trace files called "pcaps" (Wireshark may be use to open this file extension) containing all the packets shared among the peers in the network. Then we use these trace files to do feature extraction by a tool called "NETMATE" (Network Measurement and Accounting Meter) which is an Open-source tool [22]. The NETMATE will generate the data in the form of 'flows', each individual flow contains the information of protocol, Source IP, Destination IP, Source port, Destination port and the other forty-five features. The features values are represented as integers unless otherwise stated.

3.3   Feature Selection

Among these features, there're some features that have little effect on result of clustering. Thus choosing effective subset of features is important. The more accurate the set of features is, the better the performance results. In contrast, choosing less effective features reduces the performance and accuracy when doing the clustering.

For starter, we removed two features of the flows, namely SIP and DIP. The reason is because IP differs from device to device, In other words, these features depend on the network configuration. However, we need to cluster the flows based on behavior. Next, in order to find the most influential features of the entire set, we evaluate the importance of attributes (features) by computing their Information Gain (IG) using the Ranker method Algorithm, where Information Gain (IG) is described by the following equation:

$$IG(Class, Attribute) = H(Class) - H(Class|Attribute) \qquad (1)$$

Where H is the Entropy before and after comparing it to the class. Table 1 contains the first 20 features ranked according to their Information Gain value, which is shown in the first column.

Table 1. Ranked features according to information gain value

Ranking value	Feature
0.27419	srcport
0.22435	min_fpktl
0.20271	max_fpktl
0.19177	total_fvolume
0.18125	sflow_fbytes
0.18009	mean_fpktl
0.16425	dstport
0.14165	max_bpktl
0.1184	min_bpktl
0.1152	total_bvolume
0.11032	mean_bpktl
0.1025	duration
0.09241	min_active
0.08748	mean_active
0.08134	max_active
0.06864	sflow_bbytes
0.06374	std_bpktl
0.06278	total_bhlen
0.0519	std_fpktl
0.0464	min_biat

3.4 Clusters Evaluation Measurements

The data is divided later into subsets for evaluating the machine learning algorithms, the sets contain P2P traffic from both kind's malicious flows and Non-Malicious flows.

Performance statistics are calculated for all the trials. The class is ignored during all the evaluations for the clusters.

The validation measures are used to evaluate the credibility of the clusters, in this case due to the high imbalance between classes, to keep the real scenario of a network, we can't only rely on the accuracy measurement of the clusterer.

The Classification Oriented Measures of Cluster Validity are described below:

- The number of correctly classified instances as malicious is referred to as the True Positive (TP).
- The number of instances classified as malicious but should be normal and therefore rejected, is referred to as the False Positive (FP).
- The number of instances classified as normal but are actually malicious, is referred to as the False Negative (FN).
- The number of normal instances from a class correctly rejected is referred to as True Negative (TN).

The effectiveness of the clusterer is measured by the overall accuracy, denoted as:

$$Accuracy = (TP + TN)/((TP + TN + FP + FN)) \qquad (2)$$

The fraction of a cluster that consists of objects of a specified class is the precision. It is given by:

$$Precision = (TP)/((TP + FP)) \qquad (3)$$

Recall (or true positive rate): The extent to which a cluster contains all objects of a specified class. It is given by:

$$Recall = TP/((TP + FN)) \qquad (4)$$

F-measure: A combination of both precision and recall that measures the extent to which a cluster contains only objects of a particular class and all objects of that class:

$$F = 2 * ((Precision(i,j) * Recall(i,j)))/((Precision(i,j) + Recall(i,j))) \qquad (5)$$

4 Results and Discussion

We need to utilize clustering algorithms which can handle such differences in the data along with the first 10 top ranked attributes from the feature extraction. As mention before we used WEKA (Waikato Environment for Knowledge Analysis, WEKA) [23] to run this three clusterers.

4.1   Dataset Assemble for Testing

The dataset used in this paper was obtained from a previous research group that made the datasets publicly available, their paper is about a P2P traffic categorization system called "PeerRush" [21]. The labeled data of all four P2P applications (Emule, UTorrent, Vuze and FrostWire) along with Zeus and Waledac were used for testing purposes. The dataset was divided into 3 combinations of subsets, containing both kinds of flows that are labeled either Non-Malicious or Malicious:

– Dataset 1: This dataset contains network flows from all the four P2P applications along with Zeus traces. A total of 17,940 flows are contained in the dataset, 95% non-malicious traces and 5% malicious traces.
– Dataset 2: This dataset contains network flows from all the four P2P applications along with Waledac Traces. A total of 12,310 flows are contained in the dataset, 93% non-malicious traces and 7% malicious traces.
– Dataset 3: This dataset contains network flows from all the four P2P applications along with Zeus and Waledac traces combined, the selected malicious flows are different from the previous subsets mention above. A total of 12,334 flows are contained in the dataset, 92% non-malicious traces and 8% malicious traces.

4.2   Unsupervised Learning Method: Cluster Analysis

Results Dataset 1: Zeus Botnet. For dataset 1 we evaluate the performance of each clustering algorithm and those results are presented in Table 2. Let's keep in mind we are evaluating the accuracy of how good the malicious traces from the Zeus botnet are group together and separated away from the normal traffic.

Table 2. Performance of algorithms for Zeus botnet dataset

Algorithm	Incorrectly clustered	TP	FP	FN	TN
Farthest First	373	888	372	1	16679
DB Scan	241	649	1	240	17043
K-means	5388	889	5388	0	11663

The evaluation measurement for each algorithm are presented in Table 3. Farthest First retrieve almost all of the malicious flows but with a fair precision, i.e. it identified the malicious flows but also confused 372 benign flows as malicious.

Results Dataset 2: Waledac Botnet. For dataset 2 we evaluate the performance of each clustering algorithm and those results are presented in Table 4.

The evaluation measurement for each algorithm are presented in Table 5. DB Scan didn't retrieve all of the malicious flows again into the cluster labeled as malicious, but had a better performance overall with a precision over 85% keeping the count of FP low.

Table 3. Evaluation measurement for Zeus botnet dataset

Algorithm	Accuracy	Precision	Recall	F-measure
Farthest First	97.92%	0.705	0.999	0.826
DB Scan	98.66%	0.998	0.730	0.843
K-means	69.97%	0.142	1	0.248

Table 4. Performance of algorithms for Waledac botnet dataset

Algorithm	Incorrectly clustered	TP	FP	FN	TN
Farthest First	3544	254	2823	721	8512
DB Scan	351	714	90	261	11206
K-means	5235	818	5078	157	6257

Table 5. Evaluation measurement for Waledac botnet dataset

Algorithm	Accuracy	Precision	Recall	F-measure
Farthest First	71.21%	0.083	0.260	0.125
DB Scan	97.14%	0.888	0.732	0.803
K-means	57.47%	0.139	0.839	0.238

Results Dataset 3: Zeus & Waledac Botnet. For dataset 3 the results are presented in Table 6. Farthest First mixed for both clusters the normal flows (TN) with malicious flows (TP), correctly labeling the cluster containing the most count of malicious instances but along with a high count of normal traffic.

The evaluation measurements are presented in Table 7. Simple K-means retrieved all the malicious flows (TP) successfully but also mixed a high count of benign flows in the same malicious cluster.

4.3   Unsupervised Learning Comparison

The accuracy of all the algorithms is shown in Fig. 6. DB Scan performed significantly well for all the situations assigned, each change of dataset diminish slightly the accuracy, but in general it maintains the highest percentage. Simple K-means improved in the last test but still had some imbalance in the number of correctly classified malicious instances that were retrieved.

The most prominent features that helped the algorithms group the two types of P2P flows were srcport and dstport. Also features related to how much time the exchange of packets changed to idle or active helped distinguish the behavior from normal P2P flows.

In Table 8 we can compare the real performance values of each algorithm. DB Scan proves to be a worthy contender against P2P botnets flows and may be use to improve the precision of detection systems along with other security

Table 6. Performance of algorithms for Zeus & Waledac botnet dataset

Algorithm	Incorrectly clustered	TP	FP	FN	TN
Farthest First	3560	117	2678	882	8657
DB Scan	450	554	5	445	11325
K-means	1331	999	1331	0	10004

Table 7. Evaluation measurement for Zeus & Waledac botnet dataset

Algorithm	Accuracy	Precision	Recall	F-measure
Farthest First	71.14%	0.042	0.117	0.062
DB Scan	96.35%	0.991	0.554	0.711
K-means	89.21%	0.429	1	0.600

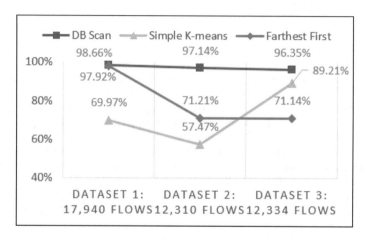

Fig. 6. Accuracy of unsupervised algorithms for all datasets

Table 8. Average measurement values for each algorithm

Algorithms	Average accuracy	Average precision	Average recall	Average F-measure
Farthest First	80.09%	0.2763	0.4588	0.3378
DB Scan	97.38%	0.9592	0.6723	0.7857
Simple K-means	72.22%	0.2364	0.9463	0.3621

tools. We believe that each algorithm performs depending on the quality of the data and the previous preprocessing of it.

5   Conclusion

The use of unsupervised learning was propose for classifying P2P traffic flows in comparison to the previous methods using supervised learning. The results lead us to believe the data is suitable for a density based clusterer, since DB Scan algorithm performed well on every situation, obtaining high precision classifying P2P botnet flows and retrieving most of these malicious flows from the normal P2P Traffic.

5.1 Future Work

The following items are things we wish to improve in the future:

- Enhance the performance of the algorithms with huge datasets.
- Testing with a completely different set of features the performance.

Acknowledgment. This research was partially supported by the Ministry of Science and Technology of the Republic of China under the Grants MOST 108-2218-E-007-053 and MOST 108-2218-E001-001.

References

1. Price, R.: Business Insider, 28 May 2015. http://www.businessinsider.com/hola-used-for-botnet-on-chrome-2015-5. Accessed 6 June 2015
2. Kaspersky Lab: SecureList. Kaspersky, 29 May 2015. https://securelist.com/blog/research/70071/statistics-on-botnet-assisted-ddos-attacks-in-q1-2015/. Accessed 6 June 2015
3. Barthakur, P., Dahal, M., Ghose, M.K.: A framework for P2P botnet detection using SVM. In: International Conference on Cyber-Enabled Distributed Computing and Knowledge Discover, Sanya (2012)
4. Liao, W.H., Chang, C.C.: Peer to peer botnet detection using data mining scheme. In: 2010 International Conference on Internet Technology and Applications, Wuhan (2010)
5. eMule: eMule Project. http://www.emule-project.net/home/perl/general.cgi?l=1. Accessed 7 June 2015
6. Skype: Skype. Microsoft. https://support.skype.com/en/faq/FA10983/what-are-p2p-communications. Accessed 7 June 2015
7. BitTorrent: BitTorrent. BitTorrent Inc. http://www.bittorrent.com/. Accessed 7 June 2015
8. Wang, P., Aslam, B., Zou, C.C.: Peer-to-peer botnets. In: Handbook of Information and Communication Security, pp. 335–350. Springer, Heidelberg (2010)
9. Vacca, J.: Computer and Information Security Handbook. Morgan Kaufmann, Burlington (2009)
10. Narang, P.: PeerShark: flow-clustering and conversation-generation for malicious peer-to-peer traffic identification. EURASIP J. Inf. Secur. (2014)
11. Symantec Security Response: Zeus Now Setting its Sights on Japanese Online Banking Customers. Symantec, 11 February 2013. http://www.symantec.com/connect/blogs/zeus-now-setting-its-sights-japanese-online-banking-customers. Accessed 7 June 2015
12. Lelli, A.: Symantec Official Blog. Symantec, 12 January 2011. http://www.symantec.com/connect/blogs/return-dead-waledacstorm-botnet-back-rise. Accessed 7 June 2015

13. Neville, A.: Symantec Official Blog. Symantec, 4 June 2013. http://www.symantec. com/connect/blogs/waledac-reloaded-trojanrloaderb. Accessed 7 June 2015

14. Ghahramani, Z.: Unsupervised learning. In: Advanced Lectures on Machine Learning, pp. 72–112. Springer, Berlin (2004)

15. Hochbaum, D.S., Shmoys, D.B.: A best possible heuristic for the k-center problem. Math. Oper. Res. **10**(2), 180–184 (1985)

16. Tan, P.N., Steinbach, M., Kumar, V.: Cluster analysis: basic concepts and algorithms. In: Introduction to Data Mining, pp. 487–559. Pearson (2005)

17. Li, J., Zhang, S., Lu, Y., Yan, J.: Real-time P2P traffic identification. In: Global Telecommunications Conference, New Orleans (2008)

18. Gu, G., Perdisci, R., Zhang, J., Lee, W.: Botminer: clustering analysis of network traffic for protocol-and structure-independent botnet detection. In: USENIX Security Symposium (2008)

19. Garg, S., Singh, A.K., Sarje, A.K., Peddoju, S.K.: Behaviour analysis of machine learning algorithms. In: 15th International Conference on Advanced Computing Technologies (ICACT) (2013)

20. Guntuku, S.C., Narang, P., Hota, C.: Real-time peer-to-peer botnet detection framework based on Bayesian regularized neural network (2013)

21. Rahbarinia, B., Perdisci, R., Lanzi, A., Li, K.: PeerRush: mining for unwanted P2P traffic. In: Detection of Intrusions and Malware, and Vulnerability Assessment, vol. 7967, pp. 62–82. Springer, Heidelberg (2013)

22. NETMATE: netmate-flowcalc. https://code.google.com/p/netmate-flowcalc/. Accessed 1 May 2015

23. Hall, M., Frank, E., Holmes, G., Pfahringer, B., Reutemann, P., Witten, I.H.: The WEKA data mining software: an update. In: SIGKDD Explorations (2009)

Intrusion Detection, Privacy and Cryptography

Achieve Web Search Privacy by Obfuscation

Hao Wang$^{1(\boxtimes)}$ ⓘ, Wenyan Liu^2, and Jiangtao Wang^1

^{1} National Trusted Embedded Software Engineering Technology Research Center,
East China Normal University, Shanghai, China
wanghao@stu.ecnu.edu.cn
^{2} East China Normal University, Shanghai, China
wenyanLiu@stu.ecnu.edu.cn

Abstract. After personal information of up to 87 million users of Facebook may be leaked to the consulting firm Cambridge Analysis in March 2018, some hackers sold 870 million user information stolen from 38 popular websites on the Darkweb in February 2019. The issue of privacy protection has once again received intense attention from people. One significant privacy risk that we encounter every day is searching across various search engines. The search engines can be curious and save users' query histories. Then the search company and adversary can dig out the users' intentions and points of interest by some technology such as machine learning. They may deliver advertisements to the user accurately. Even some malicious attackers will threaten users' personal and property safety.

A popular tool named TrackMeNot can help to protect users' search privacy to a certain extent. TrackMeNot can maintain a dynamic list of popular queries and send dummy queries to obscure users' real intentions. It is said they can significantly increase the difficulty for 3rd-parties to filter out users' real queries. The work by [2] tries to find the deficiency of TrackMeNot. They use clustering or classifying method to filter out the real queries from the queries. In this paper, we make two useful improvements to TrackMeNot. By improving the generation strategy of dummy queries, we can effectively reduce the filtering effect of clustering methods and common classifiers on dummy queries. Experiments with the real query log dataset AOL show the effectiveness of our work.

Keywords: Web search privacy · Query obfuscation · TrackMeNot

1 Introduction

With the advent of the era of data explosion and the rapid development of artificial intelligence, the security of personal information is also facing many

Supported by East China Normal University, National Trusted Embedded Software Engineering Technology Research Center.

challenges. In the first half of 2018, Facebook's privacy leaks caught the attention of the world. Facebook admitted to the media that the personal information of up to 87 million users might be leaked to the consulting firm Cambridge Analysis. Facebook has 2 billion users worldwide and 1.4 billion active users per day. This incident has once again triggered a heated discussion about security and privacy. The privacy risks faced by users have once again surfaced, and we have to worry about user privacy issues.

There is no doubt that web search has become the most critical tool for finding information online. Because billions of users generate thousands of queries per second, search engines have many data, an unparalleled potential for describing personal interests, thoughts, and behaviors around the world. One of the most common things search engine companies do is to build user profiles through user queries and then provide relevant ads to users. When we open a web page, we are often familiar with the ads on the page. Because once we search, the service provider records our operations. The DNT (do not track) settings in the browser are different, it is a http header field request the web application do not track the information of user if the companies follow the DNT protocol. But We can't ask search engine companies not to store our search records. Search engine companies are more inclined to store user search records, which could help improve the quality and efficiency of their search services and effectively improve the operation and revenue of Online behavioral advertising (OBA). However, the risks are apparent. With these user queries, search engine companies can use the related technologies of machine learning and data mining to obtain more sensitive privacy information, such as whether users have serious diseases, how about users' economic conditions, and so on. The users' true profile is exposed, leading to more targeted advertising, and the search engine does not guarantee the user's authenticity and reliability of the advertising content. These advertisements may be preferentially pushed to the user just because of the bidding ranking. But the losses caused by these ads are not easy to claim from search engine companies. What is more serious is that the users' information is maliciously exploited, and the user may be exposed to more deceptive and dangerous cybercrime.

In the past, some methods have been proposed to solve the problem of network search privacy. One of the methods is to use an anonymous network, through the proxy server or the Tor network [9], so that the user's real information such as IP address and hardware information cannot be truly tracked. This method works well, but there are also examples that specific users can be determined only by the content of the search. The New York Times reporter easily found a 62-year-old woman based on AOL search data. The old lady confirmed that the search keywords listed were indeed hers [2]. Another type of method is private information retrieval (PIR) [8], which allows users to retrieve records from a database, and the database owner cannot determine which record was accessed. This kind of method is complicated to apply in the web search field. The method we focus on is the third type of method: a category of private web search solutions that we call obfuscation-based private web search (OB-PWS) systems [10]. This method automatically generates dummy queries when users perform real queries. These fake queries are not necessarily related to user interests. When the server receives these queries, it is difficult to determine the user's real purpose. The advantage is that we only need to deploy on the client side, no server-side cooperation. Users get the initiative to a certain extent, and users can decide how to choose the setting to protect their search privacy.

In this paper, we focus on a tool based on the principle of query obfuscation, TrackMeNot (TMN), which is a browser extension that can be deployed on Chrome or Firefox browsers. TMN uses some necessary measures to simulate user queries. When the user searches in the browser, TMN could automatically send dummy queries according to the set rules to cover the user's true intentions. In [2], they propose a method of attacking TMN. During the initial startup of TMN, the dummy query sent by the TMN lacks similarity with the real query of the user, and the real query of some users can be figured out by clustering and classifying method. They use the clustering algorithm to divide all the queries into some clusters. Then they label the largest cluster as the real queries from the user. Although this cluster of queries is few, the minimum size of the largest cluster queries is only 10, the accuracy is very high and the 10 queries are all real queries created by the user. It can be considered that most of the queries in the largest cluster are users' real queries. We think this is a cold start issue. In some other attempts to attack obfuscation-based tools [3], it is assumed that the attacker has some real queries generated by the user before the user started using the obfuscation-based tool. Then they use the real query to train the supervised learning model to classify the real and dummy queries logged after user using the obfuscation-based tool. But the availability of a set of real queries is a strong assumption, this assumption cannot be confirmed. The attacker cannot know what time the user started using the obfuscation-based tool but only knows when the user started the first query. It is difficult for obfuscation-based tools to imitate the user's real query at the beginning because it is difficult to judge the user's true intention because of insufficient user query. So the queries sent by the user at the beginning are extremely valuable. These queries, in the beginning, provide a favorable reference for further analysis of the user query.

Our contribution: We verify the effect of the user's real query being filtered out during the cold start stage of the TMN tool. We propose Two enhancements for TMN, which effectively compensates for the vulnerability that TMN may leak the user's real query during the cold start phase. The structure of this paper is as follows: In the second section, we introduced some other related work on query privacy protection. In the third section, we describe the working principle of TMN, as well as its advantages and disadvantages. In the fourth section, we discuss the methods of clustering attacks and our enhancements of defense. In the fifth section, we introduce our experimental setup and results and we discuss them. In the last two sections, we give the conclusion of the study and looks into future work.

2   Related Work

In this section, we discuss three aspects of related work. At first, we introduce various OB-PWS tools. Then we describe the attacks on obfuscation-based privacy-protection tools. At last, we discuss the evaluation of the effect of query privacy mechanism related work.

2.1   Obfuscation-Based Private Web Search (OB-PWS)

One of the most OB-PWS tools is TrackMeNot, which maintains a list of fake queries that are sent along with user queries to protect the user's real queries. The update of the dynamic list is related to the most recent popular content, and the content returned by the user's real query. We demonstrate its detail of mechanism in the next section. The second tool is GooPIR [16]. GooPIR is a tool similar to TMN that selects keywords from public dictionaries to generate fake queries. When a user initiates a query, GooPIR generates k-1 fake queries at the same time to mask the user query. GooPIR aims to offer what Domingo-Ferrer et al. call h(k)-private information retrieval ($h(k)$-PIR). This property ensures that a real query R is seen by the adversary as a random variable R whose entropy is such that $H(R) \geq h(k)$ for some function h [10]. The larger k is, the more fake queries are sent to the server, and the better the protection of user privacy. GooPIR considers the popularity of a user's query term, and the k-1 fake queries all have similar levels of popularity. The popularity is calculated from the frequency of a query appears on the network.

Plausibly Deniable Search (PDS) [14], like GooPIR, also generates $k-1$ fake queries and adds some specific rules. For a k size query set $S = \{Q1, ..., Qk\}$, of which one query is real and $k-1$ are dummies. There are three rules: the probability that any real query is required to generate such a set is equal; the queries in S correspond to different topics; all Qi in S must have the same plausibility.

A PRivAcy model for the Web (PRAW) [15] hides real user interest by generating fake web interactions. Its strategy is to generate some queries that are close to the user's real topic. The purpose is to expand the user's point of interest and protect the user's a more specific point of interest topic. Also, it can defend against clustering attacks. Optimized Query Forgery for Private Information Retrieval (OQF-PIR) [13] assumes a popular profile and expect the observed profile to be similar to the popular, and they use the KL divergence between the observed profile and the popular one to measure the similarity. Noise Injection for Search Privacy Protection (NISPP) [12] is similar to OQF. The main difference is that they want the mutual information between the observed profile and real profile to be the least.

2.2   Quantity Privacy of OB-PWS

Ero Balsa presented an abstract model and related evaluation framework to evaluate the six OB tools we mentioned above. Here he proposes a profile-based analysis and query-based analysis. The profile-based approach is designed to measure the uncertainty of an opponent's actual profile. Information entropy is used to measure the uncertainty of an adversary to the user. $Z \in \mathcal{Z}$ represents the filtered profile of the adversary. $X \in \mathcal{X}$ represents the real profile of the user. When the EZ is close to 1, the profile of the user is perfectly protected.

$$\mathcal{E}_Z = H(\mathcal{X}|\mathcal{Z}) = H(\mathcal{X}, \mathcal{Z}) - H(\mathcal{Z}) \tag{1}$$

When the $\mathcal{E}_Z$ is close to 1, the adversary is totally uncertain on real profiles $X \in \mathcal{X}$ and the profile of the user is perfectly protected.

The query-based method is used to measure whether a dummy query can be distinguished from a real query. The opponent uses a classification method to separate the dummy query from the real query. Two attributes are defined, one is unobservability, which indicates how many proportions of the user's real query are misclassified in the false query classification. Another one is deniability, which indicates how much of the user's false queries are classified in the user's real query.

Gervais [17] proposed an algorithm that evaluates the effectiveness of the query obfuscation mechanism, defining the privacy criteria for query levels (links between user queries) and semantic levels (topics of interest to users). Their definition is similar to Ero Balsa' profile-based and query-based measure. They use a machine learning method to train a linkage function to determine whether a query belongs to the user's real query or generated false query.

2.3   Attack Pattern

We focus on Rami Al-Rfou's clustering attack method for TMN [2]. He attempts to separate real and false queries using clustering methods and selects the cluster of result with the largest size as the user query. There is a high accuracy rate in the case of a small dataset size. Peddinti and Saxena [2] attempted to use a classifier to separate real and dummy queries from a query set, dummy queries are generated by using TMN. They also used the AOL dataset, using the user's 2 months of the real data training model, using one month of real data to simulate the user's search in a TMN browser for 1 month, collecting query data for 1 month, using six classification algorithms for verification. Their classification method obtained a good result. For identifying the TMN query, the error rate is close to 0.02%, and the average classification accuracy for the user query is about 48.88%. In some cases, a user query recognition rate of 80% to 100% is achieved.

The work of Chow [18] uses two simple ways to classify observed queries. The first way is based on capitalization. According to the author's observation, the query generated by TMN has more uppercase letters. According to this feature, 93% of user queries and 54% of TMN queries are correctly classified. However, according to our experimental results, TMN has already fixed this Vulnerability. The generated TMN query has no significant difference with the user's real query about the rate of uppercase letters. At the same time, by analyzing the source code of TMN, there is a 90% probability of converting the query to lowercase letters. The second way is that according to the overlapping rate, Richard Chow found that queries generated by TMN have more repeated keywords. They compare the current query with the previous and following queries for repeated keywords. If either comparison yields overlapping keywords, the current query is considered to be a TMN query. Otherwise, it is a user query. As a result, 91% of user queries and 83% of TMN queries are correctly classified. If they compare the two queries previous and following the current query, the 88% user query and 89% TMN query are correctly classified. However, according to our experimental results, the query repetition rate generated by TMN is about 15%, while in the AOL data set, the repetition rate of user queries is often above 50%. The user's repetition rate is bigger, which is more in line with the search habits of ordinary users. Background: TMN Query Generation

3   TrackMeNot Background

In this section, we explain how does TMN work as an extension in Firefox. It is the same process in Chrome. We try to discuss the mechanism of TMN through literature and TMN's source code.

3.1   Mechanism of TMN

TMN is an extension in Firefox and chrome, and it aims at enhancing the privacy of web search. TMN uses query obfuscation to protect users'real queries and then protect users' real profile and intention. TMN maintains a dynamic list of queries and uses seeds from the list to send dummy queries. The dynamic list is initialized by popular RSS feeds and publicly available recent queries. In this way, the words in the dynamic list are popular in the network. The list evolves with users making web search. As the operation continues, the words in the list are randomly marked and substituted by the content from HTTP response returned by the users' real queries. Then each user has a unique dynamic list that is relevant to the user's search history. TMN eventually generates queries as users do. TMN also has other technical mechanisms to mimic users' queries and protect users' profile. Burst-Mode Querying is a function to trigger a random number of TMN queries when monitor user sends real queries. TMN monitors users' real search requests using "Real-Time Search Awareness". By setting the regular dummy queries generating and burst-mode querying, user's search frequency can be masked. Similar Clicking is a mechanism to mimic user behavior of clicking the results list returned by the search engines. Searching without clicking one of the results make dummy queries vulnerable to be distinguished by search engines. Moreover, TMN uses regular expressions to avoid clicking on advertisements. Not only because too many clicks on the advertisement may be detected as robots, but also because TMN does not want to influence the revenue by clicking advertisements. Live Header Maps and Cookie anonymization are other useful mechanisms. Live Header Maps maintains the header fields and URLs for the last search by the user. TMN then mimic the header values with the user's search header. Cookie anonymization can block the cookies sent by the user and store the cookies. Then TMN forwards the stored cookies with its generated queries.

3.2 Weakness and Strengths

TMN provides useful protection of user's query by these above mechanisms. Of course, there are weaknesses of TMN. TMN can not mask actual identifying information, and this strategy was not energetically pursued. TMN is not designed to mask IP addresses [5]. Tor is a popular tool to hide users' IP address. Some users are concerned that if TMN automatically generates sensitive queries about pornography, politics, and terrorist activities. It is indeed a problem to make users be misunderstood in these sensitive fields. TMN provides a blacklist setting for the user, but it can not promise you never generate sensitive queries out fo the blacklist. However, most time TMN fetches queries from the recently popular search, TMN does not generate queries make the user stand out. The weakness we focus on is about the cold-start. When users firstly start to send queries on search engines with TMN, the queries created by TMN may be easy to distinguish with real queries, because the dynamic list is updated randomly and need a certain number of users'real queries to evolve dummy queries more like real queries. The work in [2] proposes a useful method to filter out users' real queries from the queries mixed with dummy queries. They define a semantic distance between each pair of queries. Moreover, use the Partitioning Around Medoids (PAM) algorithm to do clustering. Their method gets a subset of the user's real query and has a high degree of precision in small-scale query samples. As a result, part of users' real intention may be detected. In this paper, we proposed two ways to improve TMN, which can effectively avoid such attacks. We show our works in the next section.

4   Methodology

We have three parts in this section to illustrate our design. The first part is about the proper semantic distance measure we use. In the second part, we discuss the clustering method. In the last party, we clarified how to improve on top of TMN and predict possible outcomes.

4.1   Semantic Distance Measure

We use the same measure to define semantic distance as the work in [2]. They use a java package DISCO to measure the semantic distance in their work. DISCO (extracting dIstributionally related words using CO-occurrences) is a Java application that allows retrieving the semantic similarity between arbitrary words and phrases [6]. Its process is collecting an extensive set of words to build word spaces, and the words in each word space are transformed into a word vector. Then the similarity can be measured by COSINE or KOLB. In our work, we use the well-known cosine vector similarity measure.

4.2 Separate User Queries

The work in [2] tries to filter out which queries are sent by real users. We already have the definition fo the distance between each pair of queries in the above. The method to find out real queries is using clustering to identify groups of related queries and then choose the largest group as the result of classification. We use the Partitioning Around Medoids (PAM) algorithm to operate clustering. PAM is similar to the well know k-means algorithm. The PAM has better robustness and is resistant to noise. Because PAM has to calculate the dissimilarities between two points and try to minimize the sum of the dissimilarities between every two points. Naturally, it will run much slower with $O(n^2)$ time complexity to calculate the center of the cluster. An idealized clustering operation should be able to divide similar points into one cluster as much as possible, and should have a suitable number of clusters so that the distance between each two clusters is as far as possible. We used the same evaluation criteria as [2] to measure the effect of clusters generated after each iteration. For every element i assigned to cluster A, we define $a(i)$ as the average dissimilarity between i and each element $j \in A$ where $j \neq i$. Define $d(i, C)$ as the average similarity between i and all the elements in a cluster C and $C \neq A$. And use $b(i)$ denotes the minimum $d(i, C)$. Then we denote $s(i)$ to quantify the result of clustering:

$$s(i) = \frac{b(i) - a(i)}{max(a(i), b(i))} \tag{2}$$

4.3 Enhance TMN

A non-negligible problem of TMN is about the cold start. When the dummy queries generated by TMN are from the initial dynamic list, it is often far away from the user's real query in the semantic distance, which makes the user's real query face great risks. There are two common methods for solving the cold start problem. One is to let the user set the corresponding points of interest before starting, but this will increase the users' cost-in-use, and it is difficult for the user to decide how to set the point of interest and protect their initial queries usefully. If the interest points set by the user are too specific, the queries generated by the TMN may be too similar to the user's real query. Although the user's query is protected, it is easy to cause the user's true profile to be revealed. Therefore, the specific implementation of this method needs further research and exploration.

Another common way is to use the most popular content as the initial generation of TMN. According to the above description of TMN, TMN has actually done this. And the existing experimental results of [2] shows that the effect is not ideal. In our experiments, similar conclusions have been obtained. The most popular is not enough to solve the problem of TMN's cold start.

Therefore, we are trying to find a convenient and effective way to solve the cold start problem. In view of the characteristics of the clustering method used, we propose two improvements to TMN. Both methods take into account the query currently sent by the user to determine what content is generated by the next TMN query. The work of [7] proposes that when generating dummy queries, not only should the user's current query be considered, but also a series of queries when searching for the same task before, so that the dummy query can be made more reasonable. When we deal with the cold start problem, we only consider the current user's query, because the cluster attack method is hard to detect if there are relationships on the semantics of multiple queries. In our first enhancement, after the user-initiated a query, we selected the query in the dynamic list that was closest to the user's last query as the query generated by TMN. Then the key to the problem is how to define the closest.

At the same time, we should not make the generated dummy query and user query very consistent or similar, which will result in the user's true intention even if the true and false query cannot be distinguished, because the true and false query may express the same meaning.

In our first enhancement, after the user-initiated a query, we selected the query in the dynamic list that was closest to the user's last query as the query generated by TMN. Then the key to the problem is how to define the closest. We use the Levenshtein distance, which is the edit distance, to measure the proximity of the words. Generally speaking, the closer editing distance will increase the degree of semantic similarity, which can improve the confusion of the TMN generated query at startup.

Our second enhancement is to randomly extract a word from the user's current query and appending it behind the dummy query that TMN will use. This will have a significant effect on the clustering attack in [2]. Because of the clustering attack in [2], when calculating the similarity of two queries Q1 and Q2, each word of Q1 are compared with each word of Q2 and the result with the largest similarity is selected as the similarity of the query. When we joint a word in the user's current query after the TMN false query, that attack method will be invalid. Because each user's real query will necessarily match a dummy query generated by TMN. At the same time, we don't have to worry too much about whether this will reveal the user's true intentions because search queries are usually phrases and not single words. Words randomly extracted from user queries do not necessarily represent the users' true intentions, at the same time after the word is combined with TMN fake query, it will also increase the difficulty of identifying the users' true intentions.

5   Experiment

The purpose of the experiment was to verify how effective our enhancements to TMN are. TMN We use the AOL dataset and build a extension of Firefox to simulate the real query of users. We also add the module for TMN extension so that each user's real query and the dummy query generated by TMN are sent to the server we set up to collect the query data.

5.1   Dataset and Setup

We send the queries from AOL to simulate a user's real queries. AOL is suited for our experiment because it is the largest english search log dataset we can get. AOL collect search data from more than 650000 users and spanning over a reasonably long period of 3 months. Even though it is somewhat out of data (created in 2006), it does not affect our experiment a lot because we haven't taken the timestamp into account. The possible problem is that the user query in 2006 may have a big gap with the most popular used by TMN in 2018. But our goal is to reduce this gap. The gap between the user's real query and the fake query generated by TMN is larger. The more we can reflect the obvious effect of our improvement on TMN.

Selecting Users. We randomly selected 30 users from the AOL dataset and divided them into three categories. These users have more than 1000 queries in the three months. Some users have queries are more evenly distributed, and the query repetition rate is less than 20%. And some user's real queries have more concentrated distribution, these query repetition rate is More than 50%, there is a situation between the two, the query repetition rate is between 20% and 50%.

Data Collection. We build a new extension for Firefox. It can read queries from the user data we choose from AOL and then type in the search box to simulate a real user triggering a search. Every query sent by our extension triggers a post request to our server with the information of this query. At the same time, we modify the source code of TMN to add a sending module. It can send the queries generated by TMN to our server too.

We run our extension and TMN extension on Firefox with 30 users. We collect 500 real queries and about 500 queries created by TMN for each user. For each user, we experimented 5 data scales: 100, 200, 300, 500 and 1000 mixed real and fake queries.

Semantic Distance and Clustering. We build a new extension for Firefox. It can read queries from the user data we choose from AOL and then type in the search box to simulate a real user triggering a search. Every query sent by our extension will send a post request to our server with the information of this query. At the same time, we modify the source code of TMN to add a sending module. It can send the queries generated by TMN to search engine server too. We run our extension and TMN extension on Firefox with 30 users. And we collect 500 real queries and about 500 queries created by TMN for each user. For each user, we experimented 5 data scales: 100, 200, 300, 500 and 1000 mixed real and fake queries. We use PAM cluster algorithm to get the clusters of each mixed queries. Then chose the largest one as the user's queries. For PAM algorithm, we set the parameter $k = 15$ means trying to get 15 clusters. And we set the iterator $= 15$ means the algorithm will execute to calculate 15 times iterator to get new cluster.

5.2   Experiment Results

In this section, firstly we implement the way to calculate the cluster PAM algorithm and draw the plot to analyze the result. Then we try to find the relationship between the result of clustering and the repetitive rate of user queries. At last, we show the results of clustering after our enhancement of TMN.

5.3   Clustering Result of TMN

Figure 1 shows the result of clustering in [2]. Position (i, j) represents the similarity between query i and query j. The lighter the shade, the more similar the queries are. Each query's class is displayed along the diagonal. If (i, i) is red, then query i was generated by TrackMeNot. If (i, i) is blue, then the query is genuine. We see that many queries are semantically unrelated. The largest cluster, seen in the top left of the graph, has 5 user queries and 0 TrackMeNot queries.

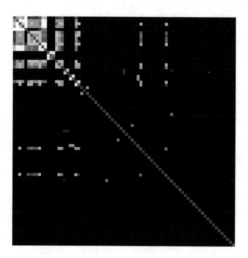

Fig. 1. A pictorial representation of the clustering in [2]

Figure 2 is our result of clustering one user queries and fake queries. It is similar to Fig. 1. Position (i, i) represents the similarity. Every yellow square means a cluster. Figure 2 has 100 queries, 50 queries are real and 50 queries are fake from TMN. We can see the fake and real queries are perfectly segmented. The largest cluster in the lower right has all the blue spot, which means all the real queries are figured out. The reason is that most of the 50 queries of users are overlapping.

We can see that when calculating 30 users, the average accuracy is between 60% and 70%, which means that we get the largest cluster by clustering algorithm, about 60% to 70% of the user's real query. Even if the adversary gets such a result, there is no guarantee that the user's true intention can be obtained, because if 50% of the true and false queries are mixed in, this is the probability

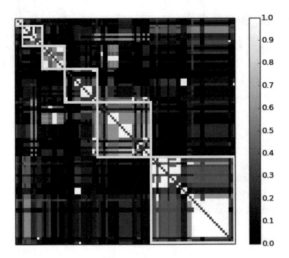

Fig. 2. A pictorial representation of the clustering of 100 queries

Table 1. Average precision and recall of 30 users.

Data size	Precision	Recall
mix100	0.689	0.629
mix200	0.714	0.553
mix300	0.678	0.525
mix500	0.659	0.502
mix1000	0.590	0.453

of a coin toss, the theoretically safest probability. At the same time, we observe that the accuracy of some users can reach 90%, which corresponds to the experimental results in [2]. If the adversary gets a 90% accurate query cluster, it is obvious that the user's true intention is exposed to a large exposure risk. The recall rate is about also 50% to 60%, because this is a clustering algorithm, and the adversary only seeks to find a subset of the user query that is as accurate as possible (Table 1).

Next we analyze the relationship between the repetition rate of the user query and the accuracy of the final result. If the user has the same keyword in either of the two queries, then we think the two queries are duplicates.

We calculate the repetitive rate of 50 user queries and compare with the mix100 data size result (see Fig. 3).

We can clearly see that there is a correlation between accuracy and repetition rate. When the user makes more repeated queries, the accuracy of the user query obtained by the clustering attack is higher, because similar queries are more likely to be divided into one same cluster.

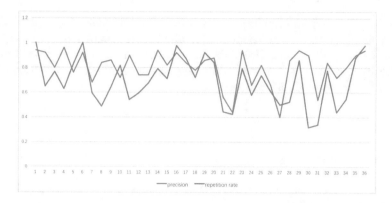

Fig. 3. Precision of user in largest cluster and the repetition rate of the user queries

Then we execute the query data generated by our enhancing TMN. We combine the user's query with the query generated by TMN as the next fake query. Table 2 gives a summary of all average precisions and recalls.

Table 2. Average precision and recall of 30 users after combining enhancement.

Data size	Precision	Recall
mix100	0.592	0.527
mix200	0.575	0.491
mix300	0.598	0.466
mix500	0.551	0.366
mix1000	0.591	0.415

When generating fake queries with user queries, the precisions of user queries in the largest cluster are falling and approaching 50%. The recall also declines. Which means that our simple enhancement has a certain effect, although the original attack method is not necessarily applicable to every user, but our defense can effectively improve the ability of all users against clustering attacks.

6   Conclusion

Web Query privacy protection is quite necessary for users, TMN made their efforts, but also proved by other methods that they have some serious vulnerabilities. We reproduce the method of clustering attacks and found that this attack does not work for all users. We analyzed the repetition rate of user queries and found out the relationship between clustering attack effects and user query repetition rates.

Meanwhile, we proposed improvement measures for clustering attacks, developed related browser extension to simulate user operations, and injected enhancement modules into TMN. Through experiments, we prove that the effect of clustering attacks is reduced after adding our enhancement method. Our method is effective against clustering attacks.

References

1. Petit, A.: Introducing Privacy in Current Web Search Engines. University of Passau, Germany (2016)
2. Al-Rfou, R., Jannen, W., Patwardhan, N.: TrackMeNot-so-good-after-all. CoRR abs/1211.0320 (2012)
3. Peddinti, S.T., Saxena, N.: On the privacy of web search based on query obfuscation: a case study of TrackMeNot. In: Privacy Enhancing Technologies, pp. 19–37 (2010)
4. Toubiana, V., Subramanian, L., Nissenbaum, H.: TrackMeNot: enhancing the privacy of Web Search. CoRR abs/1109.4677 (2011)
5. Nissenbaum, H.F., Howe, D.: TrackMeNot: resisting surveillance in web search. In: Kerr, I., Lucock, C., Steeves, V. (eds.) Lessons from the Identity Trail: Anonymity, Privacy, and Identity in a Networked Society. Oxford University Press, Oxford (2009). SSRN. https://ssrn.com/abstract=2567412
6. DIsco. Java application. http://www.linguatools.de/disco
7. Ahmad, W.U., Chang, K.W., Wang, H.: Intent-aware query obfuscation for privacy protection in personalized web search. In: SIGIR 2018, pp. 285–294 (2018)
8. Kushilevitz, E., Ostrovsky, R.: Replication is not needed: single database, computationally-private information retrieval. In: Symposium on Foundations of Computer Science, FOCS (1997)
9. Tor Anonymizing Network. http://www.torproject.org/
10. Balsa, E., Troncoso, C., Díaz, C.: OB-PWS: obfuscation-based private web search. In: IEEE Symposium on Security and Privacy, pp. 491–505 (2012)
11. TrackMeNot: Browser extension. https://cs.nyu.edu/trackmenot/
12. Ye, S., Wu, S.F., Pandey, R., Chen, H.: Noise injection for search privacy protection. In: CSE, vol. 3, pp. 1–8. IEEE Computer Society (2009)
13. Rebollo-Monedero, D., Forne, J.: Optimized query forgery for private information retrieval. IEEE Trans. Inf. Theory 56(9), 4631–4642 (2010)
14. Murugesan, M., Clifton, C.: Providing privacy through plausibly deniable search. In: SDM, pp. 768–779. SIAM (2009)
15. Elovici, Y., Shapira, B., Meshiach, A.: Cluster-analysis attack against a private web solution (PRAW). Online Inf. Rev. 30(6), 624–643 (2006)
16. Domingo-Ferrer, J., Solanas, A., Castellà-Roca, J.: h(k)-private information retrieval from privacy-uncooperative queryable databases. Online Inf. Rev. 33(4), 720–744 (2009)
17. Gervais, A., Shokri, R., Singla, A., Capkun, S., Lenders, V.: Quantifying websearch privacy. In: ACM Conference on Computer and Communications Security, pp. 966–977 (2014)
18. Chow, R., Golle, P.: Faking contextual data for fun, profit, and privacy. In: WPES, pp. 105–108 (2009)

Cryptanalysis of Proxy Re-Encryption with Keyword Search Scheme in Cloud Computing

Jen-Chieh Hsu$^{(\boxtimes)}$, Hsiang-Chen Hsu$^{(\boxtimes)}$, and Raylin Tso$^{(\boxtimes)}$

Department of Computer Science, National Chengchi University, Taipei, Taiwan
{105753501,10753036}@nccu.edu.tw, raylin@cs.nccu.edu.tw

Abstract. In this work, we review the article published by Tang et al. IN CMC 2018. We show that their construction of proxy re-encryption with keyword search exists some errors and it does not hold the correctness requirement. We point the problems out and proposed a new protocol.

Keywords: Cloud computing · Keyword search · Proxy re-encryption

1 Introduction

With the growing of Internet using rate, more and more people store their data in the cloud server. For the privacy of data, the data must be stored with encrypted. But, if other people want to retrieve your data, first, you must download your data and decrypt it, and send the data to them. It is not efficient. In 2010, Shao et al. [2] proposed a new concept called proxy re-encryption with keyword search. The scheme allows authorized user to search with the keyword and decrypt the ciphertext in the cloud server, and the cloud server doesn't know any information of the keyword and the plaintext. In 2018, Tang et al. [1] proposed a more effective proxy re-encryption keyword scheme. It is more effective in the cloud environment. However, There are some errors and inefficient parts in this paper. Therefore, we point these several problems out in our paper. First, we found that there is a parameter l in public parameter is no need. Second, we found that there is a symmetric key k in public parameter, and server S has ciphertext $Enc_k(M)$ that encrypts with symmetric key k. Therefore server S can decrypt the ciphertext, then server S can get the data of data owner A. Third, we noticed that data owner A must send lots of y to the data receiver B in the protocol, and it can be reduce. Fourth, when server S running the test algorithm, server S will find that he couldn't compute with two lacking parameters. So we suggested some methods to fix the problems in the scheme, then we proposed a new protocol to enhance the original protocol.

© Springer Nature Switzerland AG 2020
L. C. Jain et al. (Eds.): SICBS 2019, AISC 1145, pp. 329–334, 2020.
https://doi.org/10.1007/978-3-030-46828-6_28

2   Preliminaries

2.1   Bilinear Maps

Let $\mathbb{G}_1$ and $\mathbb{G}_2$ be two groups of order q, where q is large prime. A bilinear map $\hat{e} : \mathbb{G}_1 \times \mathbb{G}_1 \to \mathbb{G}_2$ satisfies these properties:

1. Bilinear: $\hat{e} : \mathbb{G}_1 \times \mathbb{G}_1 \to \mathbb{G}_2$ is *bilinear* if $\hat{e}(aP, bQ) = \hat{e}(P, Q)^{ab}$, $\forall P, Q \in \mathbb{G}_1, \forall a, b \in \mathbb{Z}$.
2. Non-degenerate: $\forall g \in \mathbb{G}_1, \hat{e}(g, g) \neq e$, where e is identity of $\mathbb{G}_2$.
3. Computable: There is an efficient algorithm to compute $\hat{e}(P, Q), \forall P, Q \in \mathbb{G}_1$.

2.2   Proxy Re-Encryption with Keyword Search Scheme

In 2018, Tang et al. [1] proposed a proxy re-encryption with keyword search in cloud computing, the scheme includes three parties: the data owner A, the cloud server S and the data receiver B. The proxy re-encryption with keyword search scheme contains eight algorithms $Setup(\lambda)$, $KeyGen(pp)$, $Enc(M, sk_a, pk_s, \omega_i, x_i)$, $ReKeyGen(H_2(sk_a), H_2(sk_b))$, $ReEnc(\tilde{\omega}, Rk_{A \to B})$, $Trapdoor(sk_b, \omega)$ $Test(T_\omega, \Gamma)$ and $Dec(FID, C, sk_b)$.

- $Setup(1^\lambda) \to pp$: Input λ is the security parameter and the output parameters is $(p, \mathbb{G}_1, \mathbb{G}_2, \hat{e}, g, H_1, H_2, \ell, k_i)$, where $\mathbb{G}_1, \mathbb{G}_2$ are two cyclic group of large prime order p, g is the generator of $\mathbb{G}_1$, and $H_1 : \{0,1\}^* \to \mathbb{G}_1$, $H_2 : \{0,1\}^{\leq l} \to \mathbb{G}_1$ are two one-way hash function, $\hat{e} : \mathbb{G}_1 \times \mathbb{G}_1 \to \mathbb{G}_2$.

- $KeyGen(pp) \to ((sk_a, pk_a), (sk_s, pk_s), (sk_b, pk_b))$: Input pp is the public parameter. Output are three key pairs. The data owner A, the server S and the data receiver B have $(sk_A = a, pk_A = g^a)$, $(sk_S = s, pk_S = g^s)$ and $(sk_B = b, pk_b = g^b)$, respectively.

- $Enc(M, sk_a, pk_s, \omega_i, x_i) \to (C, FID, \tilde{\omega})$: Input is plaintext M, the private key of data owner sk_A, the public key of server pk_S, the keyword ω. The output is ciphertext C, the index ciphertext FID and the keyword ciphertext $\tilde{\omega}$. The data owner A chooses a symmetric key k and encryption the plaintext M with key k: $C = Enc_k(M)$. Then, chooses randomly number x and then computes the index ciphertext $FID = \hat{e}(H_2(pk_S), H_2(x)) \cdot k_i$. After that, the data owner A computes the $y = Enc_{pk_B}(x)$ and sends it to the data receiver B. Next, the data owner A randomly chooses a number r and computes the keyword ciphertext $\tilde{\omega} = \hat{e}(pk_S, H_1(\omega)^r)^{H_2(sk_A)}$. Finally, the data owner A sends the tuple $(C, FID, \tilde{\omega})$ to the server S.

- $ReKeyGen(H_2(sk_a), H_2(sk_b)) \to Rk_{A \to B}$: The data receiver B sends $H_2(sk_B)$ to the data owner A. Then, the data owner A computes the re-encryption key $Rk_{A \to B} = \frac{H_2(sk_B)}{H_2(sk_A)}$ and sends it to the server S.

- $ReEnc(\tilde{\omega}, Rk_{A\rightarrow B}) \rightarrow \tilde{\omega}^{Rk_{A\rightarrow B}}$: The inputs are the keyword ciphertext $\tilde{\omega}$ and the re-encryption key $Rk_{A\rightarrow B}$. Output is the re-encrypted keyword ciphertext $\tilde{\omega}^{Rk_{A\rightarrow B}}$.

- $Trapdoor(sk_b, \omega) \rightarrow T_\omega$: The inputs are the private key of the data receiver B and the keyword ω. The output $T_\omega = [T_1, T_2] = [g^{r'}, H_1(\omega)^{1/sk_B}\dot{H}_1(pk_S^{r'})]$, where r' is random number.

- $Test(T_\omega, \tilde{\omega}) \rightarrow \{yes, no\}$: The input are the trapdoor $T_\omega = [T_1, T_2]$ and the keyword ciphertext ω. First, the server S computes the $\Gamma = \frac{T_2}{H_1(T_1^s)}$ and then checks whether the keyword ciphertext is equal $\hat{e}(pk_B^r, (\Gamma)^s)^{H_2(sk_B)}$ or not. If it is true, outputs yes, otherwise, outputs no.

- $Dec(FID, C, sk_b) \rightarrow M$: The inputs are the index ciphertext FID, the ciphertext C and the private of the data receiver B sk_B. Because the data receiver B has $y = Enc_{pk_B}(x)$ and then B can get the number x. Then B compute the symmetric key k with x and FID, $k = FID/\hat{e}(H_2(pk_S), x)$. Finally, B uses the symmetric key to decrypt the ciphertext C. The output is $M = Dec_k(C)$ (Fig. 1).

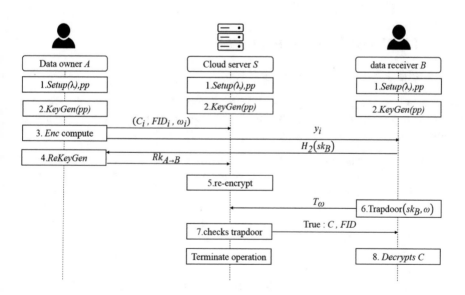

Fig. 1. The process of proxy re-encryption with keyword search scheme

The process of the proxy re-encryption with keyword search of Tang [1] is following:

1. Running the *Setup* algorithm and publishing the public parameters pp.

2. The data owner A, the server S and the data receiver B use the $KeyGen$ to generate their key pairs $(sk_A, PK_A), (sk_S, pk_s)$ and (sk_B, pk_B), respectively. And then, they publish their public key and keep their private key.

3. The data owner A uses the Enc algorithm to compute the ciphertext tuple (C_i, FID_i, ω_i) and sends these tuple to the server S. Moreover, A sends the each $y_i = Enc_{pk_B}(x_i)$ to the data receiver B.

4. The data receiver B sends the hash value of his private key $H_2(sk_B)$ to the data owner A, then A uses the $ReKeyGen$ algorithm to compute the re-encryption key $Rk_{A \to B}$. Finally, A sends the re-encryption key $Rk_{A \to B}$ to the server S.

5. The sever S uses re-encryption key to re-encrypt every keyword ciphertexts of A.

6. The data receiver B runs the algorithm $Trapdoor(sk_B, \omega)$ and gets the value $T_\omega = [T_1, T_2]$, then sends T_ω to the server S.

7. The server S checks whether the trapdoor of the data receiver B and all keyword ciphertext of the data owner A is equal or not. If it is true, S sends the index ciphertext FID and ciphertext C to the data receiver B. Otherwise, S does nothing.

8. The data receiver B first decrypts the $y = Enc_{sk_B}(x_i)$ and gets the number x_i. Then utilizes x_i and FID to get the symmetric key k. Finally, B decrypts the cipertext C with symmetric key k.

3 The Cryptanalysis Protocol

3.1 The Error of the Protocol

The proxy re-encryption keyword search scheme of Tang is more efficiency than other scheme [1]. Although, the scheme is more efficiency, but the scheme has some error.

1. **Public Parameter:** The original output of $Setup$ is $(p, \mathbb{G}_1, \mathbb{G}_2), \hat{e}, g, H_1,$ H_2, ℓ, k_i. But the ℓ is not used by other algorithm except the hash function H_2. Therefore, we remove the ℓ from the output of the $Setup$ algorithm.

2. **Symmetric Key:** If the public parameter contains the symmetric key k_i, then everyone, including server, can try to find the correct symmetric key, therefore, we also remove the symmetric key from the public parameter. Therefore, we suppose that the data owner A generates symmetric key k in encrypt phase.

3. **Unnecessary communication:** In encrypt phase, for each file, the data owner A must send a y_i to the data receiver B, if A has 1000 files, this means that A must send 1000 messages to B. But, if y_i is contained in the FID_i, then the data owner A can save these commutation.

4. **Correctness:** In the test phase, the server S must compute the equation $\hat{e}(pk_B^s, (\Gamma)^s)^{H_2(sk_B)}$, where $\Gamma = \frac{T_2}{H_1(T_1^s)} = H_1(\omega)^{1/b}$. But, the sever S doesn't have the random number r and the hash value of $B's$ private key. Therefore, the server S cannot run the test algorithm. The random r is generated by the

data owner A, we let $r = H_1(FID)$ be the new random number, since the FID is known by only the data owner A and the server S. The hash value of $B's$ private key is know by the data owner A and the data receiver B, and we don't hope that the server S know it. Therefore, we fixed the trapdoor output. The new output is $T = [T_1, T_2] = [g^{r'}, H_1(\omega)^{H_2(sk_B)/sk_B} \dot{H}_1(pk_S^{r'})]$. And then, in test phase, the server S computes the $\Gamma = \frac{T_2}{H_1(T_1^s)} = H_1(\omega)^{H_2(sk_B)/sk_B}$. Next, the server S checks the equation $\hat{e}(pk_B^r, \Gamma^s) = \hat{e}(g^{br}, H_1(\omega)^{\frac{H_2(sk_B)*s}{b}})$, where r is the hash value of the FID.

3.2   The New Protocol

The new scheme includes three parties: the data owner A, the cloud server S and the data receiver B. The proxy re-encryption with keyword search scheme contains eight algorithms $Setup(\lambda)$, $KeyGen(pp)$, $Enc(M, sk_a, pk_s, \omega_i)$, $ReKeyGen(H_2(sk_a), H_2(sk_b))$, $ReEnc(\tilde{\omega}, Rk_{A \to B})$, $Trapdoor(sk_b, \omega)$ $Test$ (T_ω, Γ) and $Dec(FID, C, sk_b)$.

- $Setup(1^\lambda) \to pp$: Input λ is the security parameter and the output parameters is $(p, \mathbb{G}_1, \mathbb{G}_2, \hat{e}, g, H_1, H_2)$, where $\mathbb{G}_1, \mathbb{G}_2$ are two cyclic group of large prime order p, g is the generator of $\mathbb{G}_1$, and $H_1 : \{0,1\}^* \to \mathbb{G}_1$, $H_2 : \{0,1\}^{\le l} \to \mathbb{G}_1$ are two one-way hash function, $\hat{e} : \mathbb{G}_1 \times \mathbb{G}_1 \to \mathbb{G}_2$.

- $KeyGen(pp) \to ((sk_a, pk_a), (sk_s, pk_s), (sk_b, pk_b))$: Input pp is the public parameter. Output are three key pairs. The data owner A, the server S and the data receiver B have $(sk_A = a, pk_A = g^a)$, $(sk_S = s, pk_S = g^s)$ and $(sk_B = b, pk_b = g^b)$, respectively.

- $Enc(M, sk_s, pk_s, \omega_i, x_i) \to (C, FID, \tilde{\omega})$: Input is plaintext M, the private key of data owner sk_A, the public key of server pk_S, the keyword ω. The output is ciphertext C, the index ciphertext FID and the keyword ciphertext $\tilde{\omega}$. The data owner A chooses a symmetric key k and encryption the plaintext M with key k: $C = Enc_k(M)$. Then, chooses randomly number x and then computes $f_i = \hat{e}(H_2(pk_S), H_2(x)) \cdot k_i$. After that, the data owner A computes the $y = Enc_{pk_B}(x)$ and $FID = (f_i, y)$. Next, the data owner A computes $r = H_1(FID)$ and computes the keyword ciphertext $\tilde{\omega} = \hat{e}(pk_S, H_1(\omega)^r)^{H_2(sk_A)}$. Finally, the data owner A sends the tuple $(C, FID, \tilde{\omega})$ to the server S.

- $ReKeyGen(H_2(sk_a), H_2(sk_b)) \to Rk_{A \to B}$: The data receiver B sends $H_2(sk_B)$ to the data owner A. Then, the data owner A computes the re-encryption key $Rk_{A \to B} = \frac{H_2(sk_B)}{H_2(sk_A)}$ and sends it to the server S.

- $ReEnc(\tilde{\omega}, Rk_{A \to B}) \to \tilde{\omega}^{Rk_{A \to B}}$: The inputs are the keyword ciphertext $\tilde{\omega}$ and the re-encryption key $Rk_{A \to B}$. Output is the re-encrypted keyword ciphertext $\tilde{\omega}^{Rk_{A \to B}}$.

- $Trapdoor(sk_b, \omega) \rightarrow T_\omega$: The inputs are the private key of the data receiver B and the keyword ω. The output $T_\omega = [T_1, T_2] = [g^{r'}, H_1(\omega)^{H_2(sk_B)/sk_B} \dot{H}_1(pk_S^{r'})]$, where r' is random number.

- $Test(T_\omega, \tilde{\omega}) \rightarrow \{yes, no\}$: The input are the trapdoor $T_\omega = [T_1, T_2]$ and the keyword ciphertext ω. First, the server S computes the $\Gamma = \frac{T_2}{H_1(T_1^s)}$ and then checks whether the keyword ciphertext is equal $\hat{e}(pk_B^r, (\Gamma)^s)$ or not. If it is true, outputs yes, otherwise, outputs no.

- $Dec(FID, C, sk_b) \rightarrow M$: The inputs are the index ciphertext FID, the ciphertext C and the private of the data receiver B sk_B. Because the data receiver B has $y = Enc_{pk_B}(x)$ and then B can get the number x. Then B compute the symmetric key k with x and FID, $k = FID/\hat{e}(H_2(pk_S), x)$. Finally, B uses the symmetric key to decrypt the ciphertext C. The output is $M = Dec_k(C)$.

4 Conclusion

Tang et al's scheme is more effectively than other proxy re-encryption with keyword search schemes. The scheme can be effectively applied in the cloud environment. However, we found that this paper has some errors and some parts are not efficient. In this paper, we point these errors and inefficient part out. Then proposed a new protocol to enhance the original scheme. New protocol not only solved the errors but increase efficiency. Reducing unnecessary communication can also make the protocol more efficient in the cloud environment.

Acknowledgments. This research was supported by the Ministry of Science and Technology, Taiwan (ROC), under Project Numbers MOST 108-2218-E-004-001, and MOST 108-2218-E-011-021-.

References

1. Tang, Y., Lian, H., Zhao, Z., Yan, X.: A proxy re-encryption with keyword search scheme in cloud computing. CMC-Comput. Mater. Contin. **56**(2), 339–352 (2018)
2. Shao, J., Cao, Z., Liang, X., Lin, H.: Proxy re-encryption with keyword search. Inf. Sci. **180**(13), 2576–2587 (2010)

An Effective Target Recovery Method from a Source Location in Geo-Indistinguishable Surroundings

Chun-I. Fan[1,2,3,4], Hsin-Nan Kuo[1], Jheng-Jia Huang[2(✉)], Yi-Hui Li[1], Er-Shuo Zhuang[1], and Yu-Tse Shih[1]

[1] Department of Computer Science and Engineering,
National Sun Yat-sen University, Kaohsiung City, Taiwan
cifan@mail.cse.nsysu.edu.tw, bluedunk@gmail.com,
jhengjia.huang@gmail.com, t91p65a@gmail.com,
zhaunges@gmail.com, steven8450@gmail.com
[2] Telecom Technology Center, Kaohsiung City, Taiwan
[3] Information Security Research Center, National Sun Yat-sen University,
Kaohsiung City, Taiwan
[4] Intelligent Electronic Commerce Research Center,
National Sun Yat-sen University, Kaohsiung City, Taiwan

Abstract. Location-based services have become increasingly more pervasive in the context of Internet of Vehicles. Nowadays, vehicles can transfer their locations to the server for route planning, attractions searching, location sharing, etc. However, the data that are transmitted over the public channel may be tracked/tampered by the attacker. Therefore, location privacy is one of the major concerns to the researchers, and several mechanisms have been designed by obfuscating the user's original location prior to transmitting it to the server. The most intuitive method for finding the target, in case an event occurs and the server needs to find original location, is to ask all of the users near to the event to return their original locations in order to find the target. In such a scenario, a large number of users are required to acknowledge their locations. Hence, how to reduce the number of affected users in the process of finding the target among the users' obfuscated locations is an important issue. Therefore, we devise a method of identifying the target and reducing the number of affected users in an efficient way.

Keywords: Location-based services (LBSs) · Differential privacy · Location privacy

1 Introduction

In the last few years, location-based services become more and more popular. People carry a variety of mobile devices, equipped with a location sensor. With these sensors, people can know where they are or share the locations with the remote server via the wireless network to acquire a variety of instant services, which is known as Location-Based Services (LBSs) [1].

© Springer Nature Switzerland AG 2020
L. C. Jain et al. (Eds.): SICBS 2019, AISC 1145, pp. 335–350, 2020.
https://doi.org/10.1007/978-3-030-46828-6_29

In addition to handheld devices, vehicles are also equipped with global positioning systems and network functions. With the popularity of the Internet of Things and big data analysis, connected vehicles have become a direction of development. Business Insider Intelligence expects that connected vehicle will grow from 36 million in 2015 to 381 million by 2020 and are expected to generate $8.1 trillion between 2015 and 2020 [2]. The vehicles can transmit messages to the back-end server. On receiving such large volume of data, the server sends back the analyzed data to the respective vehicles.

However, in the process of data transmission or improper data storage of the server, users' privacy may be leaked. Attackers may trace the user's location to know the user's daily itinerary, fixed access, or other activities, so that, user's private information, such as home address, personal interest, political direction, etc., can be identified [3, 4]. Therefore, location privacy is an important issue that needs to be studied.

In order to protect users' privacy, personal locations should be protected. Suppose, the user's location is obfuscated by adding noise. Now, the user transfers the obfuscated location instead of the original location as shown in Figs. 1 and 2. In that case, the server cannot reveal the exact location of the user, track the user, or analyze the user's personal information.

Fig. 1. Requesting services with user's real location

Fig. 2. Requesting services with user's obfuscated location

In the current location-related applications, although the protection of the user's location privacy can be achieved, the issue of event target recovery is not considered. In some situations, the server may need to find a target from a source location. How the server quickly finds the closest target with the cooperation of the users is also a problem. From the above, the problem we want to solve is to quickly find the target from a source location but with fewer users returning their locations.

The most straightforward method to find the target is to require all the users near the event location to return their original locations. As long as the searching range is large enough, the target will definitely be found. However, this will cause many unnecessary users to return their locations. Such a method is easy and quick but violates the purpose of the fewest users needing to return their original locations. Therefore, how to reduce the number of users that share their locations is our most important goal.

To mitigate the above mentioned issue, Liu [5] proposed a method by dividing the searching range to grids. The server requires users in different grids to return their original locations each time until all grids are searched instead of searching a large region at one time. The choice of grids is based on the number of users in each grid. With this method, there is a higher probability to find the target before searching all grids.

Although the method mentioned above can reduce the number of users returning their locations, it is not the best. In this manuscript, we propose an effective target recovery method from an event location in geo-indistinguishable surroundings. The contributions of this manuscript are as follows:

(1) We improve Liu's method by adjusting the searching range, and the proposed method is not only more quickly but also achieves that fewer users return their locations.

(2) We conduct experiments to implement the proposed methods, and the experimental results show that our work is better than other methods.

In Sect. 2, we describe the formal definition of the mechanisms used in the manuscript. In Sect. 3, we describe the proposed method in detail. The experimental results and the conclusion are shown in Sect. 4 and Sect. 5, respectively.

2 Preliminary and Background

In this section, we introduce the definition of differential privacy and geo-indistinguishability we will use in this manuscript, introduce the background and the mechanisms proposed for location privacy.

2.1 Differential Privacy

The concept of differential privacy was proposed by Dwork et al., using data statistics to conduct research. In recent years, attention has arisen in the rise of big data analytics and data privacy.

***Definition* 2.1:** (ε- Differential Privacy): A randomized mechanism K gives ε- differential Privacy for any two databases D_1 and D_2 differing on at most one record, and for every $S \subseteq Range(K)$, $Pr[M(D) = S] \leq e^{\varepsilon} \times Pr[M(D') = S]$, where ε is, the stronger the privacy guarantee will be. The two important properties of differential privacy are described below.

***Theorem* 2.1:** (Sequential composition): Let the randomized mechanisms K_i satisfy ε_i-differential Privacy on the database D. It will be satisfying $\left(\sum_i \varepsilon_i \right)$- differential privacy. It can be said that applying the mechanism on two distinct data will get a weaker privacy guarantee, and the two distinct data are in the associated database.

***Theorem* 2.2:** (Parallel composition): Let the randomized mechanisms K_i satisfy ε_i-differential Privacy on different databases D_i. It will be satisfying $(max_i \varepsilon_i)$ – differential privacy. Applying the mechanism on two distinct data in disjoint databases, the privacy guarantee provides from the worst privacy mechanism.

***Theorem* 2.3:** (Post-Processing): Let the randomized algorithm $K : D \rightarrow H_1$ is ε-differential Privacy. Let $g : H_1 \rightarrow H_2$ be an arbitrary randomized mapping. Then $g \circ K : D \rightarrow H_2$ is ε- differential Privacy.

2.2   Geo-Indistinguishability

Geo-Indistinguishability is to generalize the definition of differential privacy for location privacy. Where X is a set of locations and a user reports a location inside a set Z with the probabilities $P(Z)$ instead of his real location. We formally as follows:

***Definition* 2.2:** (Geo-Indistinguishability): A randomized mechanism $M : X \rightarrow P(Z)$ satisfies ε- geo indistinguishability if for all $x, x' \in X$ and $z \in Z$: $M(x)(z) \leq e^{\varepsilon d_2(x,x')} M (x')(z)$. The explanation for this definition is that for any two close points (x, x'), the geographical indistinguishability requires that the two locations should have the same probability of returning the same position z In this definition, εr represents the user's privacy level. Given r, the smaller the εr is, the stronger the privacy guarantee will be. This means that if the user wants a higher level of privacy in a larger range (i.e., r is larger), then a smaller ε should be set. At the same time, the author also proposed a mechanism to provide geo-indistinguishability by generating noise from two-dimensional Laplace distribution.

Planar Laplacian: The idea of planar laplacian is to replace the return to the real position, returning another location near the real position according to the distribution. Given the privacy parameter $\varepsilon \in R^+$, and the user's real location $x_0 \in R^2$, on any other location $x \in R^2$, the probability density function (pdf) of choosing x as the reported location is: $D_E(x_0)(x) = \frac{E^2}{2\pi} e^{-\varepsilon d(x_0,x)}$.

In order to facilitate of sampling the position according to the distribution, the plane Laplacian will be converted to a polar coordinate system with origin in x_0. The point x will be expressed as (r, θ), where r represents the distance from x to x_0 and θ represents the angle between the line xx_0 and the horizontal axis. The pdf of the polar Laplacian centered in x_0 is: $D_\varepsilon(x_0)(r, \theta) = \frac{\varepsilon^2}{2\pi} r e^{-\varepsilon r}$.

Thus, a noise (r, θ) can be generated by the following steps:

(1) Draw θ uniformly in $[0, 2\pi$
(2) Draw P uniformly in $[0, 1$ and set $r = C_\varepsilon^{-1}(P)$

First, generate θ as a random number in $[0, 2\pi$ uniformly. Second, generate P in $[0, 1$ uniformly and set $r = C_\varepsilon^{-1}(P)$. where $C_\varepsilon(r) = 1 - (1 + \varepsilon r)r^{-\varepsilon r}$. After generating the noise, the user reports the perturbation location $x =)$.

2.3   Location Privacy

In this section, the privacy issues in location privacy and the types of privacy protection are introduced.

(1) *The Goal of Protection*: There are two types of LBS privacy protection here, query privacy protection and location privacy protection [6]. Query privacy refers to the association with the user's private information and the content queries to LBS.

(2) *The Protection Model*: The privacy protection mechanism can be divided into two categories: centralized model and local model [7]. In the centralized model, there exists a trusted data curator that is trusted by all users and can provide protection. Before querying data from the service provider, users send the request to the data curator first. Then the data curator transfers the query to the service provider after sanitizing. The service provider passes the query result back to the data curator, and the data curator filters the message and sends it back to the user. The user passes sensitive information to the data curator, the attacker will regard it as the target of the attack, therefore the risk of being attacked will increase. In addition, the data curator needs to process queries from all users, which causes the system to be overloaded. Also, with the increasing awareness of personal privacy, users are more willing to sanitize sensitive information by themselves. Local models are proposed to be safer to avoid security leaks. There are no trusted third parties in this model that can access users' sensitive information. Users obfuscate their data (e.g., adding noise to the data) by themselves, and send them to the service provider. Therefore, no one other than the user himself gets the user's private information.

2.4   Location Privacy Protection Techniques

In terms of location protection location privacy, many methods have been proposed. Four countermeasures for location privacy have summarized [8]: regulatory strategy, privacy policy, anonymity, confusion. The first two are biased towards policy or regulatory discussions, while the latter two are computational countermeasures. The idea of anonymization is to hide the user's identification so that the attacker cannot connect to the corresponding user by leaking location information. For example, we can use a pseudonym instead of a user. However, Bettini et al. showed that the position information with a specific movement trajectory can form an identifier to identify the user, thereby causing the relationship between the user's pseudonym and the real person to leak [9].

Another way to protect location privacy is obfuscation, first proposed by Duckham et al. [10]. They formalized the concept of inaccuracy and imprecision. Inaccuracy refers to the lack of correspondence between the information provided and the real message. For example, returning a location different from the real location. Imprecision refers to the lack of specificity of information. For example, returning a larger area instead of a real position (it provides less detail about the actual location). The purpose of obfuscation is to deliberately reduce the quality of information to protect the privacy of the information.

(1) *Anonymization*: One method widely used to protect location privacy is k-anonymity, which was first proposed by Gruteser and Grunwald [11], and many improved methods have been proposed. The k-anonymity approach is to generate an anonymous dataset that returns the anonymous dataset rather than the user's location.

(2) *Location Perturbation*: The obfuscation mechanism is a random variable function that maps the actual location to the obfuscated location and covers the following

methods for reducing the accuracy/precision of the user's actual location: perturbation (adding noise), adding dummy locations, reducing precision (merging
regions), location hiding [12].

- **Differential Privacy:** Differential privacy was proposed by Dwork et al. in
  2006. The key concept is that the presence or absence of a piece of data in a
  database does not significantly affect the results of the data query.
- **Location Privacy:** In recent years, differential privacy has also been used for
  location protection of LBS. Andrés et al. introduced location privacy protection
  based on the concept of differential privacy, called geo-indistinguishability, and
  proposed a mechanism that complies with this privacy standard, planar laplacian. [13] proposed a new service usage framework based on differential privacy to protect the location, which is deployed on edge nodes and aims to
  provide an adjustable privacy protection solution to balance usability and
  privacy.

2.5 Points of Interest Retrieve

Intuitively, the server requires users near the source location to return their locations.
The larger the scope of the search is, the greater the chance of finding the target will be.
Just as in the operation of location-based services [14], the service provider retrieves
the points of interest (POI) in the area of retrieval (AOR) based on the location
proposed by the user and returns the results to the user, who then filters the POIs in the
area of interest (AOI). The ideal state is that the AOR completely contains the AOI, and
the user can get the information they need completely. Based on the concept of AOR
and AOI, there is a definition of its accuracy:

Definition 2.3: (LBS Accuracy): LBS application $(M, rAOR)$ is $(c, rAOI)$- accurate if
and only if for any location $x \in R^2$, we have that $Pr[C(x, rAOI) \subseteq C(M(x), rAOR)] \geq c$,
where $C(x, r)$ denotes a circle with a radius r and a center at x and c is a confidence
factor. $rAOR$ and $rAOI$ represent the radius of AOR and AOI, respectively. If the
probability that the AOI is completely contained in the AOR is bounded by c, then the
application of the LBS $(M, rAOR)$ is $(c, rAOI)$-accurate.

According to this definition, in our environment, it can be seen that as long as we
use a large enough area of retrieval, the chance of finding a target will increase.

2.6 Advanced Centralized Cloaking

In [5], the privacy of the user is protected by reporting an area instead of the user's real
location. The area that is used to protect the user called the cloaking area. Advanced
centralized cloaking is a method proposed in this manuscript to create a cloaking area
in k-anonymity.

For example, if the user Q does not reach the hidden condition as in Fig. 3(a). The
search procedure is as follows: the number of users of regions 1, 2, 3, and 4 is 3, 1, 0,
and 2, respectively. The method is to extend the cloaking area to the area with the most

hidden number, so this method looks for region 1 as an extended object, such as Fig. 3(b). If the user number is still insufficient, it continues to extend outward, such as Fig. 3(c). When the hidden demand is reached, the area can be used as a cloaking area.

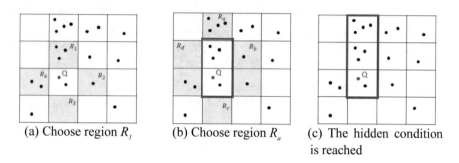

(a) Choose region R_l (b) Choose region R_a (c) The hidden condition
                                                            is reached

Fig. 3. Advanced centralized cloaking

To use the method in this manuscript, we renamed as multi-grid extension method.

3   The Proposed Scheme

In this section, we set the basic search range referred as discussed in Sect. 2.4. According to Sect. 2.5, we sort out three methods to recover the target from a source location and introduce how these methods work in our environment. We also discuss what problems such methods will encounter in actual use, and further, propose an effective solution to meet the goal of recovering the target from a source location in this manuscript.

3.1   System Architecture

In our environment, there are two main entities, users and a server. We describe the entities in our system environment as follows:

- **User**

  (1) A user obfuscates the original location and submits the obfuscated location to the server. The original location is stored in the user side.
  (2) When the server asks the user to return the original location, the user will honestly return it.

  - Server

    (1) The server collects all users' obfuscated locations.
    (2) When an accident occurs, the server will ask the user to return the original location.

3.2   System Environment

There are two situations may occur in our system, namely the general situation and event occurrence situation, where the former represents the situation in which a normal user moves on a road, and in the latter one, two users have a car accident, respectively.

(1) *General Situation*: Under the general situation, the user obfuscates the original location and submits the result to the server, or requests for the location service to the server. As shown in Fig. 4, the steps are as follows:

> Step 1: The user obfuscates his original location.
> Step 2: The user transmits the obfuscated location and requests location services to the server.
> Step 3: The server collects the locations of all users to analyze and utilize the data.
> Step 4: The server responds the corresponding results to the user.

(2) *Event Occurrence Situation*: Suppose that users A and B have an event in the same position and A wants to find B. As in Fig. 5, the steps of our scheme are as follows:

> Step 0: An event occurs.
> Step 1: A returns the event location to the server and asks the server to find B.
> Step 2: With the event location, the server sets the searching range.
> Step 3: The server asks the users in the searching range to return their original locations.

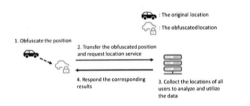

Fig. 4. General situation

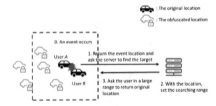

Fig. 5. Event occurrence situation

3.3   Target Recovery Methods

Referring to Sect. 2.4, we propose two methods to recover the target from the event location, which will reduce the size of the searched area. Let the event location be Q, and the target T be not in the event location area. We describe the proposed methods as follows:

(1) Circle Extension

The algorithm of the circle extension method is as follows:

> Step 0: Search the event area, if find the target then return the target. Else, set the event area as the base area.

Step 1: Search the circle around the base area, if find the target then return the target. Else, the searched area is added to the base area.

Step 2: Repeat Step 1 until the target is found.

For example, Fig. 6(a) shows the first round, the server searches for eight areas around the base area. Figure 6(b) shows that if the target is not found, the area which has already been searched will be used as the new base area, and then it will be extended outward. Figure 6(c) shows the final extension result.

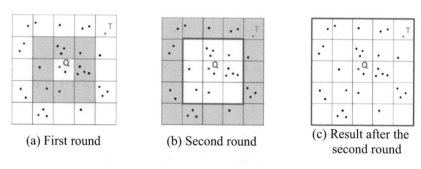

(a) First round   (b) Second round   (c) Result after the second round

Fig. 6. Circle extension method

The advantage of this approach is that the search area extends faster and the direction of extension is not affected by the number of users in each grid. Therefore, the target can be found more quickly.

(2) Single-Grid Extension

The algorithm of the single-grid extension method is as follows:

Step 0: Search the event area, if find the target then return the target. Else, set the event area as the base area.

Step 1: Set the adjacent areas of base area in the east, south, west, and north directions as the candidate areas.

Step 2: Search one of the candidate areas which has the most users in the candidate areas. If find the target, then return the target. Else, the searched area is added to the base area.

Step 3: Repeat Step 1 and Step 2 until the target is found.

For example, Fig. 7(a) shows that the target is not in the event location area. In the first round, the adjacent areas are R_1, R_2, R_3, and R_4. The number of users of each area is 3, 4, 1, and 1, respectively. This method extends to the area with the most number of users, so it sets region R_2 as the extension area. Figure 7(b) shows that if the target is still not found, the area that has already been searched will be used as the new base area, and then it will be extended outward. Figure 7(c) shows the result after the second round.

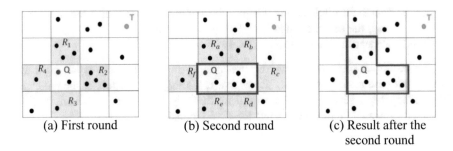

(a) First round   (b) Second round   (c) Result after the
second round

Fig. 7. Single-grid extension method

The advantage of this approach is that in areas with a large number of users, the server has a higher probability of finding the target and a lower probability of searching in insignificant areas.

3.4 Effective Target Recovery Method

We improve the method proposed in [5] as our effective target recovery method by considering following two steps. First, according to our experimental results in Session 4, we find the average distances between the obfuscated location and the original location under different privacy levels. Second, we narrowed the searching range under different privacy levels. The details of narrowing the searching range will be introduced in the next paragraphs. Finally, we use Liu's method to extend the search area within the narrowed searching range. The algorithm of the target recovery method is as follows:

Step 0: Search the event area, if find the target then return the target. Else, set the base area and narrow the searching range by the privacy level.
Step 1: Set the four sides of the rectangle as four adjacent areas of the base area as the candidate areas. If any candidate area is out of the searching range, delete the candidate area.
Step 2: Search one of the candidate areas which has the most users in the candidate areas. If find the target, then return the target. Else, the searched area is added to the base area.
Step 3: Repeat Step 1 and Step 2 until all areas in the searching range are searched. Set the searching range as the original range, and then repeat Step 1 and Step 2 until the target is found.

We introduce the details of narrowing the searching range in two parts. The first part is to reset the base area before searching. We can extend the base area just from the area where the obfuscated locations may be rather than extending the base area from the event area. For example, as shown in Fig. 8(a), we know that the obfuscated target location is outside with a certain distance, so we can exclude the area where the target does not exist before search taken place. Besides, if we know the target will far away from the original location at least three grids, we can reset the base area as shown in Fig. 8(b).

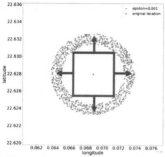
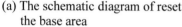

(a) The schematic diagram of reset
the base area

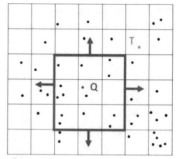

(b) Implement reset the base area

Fig. 8. Reset the base area

The second part is to narrow the maximum searching range under different privacy levels. For detail, the server narrows the searching range, colored as blue circle shown in Fig. 9, where the green areas are originally the candidate areas. Further, Fig. 9(a) shows the second round, where region R_1, R_2, and R_3 are used as candidate areas and the event location is extended by adding the area with the largest number of users. Figure 9(b) shows that if the target is still not found, the searched area will be used as the new base area and it will be extended outward. Because of narrowing the searching range, the candidate areas only leave R_a and R_b. Therefore, the server only uses three rounds to find the target. We further show the performance of the proposed method is better in Sect. 4.

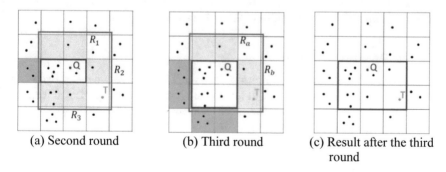

(a) Second round   (b) Third round   (c) Result after the third
                                                           round

Fig. 9. Reduction of the searching range

Combining the above two parts, we show the improved results as follows. In Fig. 10(a), the gray region represents the new base area. The users in the gray region will not be searched. Starting from the base area, the adjacent areas are R_1, R_2, R_3, and R_4. The number of users in region R_2 is the most, so the server sets R_2 as an extension area. Figure 10(b) shows that if the target is still not found, the area that has already been searched will be used as the new base area, and then it will be extended outward. Figure 10(c) shows the result after the second round.

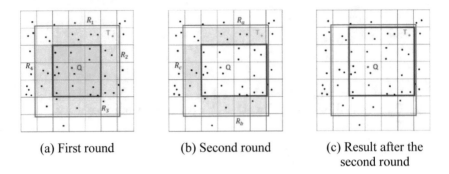

(a) First round   (b) Second round   (c) Result after the
                                                                         second round

Fig. 10. The effective target recovery method

Overall, we improve the method proposed in [5] by adjusting the searching range at different privacy levels as our effective target recovery method. By resetting the base area and narrowing the maximum searching range, the number of users being searched can be greatly reduced. Furthermore, the proposed method only needs the event location to effectively find the target without additional information.

4   Performance Evaluation

In this section, discuss the usability of the proposed method by illustrating the environmental setup followed by implementation issues and the results.

4.1   Experimental Environment

First, we introduce the basic environment settings in Table 1. We use two different data sets to compare the results. The first is to simulate a more square street, and there will be more dense vehicles representing more vehicles in a region. The second is to simulate the vehicles on the actual street of Kaohsiung City [15].

Table 1. Simulation setup

Maximum searching region	$4000 * 4000 \text{ m}^2$
The size of each grid	$30 * 30 \text{ m}^2$

4.2   Obfuscation Result

In our environment, users obfuscate their locations before sending them to the server. We use the Planar Laplace mechanism [16] as our obfuscation method to obfuscate the locations. With different levels of privacy, user's obfuscated locations can vary greatly from the original location. Figure 11 shows the distribution of the obfuscated results at ten different privacy levels. It shows that the obfuscated locations with a higher privacy level will be farther from the original location.

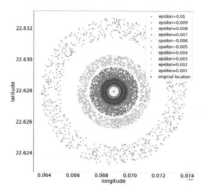

Fig. 11. Obfuscation results under different privacy levels

4.3 Number of Search Rounds

The number of search rounds is related to the times that each method performs the extension processes. Each round before asking users to return their locations has fixed overhead, such as packet header or initialization of a protocol. So, the higher the search round is, the higher the overhead will be. Figures 12(a) and (b) show the comparison of the number of search rounds of the four methods with different data sets.

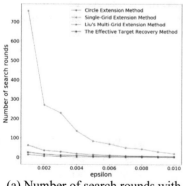

(a) Number of search rounds with data set 1

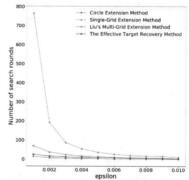

(b) Number of search rounds with data set 2

Fig. 12. Number of search rounds

4.4 Number of Searched Users

The number of searched users refers to how many users requested by the server to return their original locations. Our most important goal is to reduce the number of searched users as many as possible. Figures 13(a) and (b) show the comparison of the number of searched users of the four methods with different data sets.

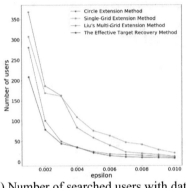
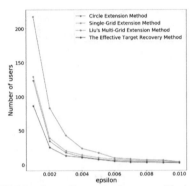

(a) Number of searched users with data set 1

(b) Number of searched users with data set 2

Fig. 13. Number of searched users

In the analysis of different methods, we use the number of search rounds and the number of searched users to compare the performance. In terms of performance considerations, we should consider both the search rounds and the searched users at the same time. And according to different needs, the priority of item considerations will be different. Such as in our situation, we want to affect the fewer number of users, so we consider the number of searched users to be less first, and secondly consider the number of the search round. In the above experimental results, the effective target recovery method has the best performance in the comparison of the number of searched users. Compared with other methods, although the effective target recovery method needs more search rounds, it reduces the number of searched users. Therefore, the effective target recovery method is more suitable for our needs.

5 Conclusion

We proposed an efficient method which affects fewer users. We improved Liu's method by adjusting the searching range at different privacy levels. To improve the efficiency, we introduced an effective target recovery strategy by adjusting the searching range at different privacy levels. The advantages of the effective target recovery method are as follows:

(1) Compared with the previous work, the effective target recovery method can reduce the number of searched users significantly.

(2) We implemented the proposed methods and demonstrated that the proposed method can achieve the goal that affected fewer users.

(3) In the effective target recovery method, the server only needs the event location to find the target effectively without auxiliary information.

To the best of our knowledge, this manuscript is the first one to consider the target recovery problem. Although the proposed method needs more search rounds, it is more suitable for reducing the number of users who need to return their locations. In the future, we will try to consider different methods to extend the search area to achieve a minimal number of searched users and reduce search rounds, such as extending the next search area with the same size of base area.

Acknowledgement. This work was partially supported by Taiwan Information Security Center at National Sun Yat-sen University (TWISC@NSYSU) and the Ministry of Science and Technology of Taiwan under grant MOST 108-2221-E-110-032. It also was financially supported by the Information Security Research Center at National Sun Yat-sen University in Taiwan and the Intelligent Electronic Commerce Research Center from The Featured Areas Research Center Program within the framework of the Higher Education Sprout Project by the Ministry of Education (MOE) in Taiwan.

References

1. Primault, V., Mokhtar, S.B., Lauradoux, C., Brunie, L.: Differentially private location privacy in practice. arXiv preprint arXiv:1410.7744 (2014)
2. Meola, A.: Automotive industry trends: IoT connected smart cars & vehicles. Business Insider (2016)
3. Gambs, S., Killijian, M.-O., del Prado Cortez, M.N.: Show me how you move and i will tell you who you are. In: Proceedings of the 3rd ACM SIGSPATIAL International Workshop on Security and Privacy in GIS and LBS, pp. 34–41. ACM (2010)
4. Krumm, J.: Inference attacks on location tracks. In: International Conference on Pervasive Computing, pp. 127–143. Springer, Heidelberg (2007)
5. Liu, C.-C.: A new hybrid cloaking system for location privacy. Master's thesis, National Taipei University of Technology (2010)
6. Shin, K.G., Ju, X., Chen, Z., Hu, X.: Privacy protection for users of location-based services. IEEE Wirel. Commun. **19**(1), 30–39 (2012)
7. Murakami, T., Hino, H., Sakuma, J.: Toward distribution estimation under local differential privacy with small samples. Proc. Privacy Enhancing Technol. **2018**(3), 84–104 (2018)
8. Krumm, J.: A survey of computational location privacy. Pers. Ubiquitous Comput. **13**(6), 391–399 (2009)
9. Bettini, C., Wang, X.S., Jajodia, S.: Protecting privacy against location-based personal identification. In: Workshop on Secure Data Management, pp. 185–199. Springer, Heidelberg (2005)
10. Duckham, M., Kulik, L.: A formal model of obfuscation and negotiation for location privacy. In: International Conference on Pervasive Computing, pp. 152–170. Springer, Heidelberg (2005)
11. Gruteser, M., Grunwald, D.: Anonymous usage of location-based services through spatial and temporal cloaking. In: Proceedings of the 1st International Conference on Mobile Systems, Applications and Services, pp. 31–42. ACM (2003)

12. Shokri, R., Theodorakopoulos, G., Le Boudec, J.-Y., Hubaux, J.-P.: Quantifying location privacy. In: 2011 IEEE Symposium on Security and Privacy, pp. 247–262. IEEE (2011)
13. Zhou, L., Yu, L., Du, S., Zhu, H., Chen, C.: Achieving differentially private location privacy in edge-assistant connected vehicles. IEEE Internet Things J. **6**(3), 4472–4481 (2019)
14. Hua, J., Tong, W., Xu, F., Zhong, S.: A geo-indistinguishable location perturbation mechanism for location-based services supporting frequent queries. IEEE Trans. Inf. Forensics Secur. **13** (5), 1155–1168 (2017)
15. F. B. of Kaohsiung City, "Fire hydrant location" (2018). https://data.kcg.gov.tw/dataset/fire-hydrant-up
16. Andrés, M.E., Bordenabe, N.E., Chatzikokolakis, K., Palamidessi, C.: Geo-indistinguishability: Differential privacy for location-based systems. arXiv preprint arXiv:1212.1984 (2012)

Practical Authentication Scheme Preserving Privacy Protection for the Internet of Things

Kuo-Yu Tsai[1], Ching-Wei Lai[2], Chen-Hua Fu[2], Chi-Hung Wang[3(✉)], and Yuh-Shyan Hwang[3]

[1] Department of Information Engineering and Computer Science,
Feng Chia University, No. 100 Wenhwa Road, Seatwen, Taichung 407, Taiwan
[2] Department of Information Management, National Defense University,
No. 70, Sec. 2, Zhongyang N. Road, Beitou District, Taipei 112, Taiwan
[3] Department of Electronic Engineering, National Taipei University
of Technology, No. 1, Sec. 3 Zhongxiao E. Road, Taipei 106, Taiwan
wang073000@yahoo.com.tw

Abstract. Based on the authentication scheme of Wu et al. (published in the Journal of Ambient Intelligence and Humanized Computing, Volume 8, Issue 1, pp 101–116, 2017), we propose a robust privacy protection authentication scheme to prevent malicious attacks in general wireless sensor networks. In addition, we implement the proposed scheme in Rasperry Pi to simulate a home monitoring system and validate the feasibility of the proposed method.

Keywords: Internet of Things · Privacy protection · Authentication

1 Introduction

Following the development of network communication and wireless sensor device technology, large quantities of wearables and wireless sensor devices are used in everyday life, providing convenient services and applications such as physiological measurements and step counting. In 1999, Ashton [1] proposed the Internet of things (IoT). In an IoT environment, each sensor device has a different identifier, and these identifiers access one another through a communications protocol to provide application programs or services and meet user demand for information and services.

A wireless sensor network (WSN) is the most vital component of the IoT. A wireless communication computing network is composed of a large number of sensor devices that collaboratively monitor various environmental aspects such as temperature, vibrations, and sound. There are three basic parts in a WSN: user, gateway, and sensors. The user and sensors must first register to obtain legal information. The gateway is responsible for the security of the WSN, enables identity authentication between the user and sensors, and communicates to retrieve sensor information.

In an IoT environment, to easily remember their passwords, users often use simple and short or initial-based passwords as a basis for identity authentication. In addition, users often employ unencrypted transmission in a section of communications protocol. However, attackers could exploit these vulnerabilities. Benson [2] highlighted the discovery of the first IoT botnet made by Proofpoint researchers, where hundreds of

© Springer Nature Switzerland AG 2020
L. C. Jain et al. (Eds.): SICBS 2019, AISC 1145, pp. 351–360, 2020.
https://doi.org/10.1007/978-3-030-46828-6_30

thousands of malicious access sources came from IoT devices. According to a proposal made by the International Telecommunication Union in 2005 [3], ubiquitous networking-the next generation of networking-and ubiquitous computing and analysis were technological visions that could be fulfilled by the IoT. Under the development of information communication technology and hardware equipment, not only could the public's need for ubiquitous communication be met but information transfer between objects could also be achieved. Subsequently, IoT security frameworks were proposed including the use of secure encryption methods and identity authentication protocol design as well as the implementation of security protocols [4]. Additionally, some scholars have discussed the issue of user privacy protection [5–7].

In 2014, Hsieh and Leu [6] proposed an authentication scheme using dynamic identity in a WSN; the user sends his/her login information to the gateway through an encrypted authentication method to verify the validity of the identity. However, Wu *et al.* [7] indicated vulnerabilities in the scheme proposed by Hsieh and Leu [6]; attackers could perform malicious insider attacks, offline guessing attacks, user forgery attacks, and sensor capture attacks. Additionally, Hsieh and Leu's scheme [6] did not use any session key for encryption to ensure the confidentiality of the transmitted information, and the scheme lacked a mutual authentication system. Wu *et al.* [7] proposed an improvement scheme to solve the problems in the scheme. However, we found that the weakness of Wu *et al.*'s scheme [7] still had potential vulnerabilities; a malicious legitimate user (attacker) could successfully impersonate any other user. Based on the application model introduced by Hsieh and Leu [6] and the improvement scheme presented by Wu *et al.* [7], this paper proposed an improvement scheme to prevent malicious attacks in general WSN environments. The attack models are described as follows:

1. Attackers can control communication channels between gateways, sensors, and users; in other words, attackers can intercept and modify any messages through public channels;
2. Attackers can modify and resend eavesdropped messages;
3. Attackers can capture data stored on smart cards;
4. When a sensor is in an insecure environment, an attacker can execute capture attacks through the sensor (i.e., the attacker can completely control the sensor); and
5. When offline, an attacker can guess the identity information or password of a user, but cannot guess two secret parameters in polynomial time.

This study designed a lightweight authentication scheme for implementation on Android platforms using Raspberry Pi 3 to verify the feasibility of the proposed method. In addition, the proposed scheme can prevent insider attacks, offline guessing attacks, user forgery attacks, sensor capture attacks, desynchronization attacks, and replay attacks, and its security features include user anonymity, mutual authentication, session key, and forward security, described as follows:

1. Insider attack: A malicious user (attacker) uses the parameter obtained during registration to deduce the relevant secret parameters of the system for forgery and impersonation attacks;
2. Offline guessing attack: An attacker obtains the correct password by repeatedly guessing and verifying the password based on intercepted information;

3. User forgery attack: Based on known information, an attacker successfully impersonates a legitimate user, gateway, or sensor;
4. Sensor capture attack: An attacker controls a specific sensor and executes actions in that sensor's capacity but cannot affect other nodes or users;
5. Prevention of desynchronization attack: An attacker interferes with or stops a message sent between the gateway and sensor to cause inconsistency in the session keys derived by the two communicating parties;
6. Prevention of replay attack: An attacker resends intercepted information such as login details;
7. User anonymity: No one can obtain the real identity information of the user;
8. Mutual authentication: The user and sensor perform mutual authentication of the legitimacy of each other's identity with the help of the gateway;
9. Session key: During the course of identity authentication, or in other words, during the exchange of partial key information, the communicating parties derive the same session key to be used in subsequent communications to ensure the confidentiality of communications; and
10. Forward security: When an attacker obtains the private key of the communicating parties at a specific time, he or she cannot deduce new session information based on the obtained information. In other words, the session key is updated at every round of communication and has no relevance.

2 Our Proposed Privacy Protection Authentication Scheme

This study proposed an improvement scheme to solve the authentication problem of Wu *et al.* [7]. The proposed scheme consists of three phase: initialization, registration, and login and authentication. The used symbols are defined in Table 1.

Table 1. Definition of symbols

Symbols	Definition
p, q	Prime numbers
$E(F_P)$	Finite field (F_P) in elliptic curve (E)
G	Subgroup in the finite field (F_P)
P	Generator of G
GW	Gateway
x	Secret key
ID_i, PW_i	User U_i's identity and password
SID_i	Sensor S_i's identity
sk_u, sk_s	Session keys computed by the user and sensor, respectively
$h(.), h(.)_1$	One-way hash functions
$a \oplus b$	a XOR b
$a \parallel b$	a cancanates b

2.1 Initialization Phase

Step 1: Gateway GW generates a subgroup G and a generator P in the finite field (F_P) of the elliptic curve (E) with the order q, where P is the root of G.
Step 2: GW generates a secret key x and secretly saves it in the gateway.
Step 3: GW generates a secret key k, which is shared with the user.
Step 4: GW selects one-way hash function.

2.2 Registration Phase

The registration phases for users and sensors are described as follows.
U_i registration:
Step 1: User U_i selects random number r_0 and inputs ID_i and PW_i.
Step 2: U_i computes MP_i and MI_i by The following equations:

$$MP_i = h(r_0 \parallel PW_i)$$

$$MI_i = h(r_0 \parallel ID_i)$$

Step 3: U_i sends $\{MP_i, MI_i, ID_i\}$ to Gateway GW through a secure channel;
Step 4: Upon receiving the registration information, GW calculates the following equations:

$$e_i = h(ID_{GW} \parallel x \parallel MI_i) \oplus MP_i$$

$$f_i = h(MI_i \parallel x) \oplus MI_i$$

Step 5: GW sends $\{e_i, f_i, P, p, q\}$ to U_i.
Step 6: GW saves $h(ID_i \parallel r_0)$ and k in the database.
Step 7: On receiving the message sent by GW, User U_i calculates $d_i = h(ID_i \parallel PW_i) \oplus r_0$ and saves d_i.
S_j registration:
Step 1: Sensor S_j sends $\{SID_j\}$ is sent to Gateway GW.
Step 2: GW calculates

$$c_j = h(SID_j \parallel x)$$

Step 3: GW secretly sends c_j to S_j.
Step 4: Upon receiving c_j, Sensor S_j saves $\{SID_j, c_j\}$.

2.3 Login and Authentication Phase

Step 1: User U_i inputs ID_i and PW_i.

Step 2: U_i performs the following calculations:

$$r_1 = d_i \oplus h(ID_i \parallel PW_i)$$

$$MI_i = h(r_1 \parallel ID_i)$$

$$MP_i = h(r_1 \parallel PW_i)$$

$$E_1 = k \oplus r_0$$

Step 3: U_i selects a random number $\alpha \in [1, q-1]$, nonces r_2 and r_3, and calculates

$$MI_i^{new} = h(r_2 \parallel ID_i)$$

$$B_1 = e_i \oplus MP_i \oplus r_3$$

$$B_2 = \alpha P$$

$$B_3 = f_i \oplus MI_i \oplus M_i^{new} \oplus h(r_3 \parallel MI_i)$$

$$B_4 = h(r_3 \parallel MI_i^{new} \parallel B_2) \oplus ID_i$$

$$B_5 = h(ID_i \parallel MI_i \parallel MI_i^{new} \parallel SID_j)$$

Step 4: U_i sends $M_1 = \{MI_i, SID_j, B_1, B_2, B_3, B_4, B_5, E_1\}$ to GW.
Step 5: Upon receiving M_1, Gateway GW calculates

$$r_3 = B_1 \oplus h(ID_{GW} \parallel x \parallel MI_i)$$

$$MI_i^{new} = B_3 \oplus h(MI_i \parallel x) \oplus h(r_3 \parallel MI_i)$$

$$ID_i = B_4 \oplus h(r_3 \parallel MI_i^{new} \parallel B_2)$$

$$r_0 = E_1 \oplus k$$

Step 6: Upon obtaining r_0, Gateway GW calculates $h(ID_i \parallel r_0)'$ and verifies whether it is identical to $h(ID_i \parallel r_0)$ stored in the database.
Step 7: Gateway GW verifies $B_5? = h(ID_i \parallel MI_i \parallel MI_i^{new} \parallel SID_j \parallel B_2)$.
Step 8: If all verifications are cleared, GW proceeds with the following steps. Otherwise, execution is terminated;
Step 9: GW calculates the following information:

$$c_j = h(SID_j \parallel x)$$

$$D_1 = h(MI_i \parallel SID_j \parallel c_j \parallel B_2)$$

Step 10: GW sends $M_2 = \{MI_i, SID_j, B_2, D_1\}$ to S_j.

Step 11: Upon receiving M_2, S_j verifies whether SID_j and $D_1? = h(MI_i \| SID_j \| c_j \| B_2)$ are identical. If verification is cleared, S_j proceeds with the following steps. Otherwise, execution is terminated;

Step 12: S_j selects a random number $\beta \in [1, q-1]$, and calculates

$$C_1 = \beta P$$

$$C_2 = \beta B_2$$

$$sk_s = h(B_2 \| C_1 \| C_2)$$

$$C_3 = h(MI_i \| SID_i \| sk_s)$$

$$C_4 = h(c_j \| MI_i \| SID_j \| C_2)$$

Step 13: S_j sends $M_3 = \{C_1, C_3, C_4\}$ to GW;

Step 14: Upon receiving M_3, GW verifies $C_4? = h(c_j \| MI_i \| SID_j)$. If verification is cleared, GW executes the following steps, otherwise execution is terminated;

Step 15: The following information is calculated:

$$D_2 = h(ID_{GW} \| x \| MI_i^{new}) \oplus h(MI_i^{new} \| r_3)$$

$$D_3 = h(MI_i^{new} \| x) \oplus h(MI_i \| r_3)$$

$$D_4 = h(ID_i \| MI_i \| MI_i^{new} \| SID_j \| D_2 \| D_2 \| r_3)$$

Step 16: GW sends $M_4 = \{C_1, C_3, D_2, D_3, D_4\}$ to U_i;

Step 17: Upon receiving M_4, User U_i verifies $D_4? = h(ID_i \| MI_i \| MI_i^{new} \| SID_j \| D_2 \| D_2 \| r_3)$. If verification is cleared, U_i proceeds with the following steps. Otherwise, execution is terminated;

$$B_6 = \alpha C_1$$

$$sk_u = h(B_2 \| C_1 \| B_6)$$

Step 18: U_i verifies $C_3? = h(MI_i \| SID_j \| sk_u)$. If verification is cleared, U_i calculates

$$d_i^{new} = r_2 \oplus h(ID_i \| PW_i)$$

$$e_i^{new} = D_2 \oplus h(MI_i^{new} \| r_3) \oplus h(r_2 \| PW_i)$$

$$f_i^{new} = D_3 \oplus MI_i^{new} \oplus h(MI_i \| r_3);$$

and

Step 19: U_i replaces the previous (d_i, e_i, f_i) with the new $(d_i^{new}, e_i^{new}, f_i^{new})$.

3   Discussions

3.1   System Simulation and Implementation

This study built a home monitoring prototype system in the software and hardware environment described in this section and implemented the proposed authentication scheme. A mobile device and Raspberry Pi 3 are used to simulate the user end and the sensor, respectively. This study also developed the required modular application programs.

1. Hardware sensors: Raspberry Pi 3 Model B, CPU: 1.2 GHz 64-bit, RAM: 1 GB LPDDR2; Raspberry Pi lens module Camera V2 8-mega-pixel lens;
2. Mobile device: Android 5.0 or later operating system;
3. Development tool: Android Studio;
4. Operating system used by the development environment: Windows 7

If a user wants to connect to a sensor device or node to retrieve information, the user and sensor must first perform identity authentication with the gateway. A user can legally access related information only after verification. We used an Android mobile device with Raspberry Pi to build a home monitoring prototype system; users can login through the app to view the visuals captured using the Raspberry Pi lens after authentication.

3.2   Security Analysis

This section describes the security analysis of our proposed scheme and discusses whether the scheme was able to meet the security requirements. This study validated the security analysis of the proposed scheme, which is based on one-way hash function security [8] and the elliptic curve discrete logarithm problem [9] described as follows:

1. One-way hash function hypothesis: Assume that H is the one-way hash function. The input value is a message of any length m, and the output value is the hash value $H(m)$. The hypothesis has the following characteristics:
(1) Given a hashing value $H(m)$, the message m is to be deduced to be computationally infeasible; or
(2) Given two messages m and m', $H(m) = H(m')$ is to be deduced to be computationally infeasible.
2. Elliptic curve discrete logarithm problem: Assume that two points P and Q are present on an elliptic curve and $Q = kP$. Given on P and Q, parameter k through reverse engineering is to be deduced to be computationally infeasible.

Our proposed scheme could prevent insider attacks, offline guessing attacks, user forgery attacks, sensor capture attacks, desynchronization attacks, and replay attacks, and includes security features such as user anonymity, mutual authentication, session key, and forward security. The workings of the scheme are described as follows:

1. Insider attack: Assume that a malicious user has already registered and obtained secret information $e_i = h(ID_{GW} \parallel x \parallel MI_i) \oplus MP_i$ and $f_i = h(MI_i \parallel x) \oplus MI_i$,

where $MP_i = h(r_0 \parallel PW_i)$ and $MI_i = h(r_0 \parallel ID_i)$. Suppose that the malicious user attempts to forge the login information of other users or impersonate another user, according to our scheme, he or she must know secret information x to be able to compute and obtain valid information. However, the malicious user does not know x; if he or she has to deduce x from e_i and f_i, he or she would not success due to the one-way hash function hypothesis;

2. Offline guessing attack: If the attacker attempts to repeatedly guess the password based on intercepted information to gain the correct password, according to our scheme, when a user registers and obtains legal information $e_i = h(ID_{GW} \parallel x \parallel MI_i) \oplus MP_i$ and $d_i = h(ID_i \parallel PW_i) \oplus r_0$, the calculations for $r_1 = d_i \oplus h(ID_i \parallel PW_i)$ and $B_1 = e_i \oplus MP_i \oplus r_3$, where the calculation for $MP_i = h(r_0 \parallel PW_i)$ is performed at the login stage. Therefore, the password is not present in the login and authentication information that is sent, and thus the attack cannot verify the accuracy of the guessed password based on the intercepted information;

3. Prevention of user forgery attack: Suppose that an attacker attempts to forge the login message M_1 of any user, he or she must generate his/her verification information $B_1 = h(ID_{GW} \parallel x \parallel MI_i) \oplus r_3$ and $B_3 = h(MI_i \parallel x) \oplus MI_i^{new} \oplus h(r_3 \parallel MI_i)$. Based on our proposed scheme, the attacker must know the secret information x to compute and obtain the valid information. However, the attacker does not know x, and if he or she attempts to randomly set B_1 or B_3 as a random value and subsequently computes and obtains valid verification information, he or she would not success due to the one-way hash function hypothesis;

4. Prevention of sensor capture attack: Suppose that an attacker gains complete control of the sensor and executes actions that could be executed by the sensor, from the captured node, the attacker can only obtain the secret information $c_j = h(SID_j \parallel x)$ of the captured sensor. According to our proposed scheme, if the attacker attempts to calculate information of other nodes or users, he or she would need to know the secret information x to obtain the valid information. If the attacker attempts to deduce x from $c_j = h(SID_j \parallel x)$, he or she would not success due to the one-way hash function hypothesis;

5. Prevention of desynchronization attack: Assume that an attacker attempts to interfere with or stop a message transferred between the gateway and the sensor and cause the session key derived by the two parties to alter, based on our proposed scheme, the session key as computed by the user and sensor were $sk_u = h(B_2 \parallel C_1 \parallel B_6)$ and $sk_s = h(B_2 \parallel C_1 \parallel C_2)$, respectively, where $B_2 = \alpha P$, $C_1 = \beta P$, and $B_6 = \alpha C_1 = \alpha \beta P = \beta B_2$. However, in our method, message authentication codes $B_5? = h(ID_i \parallel MI_i \parallel MI_i^{new} \parallel SID_j \parallel B_2)$ and $C_4 = h(c_j \parallel MI_i \parallel SID_j \parallel C_2)$ were generated during the process of information transfer to protect B_2 and C_1, respectively;

6. Prevention of replay attack: Suppose that an attacker resends the intercepted messages M_1 and M_2 during the login and authentication stage, in our proposed scheme, M_1 and M_2 include $MI_i^{new} = h(r_2 \parallel ID_i)$, where r_2 is a nonce. Based on the one-way hash function hypothesis, the attacker cannot successfully resend messages M_1 and M_2;

7. User anonymity: When a user logs in, the one-way hash function computes MI_i as the user's identity information and uses MI_i for message sending. No one can know the user's real identity information;

8. Mutual authentication: In our proposed scheme, the user and the sensor perform mutual authentication with the help of the gateway. The gateway can verify the user's identity information from message B_5 sent by the user and the sensor's identity from message C_4 sent by the sensor. The sensor receives message D_1 sent by the gateway to verify the gateway identity. The user receives message D_4 and verifies the identity of the gateway. Thus, when all the steps have been completed, mutual authentication is achieved;

9. Generation of session key: The sensor generates the session key $sk_s = h(B_2 \| C_1 \| C_2)$ and computes and sends $C_3 = h(MI_i \| SID_i \| sk_s)$ to the user for key confirmation to verify that $C_1 = \beta P$ and $C_2 = \beta B_2$. The user generates session key $sk_u = h(B_2 \| C_1 \| B_6)$ and through verification of $C_3? = h(MI_i \| SID_j \| sk_u)$, performs key confirmation to verify that $B_2 = \alpha P$ and $B_6 = \alpha C_1$; and

10. Forward security: Following our proposed scheme, the session key creation value is based on the elliptic curve discrete logarithm problem. Supposing an attacker manages to obtain the private key of user, gateway, and sensor, if the attacker attempts to deduce the previous session key, he or she must solve the elliptic curve discrete logarithm problem, which is computationally infeasible.

4 Conclusion

We found that the scheme by Wu et al. is unable to prevent insider attacks, and thus designed a lightweight identity authentication scheme to improve the scheme of Wu et al. Furthermore, we implemented the proposed scheme to simulate a home monitoring system; a mobile device and Raspberry Pi were used to simulate the user end and the sensor node, respectively. According to the experimental results, our proposed scheme could be used in wireless sensor environments.

Acknowledgements. The authors gratefully acknowledge the support from Taiwan Information Security Center (TWISC) and Ministry of Science and Technology (MOST), under the grants MOST 108-2221-E-035 -077 - and MOST 107-2221-E-182 -006 -.

References

1. Ashton, K.: That 'Internet of Things' thing. RFID J. **22**(7), 97–114 (2009)
2. Benson, V.: Personal information security and the IoT: the changing landscape of data Aprivacy. Comput. Commun. Collab. **3**(4), 15–19 (2015)
3. International Telecommunication Union. 2005. The Internet of Things—Executive Summary. ITU Internet Reports (2005)
4. Nguyen, K., Laurent, T.M., Oualha, N.: Survey on secure communication protocols for the Internet of Things. Ad Hoc Netw. **32**, 17–31 (2015)

5. Das, M.L.: Two-factor user authentication in wireless sensor networks. IEEE Trans. Wirel. Commun. **8**(3), 1086–1090 (2009)
6. Hsieh, W.B., Leu, J.S.: A robust user authentication scheme using dynamic identity in wireless sensor networks. Wirel. Pers. Commun. **77**(2), 979–989 (2014)
7. Wu, F., Xu, L., Kumari, S., Li, X.: A privacy-preserving and provable user authentication scheme for wireless sensor networks based on Internet of Things security. J. Ambient Intell. Hum. Comput. **8**(1), 101–116 (2017)
8. Bruce, S.: Applied Cryptography: Protocols, Algorithms, and Source Code in C. John Wiley, New York (1996)
9. Koblitz, N.: Elliptic curve cryptosystems. Math. Comput. **48**(177), 203–209 (1987)

Author Index

© Springer Nature Switzerland AG 2020
L. C. Jain et al. (Eds.): SICBS 2019, AISC 1145, pp. 361–362, 2020.
https://doi.org/10.1007/978-3-030-46828-6

Printed in the United States
By Bookmasters